THE MICROBE 1984
Part I Viruses

SYMPOSIA OF THE
SOCIETY FOR GENERAL MICROBIOLOGY*

* Published by the Cambridge University Press, except for the first Symposium, which was
published by Blackwell's Scientific Publications Limited.

THE MICROBE 1984
Part I Viruses

EDITED BY

B. W. J. MAHY AND J. R. PATTISON

THIRTY-SIXTH SYMPOSIUM OF
THE SOCIETY FOR GENERAL MICROBIOLOGY
HELD AT
THE UNIVERSITY OF WARWICK
APRIL 1984

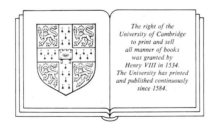

The right of the
University of Cambridge
to print and sell
all manner of books
was granted by
Henry VIII in 1534.
The University has printed
and published continuously
since 1584.

Published for the Society for General Microbiology

CAMBRIDGE UNIVERSITY PRESS

CAMBRIDGE

LONDON   NEW YORK   NEW ROCHELLE

MELBOURNE   SYDNEY

Published by the Press Syndicate of the University of Cambridge
The Pitt Building, Trumpington Street, Cambridge CB2 1RP
32 East 57th Street, New York, NY 10022, USA
296 Beaconsfield Parade, Middle Park, Melbourne 3206, Australia

First published 1984

Printed in Great Britain at The Pitman Press, Bath

Library of Congress catalogue card number: 83-19004

British Library cataloguing in publication data

Symposium of the Society for General Microbiology
(36th: 1984: Warwick)
The Microbe 1984. – (Symposia of the Society for
General Microbiology; 36)
Pt. I, Viruses
1. Micro-organisms – Congresses
I. Title II. Mahy, B.W.J. III. Pattison, J.R.
IV. Series
576   QR41.2
ISBN 0 521 26056 6

CONTRIBUTORS

BALTIMORE, D. Massachusetts Institute of Technology, Centre for Cancer Research, 77 Massachusetts Avenue, Cambridge, MA 02139, USA

BISHOP, J. M. Department of Microbiology, School of Medicine, University of California, San Francisco, CA 94143, USA

BROWN, F. Wellcome Research Laboratories, Beckenham, Kent BR3 3BS, UK

BRUNT, A. A. Glasshouse Crops Research Institute, Worthing Road, Rustington, Littlehampton, West Sussex BN16 3PU, UK

FIELDS, B. N. Department of Microbiology & Molecular Genetics, Harvard Medical School, 25 Shattuck Street, Boston, MA 02115, USA

HARRISON, S. C. Department of Biochemistry & Molecular Biology, Harvard University, 7 Divinity Avenue, Cambridge, MA, USA

McGEOCH, D. J. Institute of Virology, University of Glasgow, Church Street, Glasgow G11 5JR, UK

MAHY, B. W. J. Department of Pathology, Division of Virology, Laboratories Block, Addenbrooke's Hospital, Hills Road, Cambridge CB2 2QQ, UK

PATTISON, J. R. Medical Microbiology, Kings College Hospital Medical School, Denmark Hill, London SE5 8RX, UK

REANNEY, D. C. Department of Microbiology, LaTrobe University, Bundoora, Melbourne, Victoria 3083, Australia

SÄNGER, H. Max-Planck-Institut für Biochemie, D-8033 Martinsried bei München, Federal Republic of Germany

SYMONDS, N. D. Department of Genetics, University of Sussex, Falmer, Brighton, Sussex, UK

TYRRELL, D. A. J. Common Cold Research Unit, Harvard Hospital, Coombe Road, Salisbury, Wiltshire SP2 8BW, UK

WEISS, R. A. Institute of Cancer Research: Royal Cancer Hospital, Chester Beatty Laboratories, Fulham Road, London SW3 6JB, UK

WILDY, P. Department of Pathology, University of Cambridge, Tennis Court Road, Cambridge CB2 1QP, UK

CONTENTS

Contents: Part II Prokaryotes and Eukaryotes

EDITORS' PREFACE

This symposium celebrates the one-hundredth meeting of the Society for General Microbiology. The Orwellian connection led us to bring together in this volume a series of papers which review key areas in virology. To us, 1984 seems an entirely appropriate time for such a review, as well as for a consideration of likely future developments in the subject. Thus we can commence with a chapter which traces the development of virology from an embryonic to a fully established and diverse science. In the course of this development, firm concepts about viruses themselves have emerged. We have therefore included chapters on the structure of viruses, aspects of their evolution, and the nature and strategy for expression and recombination of their genetic material. Viruses were first detected as agents of disease, and this association is reflected by a chapter describing the continued emergence of virus diseases in animals, plants and man. There are also two chapters on the key role played by viruses in the investigation as well as the causation of neoplasia, and one on the mechanisms of virus–host interactions, an aspect of virology which has proved difficult to elucidate. To date there have been some notable successes in the control of virus diseases, and one, smallpox, has been eradicated: we include a chapter on future prospects of this area. Finally, there is a fascinating account of the viroids, small infectious nucleic acids which subvert the host cell genetic machinery to cause some devastating diseases in higher plants.

In compiling the twelve chapters we hoped to provide an overview of virology in 1984. Inevitably, however, many aspects of the subject could not be fully developed in the space available to us. Knowledge and understanding of the subject are currently experiencing a phase of rapid advancement from the application of techniques of rapid DNA sequencing and genetic manipulation. Our hope is that this volume will provide a baseline to which future advances can be related when another suitable opportunity arises to review the subject as a whole.

We wish to thank the contributing authors who met the necessary deadlines and also Roger Berkeley of the SGM, who first had the idea for this celebratory meeting and saw it through.

Division of Virology,   B.W.J. Mahy
Department of Pathology,
University of Cambridge

Department of Medical Microbiology,   J.R. Pattison
King's College Hospital Medical School,
University of London

AN ANALYSIS OF VIROLOGY

PETER WILDY

Department of Pathology, University of Cambridge, Tennis Court Road, Cambridge CB2 1QP, UK

All the world's a stage,
And all the men and women merely players:
They have their exits and their entrances;
And one man in his time plays many parts,
His acts being seven ages.

Shakespeare, *As you like it*

INTRODUCTION

Virology is a modern term which does not grace the pages of the *Shorter Oxford English Dictionary*. It is a bastard, deriving from 'virus' (a Latin term connoting a slimy liquid, slime, poison, venom, bitterness and possibly stink), coupled with the Greek -λογία (indicating something like scientific thought). Today virology is a science in its own right which continually contributes to the orderly advancement of biology and to the well-being of mankind. In this paper I wish to address the worthwhileness of virology, to examine the stuff of which it is made and to analyse why it was necessary to introduce it, and whether it will endure.

The word virology became current in the immediate post-war period. Thus we note the publication of *General Virology* by Luria in 1953 and, two years later, the introduction of George Hirst's journal *Virology* and Burnet's *Principles of Animal Virology*. This was an exciting era for biology in general. Our knowledge of the infectious agents of disease was already well advanced and the understanding of very large molecules was developing. It was becoming possible to predict mechanisms to explain established genetic facts and the necessary chemistry was being developed to establish such predictions. In retrospect one can identify several conceptual foci that were ripe for attention. There was the elucidation of the genetic code, the means by which this was expressed and how the transfer of genetic information was controlled. A number of unexplained phenomena (e.g., the nature of the transforming principle, the nature of antibody production, and the nature of viruses), had to be accounted for by any general theory that was advanced. It is not surprising that various interrelated new branches of biology (e.g., molecular biology, immunology and virology) sprouted at this time.

Turning now specifically to virology, its designation as a special pocket of biology antedated the crystallization of the modern concept of what a virus is; this topic still had to be addressed. Then, if it was to be organized as a discipline, the unity of and variety among viruses had to be more or less defined and some sort of classificatory scheme had to be developed. Since the science necessarily involves viruses and ~the people who work with them, the evaluation of virology depends as much on these people as upon the material they worked with. Finally, virology has depended very greatly on and has contributed to the development of related sciences and the techniques they offer – thus the development of biochemistry, molecular biology, immunology and virology has greatly influenced the general way things have gone in each discipline.

This volume is being published primarily to celebrate the 100th scientific meeting of the Society for General Microbiology, so it is natural also to discuss the fortunes of virology in relation to those of the Society.

EVOLUTION OF THE CONCEPT OF VIRUS

I have behaved improperly and as a punishment, the conclusion of this lecture will be prosy, coarse and vulgar! Viruses should be considered as viruses because viruses are viruses.

A. Lwoff (1957)

Central to the development of virology has been the evolution of the concept of what a virus is; until this was settled virology could never have become a discipline nor would it have been necessary. Lwoff's (1957) working definition has endured tolerably well; in the main it congealed about issues that were familiar to us all but had been obfuscated by a number of red herrings and irrelevancies which Lwoff resolutely swept out of the way.

The events that led up to the concept encapsulated by Lwoff will be familiar to all; they have been described in numerous texts and have been dealt with in detail by Waterson & Wilkinson (1978). I shall briefly recall some of the findings that drove the evolution of the central virus concept and the diversity it permits in terms of the main attributes of virus. I shall attempt to distinguish various factors which hampered progress.

The concept of the virion

When the discipline of bacteriology was young, the word virus served to describe any infective agent of disease. It is important to understand the impact of the catastrophic epidemic diseases such as smallpox and plague upon the mid-Victorians – to comprehend the implications of the term. When bacteriological methods were applied rigorously, the aetiologies of more and more bacteriological diseases became established. There remained a rump of diseases, some lurid, to which no microbe could be assigned by Koch's methodology. These failures became uncomfortable misfits in an otherwise successful scheme of infectious agents. These were agents infective for susceptible hosts but shared the negative characteristics of invisibility and inability to grow on inert media.

That the infective principle of some viruses was small in size was unequivocally established for tobacco mosaic disease by Beijerinck (1898) and for foot-and-mouth disease by Loeffler & Frosch (1898) and, within a few years, a number of other viruses were also shown to be filterable and diffusible and unable to sediment in the centrifuge. Such anomalous behaviour led to various controversial hypotheses, some of which endured for decades. Beijerinck believed in soluble life – the *contagium vivum fluidum*. He believed that he was dealing with infective molecules and conjectured means by which such molecules could infect, cause disease and make more molecules, thus foreshadowing our present day ideas.

We find this idea occurring over and over again, often being reinforced by fresh evidence as, for example, when Stanley (1935) purified tobacco mosaic virus and reported a crystallizable protein that was infective. Likewise Gratia (1921) and Bordet (1925) had interpreted bacteriophage as a lytic enzyme. On the other hand Loeffler and Frosch thought of their filterable agent as a tenuous microorganism, a view that similarly endured, viz., Burnet's (1945) *Virus as Organism*. On the whole, the molecular theory developed on account of the attributes of the infective entity and the microbial theory depended upon many other attributes as well.

From the beginning up until the 1950s one must recognize that the chief attributes of diversity among viruses related to their hosts. To some extent, this reflected the differences naturally imposed by the different nature of the host material which has, until recently, been the principal factor determining the feasibility of experiment. But one must also recognize that the 'virologists' of the day had heterogeneous backgrounds and interrelated more with others working in their field than with those working on other viruses. What could not be compared was not compared and the apparent diversity among viruses seemed the greater because of this.

The concept of the virion – or virus particle – followed most naturally from the microbial theory. One might expect eventually to see the particles and to measure their size by improved technique, a prediction that held. It is assumed that the first publication of pictures of virus particles was by Buist (1886) who visualized the 'spores of micrococci' in vaccine lymph. In course of time, the improvement of optical systems, and the introduction of special staining allowed detection of the 'elementary bodies' of a number of viruses but told us little else about them. It was not until the electron microscope was developed that particles were seen in detail and some of the bacteriophages were plainly demonstrated to have a complex structure (Ruska, 1940), as might be expected of micro-organisms. Little detail could be made out until the techniques of specimen preparation for the electron microscope had improved. Meanwhile, the sizes of virus particles were measured in the electron microscope and these were compared with estimates of functional size by various physicochemical techniques and radio-logical methods.

Thus, not only were most virus particles found to be smaller than bacteria but the estimated sizes differed over a considerable range. The larger viruses were only a little smaller than the smallest bacteria (some of them we regard today as mycoplasmas, rickettsia and chlamydia) and, more important, there seemed no discernible discontinuity in the size range from the bacteria to the smallest viruses. This, together with other attributes, seemed to support the notion of a continuous development from a free-living state to one of increasing parasitism. A second inference developed as a result of this distribution of size: namely that the 'larger viruses' might be akin to microbes while the smallest were molecules. This uneasy thought probably did much to disturb the evolution of a unified virus concept.

Some viruses were found to be very stable in the environment whereas others were not. Much was learned about thermal death points and resistance to various chemical agents. Particularly among the viruses of vertebrates, diversity was found in the sensitivity of infectivity to diethyl ether (Andrewes & Horstmann, 1949). This became the basis of a useful diagnostic test and served as a criterion for classification.

The next steps concerning the nature of virus particles had to await technical advance. Mostly they depended upon the ability to provide more and more virus and the introduction of techniques that could be applied to smaller and smaller amounts of material. Where these requirements could be met, the purification of infective particles was achieved and their nature determined by conventional chemical methods. The studies of Stanley (1935), demonstrated that tobacco mosaic virus was a protein (using bucket and spade chemistry); these results had great impact on the developing concept of virus. More important was the demonstration that viruses contain also a nucleic acid. (Schlesinger, 1935; Bawden *et al.*, 1936; Hoagland *et al.*, 1940; Ada & Perry, 1956). During this period, viruses were beginning to be considered as different from microorganisms; for instance Cowdry (1934) stated 'Viruses are carried by substances whose general nature has not been proved. As soon as any virus is found to be a microorganism it is no longer considered to be a member of the group'. However, what they were remained a mystery. Cowdry went on to say 'No proof is forthcoming that the particles (i.e., elementary bodies) are anything more than masses of nucleoprotein'.

Somewhat later, many other viruses had been shown to possess either DNA or RNA in their particles and (as methods became more critical) it became clear that, in general, only one kind of nucleic acid was to be found. In the 1950s it seemed that, whereas viruses of vertebrates might contain either DNA or RNA, bacteriophages and the viruses of invertebrates all contained DNA and plant viruses all contained RNA. For a time this became a dogma only to be shattered when the first RNA phage was demonstrated and later the first DNA plant virus.

The crucial role of the nucleic acid became evident as a result of a number of varied experiments on widely differing viruses (Markham, Matthews & Smith, 1948; Hershey & Chase, 1952; Fraenkel-Conrat & Williams, 1955; Gierer & Schramm, 1956; Colter, Bird & Brown, 1957; DiMayorca *et al.*, 1959). These findings underlined the crucial role of nucleic acids (both DNA and RNA) as genetic blueprints and were thus of general biological importance. In the enthusiasm of the day we believed that the nucleic acid of all viruses would be found to be infective. When difficulties were observed with poxviruses and myxovirus, for example, we reiterated the explanation mentioned earlier that such negative results were likely to have arisen for some technical reason. We had to wait for later, more sophisticated, studies before the question was resolved and its meaning comprehended.

In the meantime, studies on the structure of virus particles became more and more penetrating; this again was only possible through improvements to techniques such as radio-labelling and gel electrophoresis for handling large biological molecules. In the 1960s there was a flush of experiments on the nature of virus products. Within a few years it became plain that small viruses might contain few polypeptides (some as few as one) whilst large viruses contained very many in their particles. On the whole, this related well to the sizes of the virus but more importantly to genomic sizes which were also the subject of investigation.

At this time also, there was rapid development in the field of biochemical activities of virions. In the mid-1950s, virus particles had come to be thought of as packets of nucleic acid enclosed in a protein box. With the exception of neuraminidase in the myxovirus virion and lysozyme-like enzymes in the bacterophage tail, virions had been thought of as biochemically inert. Between 1960 and 1970 the virions of many of the larger viruses were found to contain enzymes which were essential for virus function. The most striking of these discoveries, which set the biological world buzzing, was the demonstration of reverse transcriptase in the retroviruses (Baltimore, 1970; Temin & Mitzutani, 1970).

Other chemical studies on purified materials revealed further diversity among different viruses: some virions contained lipid, others did not; some contained carbohydrate, others did not; some appeared to contain trace metals, others did not.

Turning now to how the virus components fit together in the virion, we note a considerable effort in the post-war period. Beginning with the studies of Bernal & Fankuchen (1941), a whole series of crystallographic studies were made which resulted in a most precise accounting of the fine structure of tobacco mosaic virus and other small crystallizable virus particles (cf. Franklin, Caspar &

Klug, 1959). Crick & Watson (1956) produced arguments to explain why small viruses at least were almost certain to assemble with a high degree of symmetry. At that time, electron microscopists were striving hard to contribute to the area of virus structure but were severely hampered by the need to use purified material, the need to desiccate it without distortion and the lack of electron scattering by virus structures. The conventional methods of this time contributed little on the matter beyond showing that Tipula iridescent virus was icosahedral in shape (Williams & Smith, 1958) and that some viruses appeared hexagonal in section. The exploitation of the negative-contrast method (Brenner & Horne, 1959) proved an important technical advance since, not only could unfixed material be visualized embedded in phosphotungstic acid, but the method was compliant enough to accept crude virus suspensions. Between 1959 and 1962 sufficient observations had been made to conclude that not only small viruses, but large ones too, had strong symmetrical elements in their structure; helical and icosahedral symmetry were common features of capsids.

The ability of the negative-contrast method to reveal surface structure also allowed fine details of surfaces of virions to be disclosed e.g., the spikes of influenza virus, the complexities of the T-even phage tail and the fibres of the adenovirus particle. The principal differentiation afforded by the method was the distinction between naked and enveloped particles. It became obvious that the possession of an envelope correlated with the presence of lipid, low buoyant density and lability of the infectious unit to lipid solvents.

As so often happens with versatile techniques, negative contrast led to a number of uncritical studies. It, nevertheless, did allow the study of otherwise impossible subjects (such as coronaviruses), it enabled the detection of virus-like particles in crude and frequently unwholesome specimens and, more important, it rapidly extended the range of viruses whose anatomy could be considered in terms of self assembly (Horne & Wildy, 1962). A general theory of how virus capsids are put together was given by Caspar & Klug (1962).

Returning to what has been deduced about the functional aspects of virions we should note the powerful influence of the Cold Spring Harbor group. The discovery of bacteriophage plaquing by d'Herelle (1921) had added a useful tool for quantifying viruses and enabled statistical treatment of what constituted the plaque forming unit (Ellis & Delbrück, 1939). The well known result was that this functional unit corresponded in its behaviour with expectation for independently acting particles; as a corollary all the apparatus necessary for causing a plaque on appropriate host cells must be housed in a single particle. With improvements in cell culture technique, focal assays became available for many viruses of vertebrates and similar results were found (Dulbecco, 1952). The dogma grew that the law of independent action might be applicable to all viruses. It was modified only with the discovery of multi-component viruses, satellites and defective viruses (of which a good number of examples are now known).

The concept of vegetative virus

It is uncertain when the failure to grow certain viruses ceased to be regarded as a mere technical failure and became accepted as true. Beijerinck had general ideas on the matter but Waterson & Wilkinson (1978) ascribed to M'Fadyean (1908) the first statement that the 'ultravisible organisms appear to be obligatory parasites'. Such an opinion was fed by the observation of inclusion bodies in the cells of tissue infected with many viruses from tobacco mosaic disease (Iwanowsky, 1903) to vaccinia (Guanieri, 1892). Notwith-standing the acceptance of a filterable infective phase, such bodies were often interpreted as the organisms themselves and were apt to be considered as intracellular protozoa. Hence the designa-tions *Chlamydozoa* (von Prowazek, 1907) and *Strongyloplasma* (Lipschütz, 1909). Some even went so far as to invent binomials for them e.g., *Entamoeba mortinatalium*, the inclusion of cytomegalic inclusion disease (Smith & Weidman, 1910/11).

The nature of inclusion bodies occupied attention for some 30–40 years, being interpreted variously as masses of elementary bodies, the matrix in which they developed or simply as a scar in the damaged cell. It became evident that inclusions were diverse in appearance but specific enough to characterize particular infectious diseases; groups of similar inclusion-forming agents were noted. Inclusions ceased to be important to the concept of virus when the chemical and physical constitution of individual examples showed lack of homology. Cowdry (1934) wrote of type A inclusions that they were 'the fingerprints of a special group of viruses which is small in comparison with that hodge-podge of agents to which the term "viruses" is usually applied'.

The unifying concept that viruses have a particular general

pattern of replication began with the critical studies on bacterio-phage by Ellis & Delbrück and Doermann (see Luria, 1950). The nature of the replicative process had considerable impact on the virological world. The focus of attention was on the eclipse phase of the growth cycle during which no evidence of virions could be found, though other specific virus attributes might be detected. The work of the Henles and of Hoyle provided evidence that influenza virus replicated in a similar fashion and the frequency of genetic reassortment was in line with this view (cf. Burnet, 1955). There followed a spate of papers, many of them uncritical, reporting 'eclipse phase' for many different animal viruses. These findings did not fit with the idea that viruses were microbes and were pounced upon to support a unifying concept embracing a distinct class of biological entities (Burnet, 1955). Considerable controversy en-sued, contrary views were expressed notably by S. P. Bedson who rightly objected to many of the less critical studies and whose own studies with the psittacosis 'virus' did not accord with the bacter-iophage pattern of replication. The question was resolved only when, in the mid-1950s, improved cell culture techniques allowed more critical experiments to be done.

Throughout the 1960s and 1970s there was steady progress in analytical methods for characterizing macromolecules and the wide-spread adoption of labelling techniques. The processes of nucleic acid replication and expression began to unfold and wide differences in the way viruses manage their economy were exposed. The armamentarium was supplemented by the availability of specific metabolic inhibitors so that the vegetative process could be inhi-bited at known points in the replicative cycle. The bacterial and bacteriophage geneticists introduced conditional lethal mutants as tools by which genetic functions could be explored. Libraries of these, principally temperature sensitive(ts), mutants were gathered for many different viruses. These mutants have been powerful tools for relating structure and function and, in particular, the replicating mechanisms of many viruses.

The discovery of reverse transcriptase in the retroviruses must be the most momentous latter-day event in virology. The finding that an RNA virus was inhibited by DNA inhibitors led Temin to commit a heresy challenging the central contemporary dogma on the storage and transfer of genetic information. The effect of the discovery of the enzyme has revolutionized biological thinking and constitutes one of the greatest single gifts virology has made to science as a whole, especially now that this enzyme is an every-day tool for engineering genes.

The enormous effort expended in the pursuit of vegetative virus has exposed an overall homology which is central to the concept of virus. It has also uncovered extraordinary diversity in the way viruses manage their intracellular economy. This diversity was at one time bewildering until certain principal patterns began to emerge and were encapsulated as the well known 'strategies' of virus replication (Baltimore, 1971).

The concept of provirus

The existence of bacteria that had the ability to generate bacteriophage *de novo* even after cloning was established by Bordet (1925) and Bail (1925). For some time the nature of the element resident in lysogenic microbes was not clear. Burnet & McKie (1929) established that they contained no infective material but that a principle – 'anlage' – must be carried in 'intimate symbiosis' and 'multiply by binary fission *pari passu* with the bacteria'. The concept was later extended by the Wollmans and subsequently crystallized by Lwoff and his colleagues at the Institut Pasteur as the modern concept of prophage (cf. Lwoff, 1966). The study of lysogeny became inextricably linked with microbial and molecular genetics, generating a fertile discipline at the heart of modern biology. From our present viewpoint the essential features of importance are that certain viruses can integrate their genome in the host chromosome, that they can be induced (Lwoff, Siminovitch & Kjeldigaard, 1950) and that while integrated can express limited functions (including proteins controlling their integration (Jacob & Monod, 1961)).

The dawning recognition of lysogeny suggested to many that viruses might be generated from host organisms. The idea was applied to certain animal viruses; for example Doerr (1938) suggested that 'herpes is not an infectious agent which is maintained by a chain of infection but that it is endogenously generated in the human organism'. This notion did not persist but, because latent herpes virus can be induced in much the same way as prophage, the analogy with prophage is irresistible and the question is still unresolved.

The analogy of prophage with tumour viruses has proved more fruitful. From the first demonstrations of filterable oncogenic agents (Ellerman & Bang, 1908; Rous, 1911) controversy raged as to whether these were viruses at all. Gardner (1931), in a monograph on microbes and ultramicrobes, commented on the similarity between the behaviour of the infective fowl tumours and bacteriophagy: 'together they constitute a strong argument for the theory of externally transmissible genetic factors'. While allowing the obvious resemblances between genes and viruses he was fixed in his idea that the latter were ultramicrobes.

By the 1960s many (both DNA and RNA) tumour viruses were under study. The development of *in vitro* cell transformation techniques (Dulbecco, 1963) allowed the essential analysis of the virus–cell relationship and provided evidence of integrated virus DNA. But we had to await the discovery of reverse transcriptase before the provirus theory for retroviruses became established. The study of these viruses has proved particularly fruitful over the past ten years revealing the importance of defectives, pseudotypes, helper functions, the existence of avirulent viruses (which had been foreshadowed by Twort, 1915) and the endogeneous viruses and their apparent benign and passive role in the host genome (cf. Weiss, 1973).

The total concept, prolonged gestation

Ever since viruses caught the attention of investigators, there has been the usually tacit recognition that they behave as genetic elements. In general this was taken for granted but (whether studying the patterns of infective disease, bacteriophagy, filterable infective principles, tumours or nucleic acid molecules) the phenomena of genetic stability and variation have been noted and accepted as something normal. Seldom was much emphasis placed on the fact until Müller (1922) drew attention to the genetic attributes of phage; 'it would be rash to call them genes, and yet at present we must confess that there is no distinction between genes and them'. He foresaw the power of the bacteriophage as a genetical tool.

The genetical aspects seem so clear to us now that it is natural to speculate why they did not immediately form the central theme about which the concept of virus emerged. One can only point to the following inhibitory factors. The first is preconception; infective diseases are caused by microbes, therefore, if the infective principle is filterable and invisible one considers it an ultramicrobe. Other analogies were considered (e.g., with spermatozoa and antigens) but these were less inhibitory. The second is irrelevant or premature speculation; microbes are living, molecules are dead; are viruses alive or dead? Similarly, what are the origins of virus? Have they evolved from free-living microbes (Green, 1935; Laidlaw, 1938) or from host cells (Darlington, 1948). Third, the scientists of the day were faced with one apparently continuous spectrum of hetero-geneous agents including mycoplasma, rickettsia, chlamydia and these frequently clouded the field of vision. Clarity followed the rejection of such agents as viruses. Fourth, though reasonably certain that they were dealing with a novel biological entity, they had to build on such knowledge as they possessed and had no means of testing it. Finally, techniques were primitive; enlightenment depended upon their improvement. No other discipline had been so dependent before. There was little rigour in the experiments that were done, with notable exceptions (e.g., Beijerinck), until the development of the Cold Spring Harbor School.

Lwoff (1957) pointed out that the total concept depends on the sum of its parts. Because of this, we had to wait until the virion, vegetative virus and provirus were more or less understood before the total concept could be assembled.

The evolution of the concept of virus has been fumbling with many wrong turnings, many futile hypotheses, rejections and even-tual selections. Our present day concept represents the survival of the fittest. But we seem to have ended up with a central concept which allows us to encompass a rich diversity of entities.

UNITY AND VARIETY AMONG VIROLOGISTS

They say we are Almost as like as eggs.

Shakespeare, *Winters Tale*

When virology became established as a discipline, none of those lucky enough to join in with the enterprise were trained virologists. There were great names such as Andrewes, Bawden, Burnet, Enders, Gard, Luria, Lwoff, Markham, Rivers and K. M. Smith. These pioneers attracted various lesser known individuals round them, all as varied in temperament and training as their mentors. Enthusiasm for the new discipline drew them together.

By the 1960s it seemed right to train young graduates as virologists and a number of postgraduate courses grew up such as the one at the University of Birmingham, England. It provided a thorough grounding in fundamental virology, virological techniques and illustrated the breadth of the subject as well as possible. The interesting result was a collection of individuals, to each of whom virology meant something quite different. They went their different ways and now display their heterogeneity according to ability, opportunity and, most important, personal inclination. The point I address is that the personal motivation of individuals is a powerful force fashioning the virological scene. To some, the primary aim is to find out how viruses work. To others viruses are convenient tools to be used for other ends.

Sir Ashley Miles once confided to me that he distinguished microbiologists, whose purpose is to find out about microorganisms, from 'microbators' who exploit microorganisms as tools to solve other problems. Likewise, we can distinguish virologists from 'viropractors'.

In making this distinction there is no intention of disparaging either 'microbation' or 'viropraxy'. They are enormously important in many disparate areas such as genetic engineering, insect control and vaccine development. What is important is that the heterogeneity amongst us is such that it is doubtful if virology can endure as a discipline centred simply around the concept of virus. I sense that the number of viropractors amongst us is growing faster than that of virologists. Very probably they are already the majority and it seems increasingly unlikely that the central concept of virus can remain an appealing common purpose.

Before leaving the matter, it may be interesting to enquire into the motivation of virologists. Viropractors we can leave aside since presumably the virus is incidental to their purpose. Are the virologists motivated by avarice? Scarcely, considering the poor material rewards that society provides for us. Are they fond of solving crossword puzzles? A survey would be interesting. Do they collect stamps? Presumably taxonomists might do so. Are they fond of gardening, hang-gliding or bridge? We do not know at all; no doubt a questionnaire on such matters would reveal many embarrassments, much amusement and more importantly that virologists are heterogeneous. What of the questions concerning what draws virologists together and what makes a virologist? I can only answer Lwoff-like; virologists are virologists because virologists are virologists.

DEVELOPMENT OF VIROLOGY

'When I use a word' Humpty Dumpty said, in rather a scornful tone, 'it means just what I choose it to mean – neither more nor less'.

Lewis Carroll, *Through the Looking Glass*

Orthovirology

It is easy enough to define the term virology but difficult to convey exactly what one means by it. No doubt there are as many virologies as virologists. So while each meaning necessarily involves the study of viruses, each may reflect a unique sense of purpose, each may cover a different field of activity and each is sure to be individual. The diversity arises not because viruses are different but because virologists are. We shall deal first with main line virology, ortho-virology, remembering that this may not really exist since, like a median, it is a figment expressing central tendancy. Virology has evolved because of a series of chance observations which it has seemed urgent to explain for entirely human reasons. For instance, if the nature of infective disease had not seemed so important and if, for some reason, molecular biology had evolved precociously by another route we should have come upon viruses from quite a different direction and orthovirology might never have developed. Its development was accidental, depending upon the evolution of the virus concept and upon a great many other circumstances as well. As it was, the jumble of facts and fancies that eventually became forged into shape as orthovirology was overshone by the enticing mystery of virus. Something akin to a religion developed with virus as the principal article of faith. This entity I call protovirology.

Orthovirology is a human activity and it seems natural to compare its development with our own much as one would for, say, the 'Birth of a Nation'. Because we shall be meeting in the Forest of Arden not many miles from Shakespeare's birthplace the temptation to turn to the philosophies of the melancholy Jaques is irresistible.

The seven ages of orthovirology
'*The infant mewling and puking*'. We begin with a series of unexplained observations which do not fit in with the general scientific schemes of the day. The facts are indigestible (accounting for the mewling) and some of them are mistaken. Many are regurgitated (hence the puking) but the better ones are retained and together nourish the infant science.

'*The whining schoolboy creeping like a snail*'. The infant has assimilated the awkward facts and the child now grapples with a series of empirical 'principles'. These are unpalatable and have to be learned more or less by rote like the ABC or the two-times-table. In 1940, the virology that was purveyed to us was in this amorphous state; virus was incomprehensible, comprising a number of unrelated incoherent 'principles' crammed together in two hurried lectures.

'*The lover with woeful ballad made to his mistress' eyebrow*'. Principles have become hypotheses, not many of which have been tested and some of which are so impractical that they cannot be. A strange gasping mania, a mysterious excitement and sense of promise now enter the scene. The young science is all heart but still half-baked and half-witted.

'*The soldier full of strange oaths, jealous in humour and quick in quarrel*'. Hypotheses are put to the test and become theories. At last, discipline is introduced and the science matures. The theories are tested rigorously and everything is disputed. This is the turning point when the virus concept is agreed. Quite suddenly proto-virology becomes orthovirology.

'*The justice in fair round belly with good capon lin'd, full of wise saws and modern instances*'. This is a dangerous age; theories become laws and discipline becomes restrictive. The laws are organized and classified and we begin to notice a shift in purpose towards codification and a loss of vigour.

'*The lean and slippered pantaloon*'. Here is the involution; laws degenerate to dogma dry as dust. All libido is spent and the previously robust heart and mind turn to disinterest and feeble-mindedness. Only the religion remains.

'*Last stage of all is second childhood and mere oblivion*'. The dogmas die and become useless except perhaps to form an epitaph on the gravestone of orthovirology. The bones of the subject remain and endure only to fossilize.

This mournful sequence seems to apply to all human activities from the Dynasties of China to the British Empire. But before we become too depressed over what, at first sight, seems to confirm the colossal futility of human endeavour we should notice that we have traced only the rise and fall of orthovirology. The science has served its purpose. Many younger and more vigorous virologies have grown up and replaced it – some still in their infancy, others mature and fruitful. I do not really believe that orthovirology has yet quite reached the last stage of all but it is now weak. To some extent it seems to hold its younger offspring together though I sense that as a ligand it is steadily growing weaker.

Metavirology and paravirology

Nobody should pretend that science is divisible into watertight compartments, yet one can distinguish species of virology that seem to have developed a life of their own. For these, the concept of virus is eccentric and for some it may play no part at all.

A number of distinct subdisciplines have developed from ortho-virology. I lump them together as metavirology. The centre of gravity has shifted for one reason or another. For instance, clinical virology has great affinity with human and veterinary medicine; virus genetics with molecular biology, and genetics; and immunovirology with immunology. These are strong, rapidly developing subjects and already their impact is being felt.

Four main mechanisms seem to have contributed to the emergence of metavirology. First, the demand of 'peopledom' for the solution of current human problems has attracted young scientists to work towards particular ends; this touches on personal motivation which we have already discussed. For example, the consciousness of the world food shortage might motivate someone to seek virological solutions. Second, arising from similar pressures, governmental and philanthropic bodies favour certain areas of investigation (for instance the conquest of cancer). Because funds are made available for this, scientists are lured into the field of oncovirology. Third, it is common for virologists to recruit expertise in areas in which they feel incompetent. When the work is in progress they are astonished to find that its direction has shifted towards quite other ends. This was especially evident in the 1960s when there was a massive importation of physical scientists and biochemists into virology. Fourth, there is the satisfaction of believing one is working in a new field. The well-worn paths have lost their attraction; the new pathways seem to lead endlessly to fields of infinite fascination.

Yet other disciplines, comprising paravirology, can scarcely be claimed to have originated from orthovirology but have been greatly influenced by it. For example biochemistry and molecular biology must properly be considered to have stemmed from the study of biologically important molecules and, as we have seen, viruses may be seen simply as examples of these. This is, *par excellence*, the domain of the viropractor.

Both of these subdisciplines are cognate with orthovirology but they have now supplanted it. No doubt each will have its day and give way in future years to further sub-subdisciplines. For the present, they enrich virology enormously and, while one is happy to allow them the independence they deserve, it is clearly important that we all march together. But here is the danger – few of us are equipped to participate in, or sometimes properly comprehend, all the subdisciplines. We have to make do with digests of each other's subdiscipline to an increasing extent. The situation is not helped by the rapidity of progress and the means of communication have been too cumbersome to keep us properly informed. It is important for us all to stay linked if virology is to continue coherent and contributory to science.

Epivirology

Several virological activities do not deal directly with investigation but seem to be inescapable if the science is to flourish. I lump them together as epivirology.

Virus taxonomy

Matthews (1983) has recently produced a critical appraisal of virus taxonomy and this should be consulted for points of detail. He points out that there is a universal innate desire to classify and name natural objects. Viruses have been no exception. He divides the various attempts into four epochs. Up until 1961 there were numerous premature efforts to introduce taxonomic schemes. That is, virologists were trying to invest the schoolboy or the lover with the beard of formal cut. F. W. Andrewes (1930) stated 'in the case of creatures which we cannot see and whose very existence is, in many cases, a matter of inference only, it is idle to talk of classification in the usual sense'.

The second epoch, 1962–66, was characterized by considerable enthusiasm and much dispute. During this period, virologists began to see how all viruses could be compared using several attributes of virus particles. Various schemes were advanced (e.g., Cooper, 1961; Lwoff, Horne & Tournier, 1962; Wildy, 1962) which made use of morphological and chemical features. During the early periods much credit must go to C. H. Andrewes who was continually suggesting and improving upon the criteria available for classification and advocating a unified scheme of nomenclature principally for the viruses of vertebrates. The proposals of Lwoff et al. (1962) provoked considerable controversy, partly owing to the hierarchic nature of the proposals and partly because of the suggested introduction of startling new names. It was seen by Gibbs et al. (1966) as too fixed a scheme to allow satisfactory evolution to occur and they suggested a temporary means of designation: the 'cryptogram'.

The third epoch, from 1966 to 1971, saw the birth of the International Committee for the Nomenclature of Viruses and a cautious beginning was made. As many viruses as possible were considered and data pertaining to them reviewed. The results of four years' work were presented at Mexico City in 1970 and were subsequently published (Wildy, 1971).

The fourth epoch has seen the consolidation and expansion of the taxonomic process begun in 1966. Essentially, all useable virus attributes are taken into account and virus groups or genera are given latinized names. The process has been valuable to virology and the periodic reports embody all the essential features of virus groups and are useful to both students and scientists.

One is entitled to ask what the next epoch will bring. The dangers are that flexibility will give way to rigidity and fossilization, that too much detail will be embodied in the scheme and that the relentless urge on the part of some will lead to latinized 'species' names. That is harmless in itself but conducive to pedantry and to false values being placed on the names rather than on the viruses themselves. So far, the taxonomy of viruses has escaped hierarchical treatment; perhaps the ultimate danger is that this may one day be introduced.

Spreading the gospel
When a discipline is formative, it is usual to find personal communication, such as the correspondence between Jenner and Hunter, the most effective means of exchanging ideas and findings. Once it is established, organization becomes necessary. There develop

scientific societies which run formal meetings, publish journals and provide other means of communication. Since we are celebrating the one hundredth scientific meeting of the Society for General Microbiology a digression seems appropriate which exemplifies the process by describing how this Society has handled this in Britain.

Soon after the war, the Society for General Microbiology was established. Virologists were then considered as specialized eccentric microbiologists and had been accustomed to give papers at meetings organized by the Society, as well as others. The Society provided quite well for the virologists and even held and published symposia on virus topics e.g., The Nature of Virus Multiplication in 1952 and Virus Growth and Variation in 1959.

Virology had by then reached the fourth age and the tenor of the meetings seemed stodgy and over formal. Virologists felt diluted out by bacteriologists. An informal group grew up surrounding Sir Christopher Andrewes, and used to gather together more or less annually to listen to what was new, present their findings and to discuss them. From time to time a distinguished overseas virologist would speak. The first such meeting that I recall was held at Sheffield; about 15 people were present. Within five years, the size of this group had increased dramatically and had about 200 regular attenders. It was no longer feasible to continue without an organized structure to cope with the secretarial work and, for a few years, Sir Christopher and Alick Isaacs managed to continue running such meetings by effectively parasitizing the secretariats of several societies.

In 1960 it became obvious that this informal virus group had to decide whether to form a new independent organization or to assimilate as a group into an established scientific society. The latter course was chosen but opinion was divided on which Society was most appropriate. There was heated debate on this matter but it seemed to most of us in 1961 that the Society for General Microbiology best served our discipline. However, up to that time, the committee of that Society had been totally opposed to a group structure. Fortunately for the virologists the microbial systematists also wished to form a group; accordingly a small *ad hoc* committee was set up to formulate such a structure. This was eventually approved and the virus group was established in 1962. It had a committee comprising Sir Christopher Andrewes, B. D. Harrison, A. Isaacs, D. A. J. Tyrrell and M. G. P. Stoker. At its first meeting on 11 July 1961, it was decided to set a pattern of one annual meeting which should take the form of a symposium on a basic subject and a symposium on an applied one. The first scientific meeting of the Virus Group took place on 11 April 1963 and indeed comprised two symposia, on viral nucleic acids and the Coxsackie viruses.

It is interesting to note that the establishment of the group structure proved to be a turning point in the fortunes of the Society and is now its predominant feature; there are 10 groups reflecting the diversity within the Society which enables it to perform its prime function of furthering the microbiological sciences in the UK.

From its earliest days the Virus Group recognized the need to cover all aspects of virology and it has always insisted that its committee should reflect this. Thus it laid down that the membership should comprise medical virologists, at least one veterinary virologist and at least one plant virologist. Curiously at this time the bacteriophages did not seem to count; the first inclusion of a phage worker was not until 1968. The group has continued to flourish holding symposia on many topics, organizing open-paper meetings and, from 1973, workshop sessions to cater for superspecialized virus topics.

A curious repetition of events occurred which led to the recent formation of a second virological group within the Society. Clinical virologists began to notice the change from orthovirology to metavirology. There had been a large importation of viropractors into the virus group. The meetings no longer seemed to be directed toward virological ends and certainly not towards diagnostic, epidemiological and prophylactic objectives. Accordingly, a small informal group, this time of clinical virologists, formed itself around A. P. Waterson and held annual meetings in London. Once again, this group became too large to run on an informal basis and discussions were held as to whether a new Society of Clinical Virologists should be formed or whether the Society for General Microbiology should establish a new Clinical Virology Group. The record shows that the existing Virus Group within the Society had been much concerned over this matter and had gone out of its way to elect clinical virologists on to its committee and to organize special workshops and symposia that would please them. Eventually, with the strong support of the Virus Group, the Council of the Society approved the establishment of a second virological group to cater for this special interest.

Now that there are two virus groups within the Society we note that the interests of the two groups are converging. For instance, the molecular virologists are becoming increasingly interested in diagnosis, epidemiology, prophylaxis and chemotherapy and the clinical virologists, willy nilly, are interesting themselves in molecular approaches. The dual group system nevertheless suits the present situation admirably. Though there is a gratifying overlap of interests and joint meetings are increasingly common there still remain the extremes of interest which can be catered for separately.

By 1965 it had become plain that there was a need to increase the opportunities to publish (especially fundamental) virology. The records show that the Virus Group proposed that the Society publish a new virological journal. This was supported by the Council. The Journal of General Virology first appeared in 1967. We did not know that similar thoughts were stimulating the introduction of the Journal of Virology by the American Society of Microbiology. In the event, the two journals appeared more or less simultaneously in 1967. It is symptomatic of the strong growth of virology that both have prospered over the years.

Since it began, the Society has organized many meetings for virologists in the UK, produced a number of its symposium volumes on virological topics, backed a virological journal and yet has allowed virologists considerable independence of action. For this and the wise decisions taken in 1961 we must be grateful.

From time to time the question reappears: why not a separate British Society for Virology? There are two arguments against this at present. First, there are not enough virologists in this country (unlike Japan and the USA) to sustain the advantages that virologists now enjoy and secondly the general nature of the Society for General Microbiology allows easy relationships with other groups. Joint meetings between groups are encouraged and are frequently useful.

In saluting the Society for General Microbiology, which has served virology well in Britain, we must not forget the important work of other numerous scientific societies all over the world.

Something must be said about International Virology. Up until 1966, virologists were provided for by meetings held from time to time in different countries under the auspices of the International

Association of Microbiological Societies (IAMS). These meetings steadily grew in size and dullness and virologists became more and more disinterested. In 1966, four mutineers (ringleader: J. Melnick) decided to run an International Virology meeting independently. Though restricted mainly to viruses of vertebrates, the meeting (held in Helsinki) was successful and followed by other more comprehensive programmes in Budapest, Madrid, Amsterdam and Strasbourg. The last three were once again sponsored by IAMS (now IUMS).

Several points have become clear. First, it is important for the health of virology and for the interests of virologists that meetings be held separately from others dealing with microbiological disciplines, in order to attain the maximum chance of personal encounter. Second, the breadth of virology is great enough to sustain varied combinations of interest. Indeed, considerable ingenuity is required to ensure mutually fruitful exchanges. Third, the heterogeneity of virologists (and viropractors) is such that it has often been questioned whether it is wise to continue with such large congresses. So far, I believe they have been a powerful force of good communication. I hope that they remain so.

There is now a movement to create an independent International Virological Organization for the promotion of virology just as, for example, the immunologists have done. I see little sense in this since it would add yet another administrative compartment to international bumbledom with no advantage for the science.

Since the time of Johannes Gutenberg we have relied principally upon the printed word for the broadcast of scientific information. This is a cumbersome enough method but, what is worse, it has led to the propagation of a mass of unwanted information which clutters libraries all over the world. Waksman (1980) has commented on similar informational overload in immunology. This he ascribes largely to the scientists' need to compete for funds and status, our measurement of merit by frequency of publication and the enthusiasm of commercial publishers. Waksman's solution is to publish abstracts only and use some electronic retrieval system for the data, to publish papers without peer review and to boycott commercially motivated journals. This was startling enough to provoke discussion ranging from the iniquity of page charges to the indispensability of referees, from the potential of electronics to the prestige of the quality journal. This very obvious bottleneck in the transmission of information will have to be resolved and, provided that vested interests are controlled, it can be. There are however two more bottlenecks to consider. The limitation imposed by the amount we can absorb can be overcome by the use of memory banks and electronic means of retrieval and correlation. But machines can only do what they are told; instruction must depend upon our understanding and decisions about what priorities we wish to address. This last feature appears to form the bottleneck that may obstruct future advance.

CONCLUSIONS

Virology has contributed much to science and to mankind. By restriction of matter under examination it has enabled relatively simple genetic systems to be analysed in a way that would have been impossible some years ago. It developed a rigour which has provided a model for experimental biology as a whole. Specifically it has provided tools which have become essential for genetic engineering. On the practical side it has served us well in the control of yellow fever and poliomyelitis and it has eliminated smallpox. It has revealed new ways of developing vaccines (Bittle et al., 1982; Smith, Mackett & Moss, 1983). It has opened up approaches in the control of neoplasia. It has continually lured us on with utopian promises of chemotherapy and interferon, and it has opened up new vistas for biology (such as reverse transcription, jumping genes and oncogenes). The science is technically orientated and its development has depended greatly on borrowed and invented techniques and it has contributed techniques to other branches of virology.

Ten or fifteen years ago virology appeared to have a straightforward structure but this has now reached a critical stage of complexity. Formerly we recognized the central concept of virus holding the discipline orthovirology together, with numerous budding specialities thrusting up from the main stem. Now orthovirology has strengthened little in comparison with the accelerating development of metavirology and paravirology. These subdisciplines are taking on an increasingly independent character. This is a matter for rejoicing since the new disciplines are accomplishing much. One might well question whether it is sensible to continue using virology as the banner under which the new disciplines should be assembled. Their scope now exceeds that of orthovirology.

We have noticed the existence of epivirology. This somewhat despised handmaiden of virology has been developed with enthusiasm over the past fifteen years and looks as if it will continue for some time. It seems to me that this Cinderella is at present the principal influence binding virology as a whole together. It is as though virology has turned itself inside out; the main supporting element, formerly an endoskeleton, is now an exoskeleton. One is reminded not so much of an invertebrate as of a tortoise for, while this animal depends for its rigidity on its shell, the shell itself is still centred on the primitive vertebral system. One is tempted to speculate how long such a structure can endure because shells are restrictive; perhaps the vital material will burst out and, quatermass-fashion, overtake us all. Obviously the shell will have to expand if it is to control and still house its contents. Tortoises manage to do this, expanding their shells *pari passu* with their contents. They grow to enormous size and endure for many years, but this is at the expense of becoming more and more cumbersome. The fashioning of the shell, epivirology, is in our hands and decisions have to be made whether to preserve the intactness of virology by increasing the elasticity and scope of the epivirology or whether to allow its disruption so that the vital matter can reform in different packages. One thing is certain; if epivirology grows stronger and more rigid it will strangle our discipline. Virology will wither to mere oblivion, sans teeth, sans eyes, sans taste, sans everything.

I wish to acknowledge the usefulness of the book by A. P. Waterson and L. Wilkinson whose researches were of greatest value. I am grateful to Professor S. R. Elsden and Dr B. W. J. Mahy who provided historical details of the Society for General Microbiology and the Virus Group. I am also grateful to Miss M. F. Layton for the typing and for deciphering my handwriting.

REFERENCES

ADA, G. L. & PERRY, B. T. (1956). Influenza virus nucleic acid: relationship between biological characteristics of the virus particle and properties of the nucleic acid. *Journal of General Microbiology*, **14**, 623–33.

ANDREWES, C. H. & HORSTMANN, D. M. (1949). The susceptibility of viruses to ethyl ether. *Journal of General Microbiology*, **3**, 290–7.

ANDREWES, F. W. (1930). The nomenclature and classification of micro-organisms. IN *A System of Bacteriology*, vol. 1, ed. P. Fildes & J. C. G. Ledingham, pp. 292–310. London: H.M. Stationary Office.

BAIL, O. (1925). Der Kollistamm 88 von Gildemeister und Heizberg. *Medizinische Klink (München)*, **21**, 1271–3.

BALTIMORE, D. (1970). RNA-dependent DNA polymerase in virions of RNA tumour viruses. *Nature*, **226**, 1209–11.

BALTIMORE, D. (1971). Expression of animal virus genomes. *Bacteriological Reviews*, **35**, 235–41.

BAWDEN, F. C., PIRIE, N. W., BERNAL, J. D. & FANKUCHEN, I. (1936). Liquid crystalline substances from virus infected plants. *Nature*, **138**, 1051–2.

BEIJERINCK, M. W. (1898). Ueber ein *contagium vivum fluidum* als Ursache de Fleckenkrankheit der Tabaksblätter. *Verhandelingen der Koninklyke akademie van Wetenschappen, Amsterdam*, **65**, (2), 3–21.

BERNAL, J. D. & FANKUCHEN, I. (1941). X-ray and crystallographic studies of plant virus preparations. *Journal of General Physiology*, **25**, 111–46.

BITTLE, J. L., HOUGHTEN, R. A., ALEXANDER, H., SHINNICK, T. M., SUTCLIFFE, J. G., LERNER, R. A., ROWLANDS, D. J. & BROWN, F. (1982). Protection against foot-and-mouth disease by immunization with a chemically synthesized peptide predicted from the viral nucleotide sequence. *Nature*, **298**, 30–3.

BORDET, J. (1925). Le problème de l'autolyse microbienne transmissible ou du bacteriophage. *Annales de l'Institut Pasteur, Paris*, **39**, 711–63.

BRENNER, S. & HORNE, R. W. (1959). A negative-staining method for high resolution electron microscopy of viruses. *Biochimica, Biophysica Acta*, **34**, 103–10.

BUIST, J. B. (1886). Vaccinia and variola: a study of their life history. *Proceedings of the Royal Society of Edinburgh*, **13**, 603–20.

BURNET, SIR M. (1945). *Virus as Organism*. Cambridge (Mass.): Harvard University Press.

BURNET, F. M. (1955). *Principles of Animal Virology*. New York: Academic Press.

BURNET, F. M. & McKIE, M. (1929). Observations on a permanently lysogenic strain of *B. enteritidis gaertner*. *Australian Journal of Experimental Biology and Medical Science*, **6**, 277–84.

CASPAR, D. L. D. & KLUG, A. (1962). Physical principles in the construction of regular viruses. *Cold Spring Harbor Symposia on Quantitative Biology*, **27**, 1–24.

COLTER, J. S., BIRD, H. H. & BROWN, R. A. (1957). Infectivity of ribonucleic acid from Ehrlich ascites tumour cells infected with Mengo encephalitis. *Nature*, **179**, 859–60.

COOPER, P. D. (1961). Chemical basis for the classification of animal viruses. *Nature*, **190**, 302–5.

COWDRY, E. V. (1934). The problem of intranuclear inclusions in virus diseases. *Archives of Pathology*, **18**, 527–42.

CRICK, F. H. C. & WATSON, J. D. (1956). Structure of small viruses. *Nature*, **177**, 473–5.

DARLINGTON, C. D. (1948). The plasmagene theory of the origin of cancer. *British Journal of Cancer*, **2**, 118–26.

D'HERELLE, F. (1921). *Le Bacteriophage*. Paris: Masson.

DiMAYORCA, G. A., EDDY, B. E., STEWART, S. E., HUNTER, W. S., FRIEND, C. & BENDICH, A. (1959). Isolation of infectious deoxyribonucleic acid from SE polyoma-infected tissue cultures. *Proceedings of the National Academy of Sciences (USA)*, **45**, 1805–8.

DOERR, R. (1938). Die Entwicklung der Virusforschung. In *Handbuck der Virusforschung*, vol. 1, ed. R. Doerr & C. Hallauer. Vienna: Verlag von Julius Springer.

DULBECCO, R. (1952). Production of plaques in monolayer tissue cultures by single particles of an animal virus. *Proceedings of the National Academy of Sciences (USA)*, **38**, 747–52.

DULBECCO, R. (1963). Transformation of cells *in vitro* by viruses. *Science*, **142**, 932–36.

ELLERMAN, V. & BANG, O. (1908). Experimentelle Leukämie bei Huhnern. *Zentralblatt für Bakteriologie, Parasitenkunde, Infectionskrankheiten und Hygiene., Abt. 1. Orig.*, **46**, 595–609.

ELLIS, E. L. & DELBRÜCK, M. (1939). The growth of bacteriophage. *Journal of General Physiology*, **22**, 365–84.

FRAENKEL-CONRAT, H. & WILLIAMS, R. C. (1955). Reconstitution of active tobacco mosaic virus from its inactive protein and nucleic acid components. *Proceedings of the National Academy of Sciences (USA)*, **41**, 690–8.

FRANKLIN, R. E., CASPAR, D. L. D. & KLUG, A. (1959). The structure of viruses as determined by X-ray diffraction. In *Plant Pathology: Problems and Progress, 1908–58*, ed. C. S. Hotton, G. W. Fischer & R. W. Fulton, pp. 447–61. Madison (Wisconsin): University of Wisconsin Press.

GARDNER, A. D. (1931). *Microbes and Ultramicrobes*, ed. G. R. de Beer. London: Methuen & Co. Ltd.

GIBBS, A. J., HARRISON, B. D., WATSON, D. H. & WILDY, P. (1966). What's in a virus name? *Nature*, **209**, 450–4.

GIERER, A. & SCHRAMM, G. (1956). Infectivity of ribonucleic acid from tobacco mosaic virus. *Nature*, **177**, 702–3.

GRATIA, A. (1921). Preliminary report on a staphylococcus bacteriophage. *Proceedings of the Society for Experimental Biology and Medicine*, **18**, 217–19.

GREEN, R. G. (1935). On the nature of filtrable viruses. *Science*, **82**, 443–5.

GUANIERI, G. (1892). Richerche sulla patogenesi ed etiologia dell'infezioni vaccinica e vaiolosa. *Archivio Scientifico di Medicina Veterinaria*, **16**, 403–24.

HERSHEY, A. D. & CHASE, M. (1952). Independent functions of viral protein and nucleic acid in growth of bacteriophage. *Journal of General Physiology*, **36**, 39–56.

HOAGLAND, C. L., LAVIN, G. I., SMADEL, J. E. & RIVERS, T. M. (1940). Constituents of elementary bodies of vaccinia. II. Properties of nucleic acid obtained from vaccine virus. *Journal of Experimental Medicine*, **72**, 139–47.

HORNE, R. W. & WILDY, P. (1962). Symmetry in virus architecture. *Virology*, **15**, 348–73.

IWANOWSKY, D. I. (1903). Uber die Mosaikkrankheit der Tabakspflanze. *Zeitschrift für Pflanzenkrankheiten*, **13**, 1–14.

JACOB, F. & MONOD, J. (1961). Genetic regulatory mechanisms in the synthesis of protein. *Journal of Molecular Biology*, **3**, 318–56.

LAIDLAW, SIR P. P. (1938). *Virus Diseases and Viruses*. Cambridge University Press.

LIPSCHÜTZ, P. B. (1909). Ueber Mikrosckopisch sichtbar, filtrierbare virusarten (Strongyloplasmen). *Zentralblatt für Bakteriologie, Infectionskrankheiten und Hygiene, Abt. I Orig.*, **48**, 77–90.

LOEFFLER, F. & FROSCH, P. (1898). Berichte der Kommision zur Erforschung der Maul-und Klauenseuche bei dem Institüt für Infectionskrankheiten in Berlin. *Zentralblatt für Bakteriologie, Parasitenkunde, Infectionskrankheiten und Hygiene, Abt. 1, Orig.*, **23**, 371–91.

LURIA, S. E. (1950). Bacteriophage: an essay in virus reproduction. *Science*, **111**, 507–11.

LWOFF, A. (1957). The concept of virus. *Journal of General Microbiology*, **17**, 239–53.

LWOFF, A. (1966). The prophage and I. In *Phage and the Origins of Molecular Biology*, ed. J. Cairns, G. S. Stent & J. D. Watson, pp. 88–99. Cold Spring Harbor: Cold Spring Harbor Laboratory.

Lwoff, A., Horne, R. W. & Tournier, P. (1962). A system of viruses. *Cold Spring Harbor Symposia on Quantitative Biology*, **27**, 51–5.

Lwoff, A., Siminovitch, L. & Kjeldigaard, N. (1950). Introduction de la production de bacteriophage chez une bactérie lysogène. *Annales de l'Institut Pasteur, Paris*, **79**, 815–58.

Markham, R. M., Matthews, R. E. F. & Smith, K. M. (1948). Specific crystalline protein and nucleoprotein from a plant virus having insect vectors. *Nature*, **162**, 88–90.

Matthews, R. E. F. (1983). The history of viral taxonomy. In *A Critical Appraisal of Viral Taxonomy*, ed. R. E. F. Matthews. Florida, C.R.C. Press.

M'Fadyean, J. (1908). The ultravisible viruses. *Journal of Comparative Pathology & Therapeutics*, **21**, 232–42.

Müller, H. J. (1922). Variation due to change in the individual gene. *American Naturalist*, **56**, 32–50.

Rous, P. (1911). Transmission of a malignant new growth by means of a cell free filtrate. *Journal of the American Medical Association*, **56**, 198.

Ruska, H. (1940). Die Sichtbarmachung der Bakteriophagen Lyse in Übermikroscop. *Naturwissenschaften*, **28**, 45–6.

Schlesinger, M. (1935). The Fuelgen reaction of the bacteriophage substance. *Nature*, **138**, 508–9.

Smith, A. J. & Weidman, F. D. (1910/11). Infection of a stillborn infant by an amebiform protozoan (*Entamoeba mortinatalium* N.S.) *Medical Bulletin of the University of Pennsylvania*, **23**, 285–98.

Smith, G. L., Mackett, M. & Moss, B. (1983). Infectious vaccinia virus recombinants that express hepatitis B virus surface antigens. *Nature*, **302**, 490–5.

Stanley, W. M. (1935). Isolation of a crystalline protein possessing the properties of tobacco-mosaic virus. *Science*, **81**, 644–54.

Temin, H. M. & Mitzutani, S. (1970). RNA-dependent DNA polymerase in virions of Rous sarcoma virus. *Nature*, **226**, 1211–13.

Twort, F. W. (1915). An investigation on the nature of ultramicroscope virus. *Lancet*, ii, 1241–3.

von Prowazek, S. (1907). Chlamydozoa. *Archiv für Protistenkunde*. **10**, 336–58.

Waksman, B. H. (1980). Information overload in immunology: possible solutions to the problems of excessive publication. *Journal of Immunology*, **124**, 1009–15.

Waterson, A. P. & Wilkinson, L. (1978). *An Introduction to the History of Virology*. Cambridge University Press.

Weiss, R. (1973). Inherited oncornavirus genes and their activation. In *Viral Replication and Cancer*, ed. J. L. Melnick, S. Ochoa & J. Oró, pp. 255–67. Barcelona: S. A. Calabria.

Wildy, P. (1962). Classifying viruses at higher levels: symmetry and structure of virus particles as criteria. *Symposium of the Society for General Microbiology*, **12**, 145–63.

Wildy, P. (1971). *Classification and Nomenclature of Viruses. Monographs of Virology*, vol. 5, ed. J. L. Melnick. Basel: Karger.

Williams, R. C. & Smith, K. M. (1958). The polyhedral form of the Tipula iridescent virus. *Biochemica, Biophysica Acta*, **28**, 464–9.

STRUCTURES OF VIRUSES

STEPHEN C. HARRISON

Department of Biochemistry and Molecular Biology, Harvard University, Cambridge, MA, USA

A virus particle is a structure for transferring nucleic acid from one cell to another. The nucleic acid may be either RNA or DNA and, in both cases, particles of varying complexity are found. Observed structures reflect requirements for efficient and accurate assembly, for exit and re-entry, and for correctly localized disassembly. There is an important distinction between viruses with lipid-bilayer membranes ('enveloped' viruses) and those without membranes. The distinction corresponds to differences in the mechanism of exit and re-entry. Our knowledge of virus structures in 1984 extends to nearly atomic detail in a few cases, where X-ray crystallography has permitted such a 'close-up' view, and to the level of electron microscopic visualization in most others. Questions about regulation of assembly, antigenicity, membrane fusion, and nucleic acid packaging can be couched, at least in part, in terms of three-dimensional molecular organization. This paper describes the molecular architecture of non-enveloped and of enveloped viruses, as examples of regulated assembly and as structural 'samples' of cellular compartments.

THE NUCLEIC-ACID/PROTEIN PACKAGE

There are two basic ways in which nucleic acid is packaged: into rod-like or filamentous structures and into roughly spherical ('isometric') ones (Crick & Watson, 1956). In rod-like or filamentous particles, protein subunits bind in a periodic way along the nucleic acid, winding it into a helical path. The best-known examples of such structures are plant viruses like tobacco mosaic virus (TMV). The RNA of myxo- and paramyxoviruses and of rhabdoviruses is also incorporated into filamentous structures by association with subunits of the nucleoprotein. These 'nucleocapsids' are coiled inside the viral membrane. In isometric particles, nucleic acid is condensed within the virus in a manner geometrically independent of the organization of the shell. For example, the double-

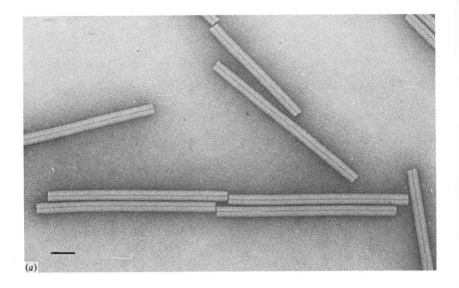

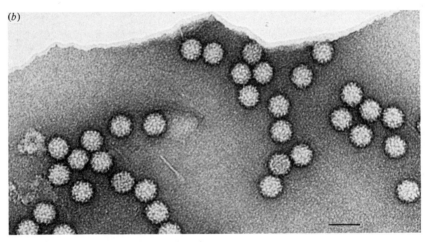

Fig. 1. Electron micrographs of (*a*) tobacco mosaic virus and (*b*) tomato bushy stunt virus, illustrating helically and icosahedrally symmetric viruses respectively. Bar = 500 Å. (Courtesy John T. Finch.)

stranded, circular DNA of papova viruses forms nucleosomes by binding cellular histones (Fey & Hirt, 1975; Griffith, 1975). In small plant viruses, the links from protein shell to RNA are sufficiently flexible that few constraints are placed on the precise way in which RNA folds.

The structures of both rod-like and isometric particles reveal biochemical properties needed for efficient and error-free assembly.

One important general feature arises from the condition that viral substructures assemble from many identical copies of one or a few kinds of protein. Repeated occurrence of similar protein–protein interfaces leads to a symmetrical arrangement of the subunits. Otherwise, the same sets of amino acid side chains would have different patterns of non-covalent bonding in different places, and ambiguities in assembly would result. Thus rod-like and filamentous viruses and the filamentous nucleocapsids of many enveloped viruses have helical symmetry, and simple isometric viruses have icosahedral symmetry (Fig. 1). If homologous sets of side chains *do* bond differently in different places, there are a limited set of alternatives, and some further switch must be present as described below.

A second significant property of larger and more complex virus structures is that particles are constructed from distinct subassemblies. The most dramatic illustration is found in the complex bacteriophages such as T4 (Fig. 2). Heads, tails and tail fibres assemble independently in pathways, each of which has a defined, sequential character (see Casjens & King, 1975). Scaffold or template structures can occur (e.g., the head core): these are required for correct assembly but are then deleted to make way for subsequent steps. Processing (e.g., cleavage) by non-structural proteins is also involved at some stages. Within a pathway, a particular intermediate serves to nucleate addition of the next component. For example, tail core subunits in T4 phage do not associate with each other unless assembly is initiated on a baseplate. Likewise, budding of Sindbis or Semliki Forest virus occurs only around pre-assembled cores (Acheson & Tamm, 1967).

A third important property of virus particles is that incorporation of viral nucleic acid is specific but independent of most of its base sequence. In the case of helical structures, each subunit may interact with a definite number of nucleotides. This is independent of the sequence of nucleotides except that there will be preferential binding to a site for initiating assembly. In the case of isometric particles, condensation of the nucleic acid is independent of details of its structure, since there are many cases where various sequences can be packaged by the same shell (Jaspars, 1974). For single-stranded RNA viruses, it therefore appears that no definite secondary or tertiary fold is needed for the RNA, aside from the restriction that it fit within the shell. What determines specific incorporation has been studied in only a few cases. The mechanism

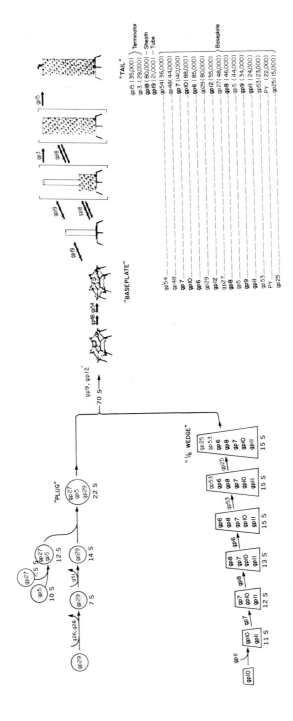

Fig. 2. Pathway of tail assembly in bacteriophage T4. Individual gene products are indicated by gp18, etc. Formation of the baseplate must precede assembly of the core, which in turn nucleates addition of the sheath. Baseplate assembly is itself a complex process: eight different polypeptides interact in a sequential manner to form the 'wedges', six of which add to an independently formed 'plug'. (Kikuchi & King, 1975.)

RNA

Protein
subunit

Fig. 3. Diagram of the TMV structure, showing 49 subunits in three turns. The pitch of the basic helix is 23 Å. The surface lattice is indicated in the lower part of the drawing. (Klug & Caspar, 1960.)

depends in part on whether nucleic acid is packed into a pre-formed shell (as in DNA phages, Casjens & King, 1975) or whether protein and nucleic acid co-assemble (as in many plant viruses, Bancroft, 1970).

Helical structures

The symmetry of a helical arrangement is conveniently described by the number of units per turn (u) and the axial rise per unit (p). The helical pitch (P) is equal to u times p. The diagram of TMV in Fig. 3 illustrates that a structure with helical symmetry can also be described by a 'surface lattice'. One can imagine generating the object from a planar net by rolling the array into a tube. Helical structures permit coordinated interwinding of single-stranded nucleic acid and protein and hence repeated interactions. The coat proteins or nucleocapsid proteins of such viruses must therefore disrupt local nucleic acid secondary structure. An exception to the presence of regular interactions between protein and nucleic acid in helical structures appears to be found in the filamentous phages, where certain strains have non-integral numbers of bases per protein subunit and where the circular single-stranded DNA cannot conform in helical symmetry to the polar symmetry of the coat (Day & Wiseman, 1978). These structures probably can be thought of as extremely elongated shells with somewhat variable DNA/protein contacts, rather than as a single nucleoprotein strand. The transformation documented by Griffith, Manning & Dunn (1981) also supports this view.

Icosahedral symmetry

Closed-shell virus particles usually have structures based on icosahedral symmetry (Caspar & Klug, 1962). This is the most efficient possible arrangement in the sense that it uses the smallest unit to build a shell of fixed size. There are exactly 60 identical elements in the surface of such a structure, and theory shows that there are no symmetries for a closed shell having more than 60 units. Most viruses have genomes larger than can be contained within 60 subunits of reasonable molecular weight, and each of the icosahedrally equivalent elements is then composed of a number of protein chains. These may be chemically distinct, as in poliovirus, where there are 60 copies each of four different species (Baltimore, 1969). In many cases, however, the viral shell is composed of some multiple of 60 chemically identical structural units, with each such unit consisting of one or a few distinct polypeptides.

The structural units thus need some flexibility in the way they interact: in a structure with $60N$ units there will be N different packing environments. This creates a non-trivial problem for regulation of assembly, since each unit must 'know' which mode of interaction to adopt as it enters the growing shell. The description of tomato bushy stunt virus (TBSV) and related structures below shows a mechanism for controlled switching among the three different modes in a 180 subunit structure. Larger particles (such as P22 and λ phage heads) appear to need additional proteins to accomplish an analogous function; these proteins are then discarded (or in P22, re-used) once the shell is formed (Casjens & King, 1975).

Most of the known icosahedral shells with $60N$ units also conform to one of a set of structures, first recognized by Caspar & Klug

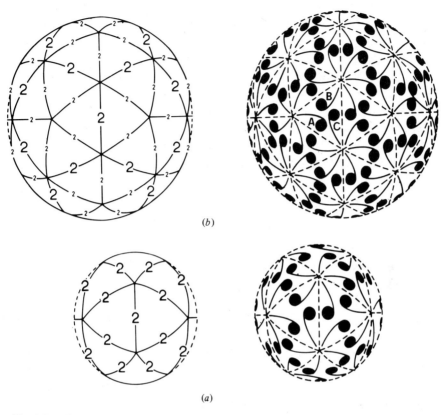

Fig. 4. Icosahedral surface lattices containing commas as structural units. (a) $T = 1$ structure, with 60 commas; positions of dyads indicated by '2' in left hand surface lattice diagram. (b) $T = 3$ structure, with 180 commas; twofold axes of the entire structure ('strict' dyads) are shown by large '2', twofold axes of only local validity ('local' dyads) by small '2'. Note that the *local* packing of units is the same in all cases, but that in the 180 unit structure there are three distinct environments for the commas (denoted A, B and C).

(1962), in which the differences among the N distinct packing environments are minimized (Fig. 4). This reduces the flexibility or adaptability required of the structural unit. Inter-subunit bonding in these shells has been called 'quasi-equivalent', to emphasize the similarity of the N symmetrically distinct locations (Caspar & Klug, 1962). Indeed, in the case of TBSV and relatives, it is clear that subunit domains are essentially invariant and that internal flexibility is built into hinges or linkers between domains (Harrison *et al.*, 1978). Moreover, the non-covalent contacts made by these domains are quite similar in all three packing modes, with a switch mechanism (mentioned above) for unambiguous selection (Olson, Bricogne & Harrison, 1983).

Designs   permitting   quasi-equivalent   bonding   have
$N = T = h^2 + hk + k^2$, where h and k are any integers (Caspar &
Klug, 1962). This relationship can be derived by sub-triangulating
the icosahedral surface lattice or by examining ways of folding up a
p6 net into a closed shell, and T has been called a 'triangulation
number'. Structures with $T = 1$ (poliovirus) and $T = 3$ (TBSV and
many other plant viruses) are known. Sindbis and Semliki Forest
virus probably have a $T = 4$ structure (von Bonsdorff & Harrison,
1975, 1978), and P22 phage head may have a $T = 7$ structure
(Casjens, 1979). Viruses with larger shells have different kinds of
subunits at the fivefold vertices and deviate in other ways from the
simple geometries derived from a p6 net. Such complex structures
need an organized assembly pathway, as in the assembly of P22
bacteriophage heads (Fig. 21), with one set of proteins acting as
framework or scaffold to position another. Thus, the assumptions
that go into deriving 'permitted' triangulation numbers are not
obeyed in larger particles. The sub-triangulations of an icosahedral
lattice do, however, indicate the possible ways of obtaining close
packing in the surface, whatever the symmetry of local interactions.
Such close packing probably explains why adenovirions appear to be
$T = 25$ structures, or papova viruses $T = 7$, even when actual local
symmetry does not turn out to correspond to a folded p6 net
(Franklin et al., 1971; Crowther & Franklin, 1972; Rayment et al.,
1982; Baker, Caspar & Murakami, 1983).

VIRAL MEMBRANES

Most enveloped viruses acquire their membrane by budding
through the cell surface (Fig. 5; cf. Compans & Klenk, 1979). The
viral membrane is therefore a differentiated segment of the plasma
membrane, and its lipid composition accurately reflects this origin
(Choppin et al., 1971). The presence of a bilayer introduces the
requirement for recognition of components on opposite sides of the
membrane. The viruses take advantage of cellular compartmentalization mechanisms to place viral glycoproteins on the cell surface
and internal proteins in the cytoplasm, and budding consists of
matching internal and external structures.

Enveloped viruses generally contain one or more surface glycoproteins on the outside of the bilayer, a matrix (M) protein on
the inside, and one or more proteins complexed with the nucleic

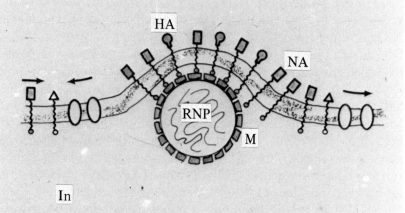

Fig. 5. Diagram showing pathway of assembly of an enveloped virus, using influenza as an example. RNA associates with N-protein to form RNP structures. Budding is a concerted interaction of RNP, of the so-called M protein that underlies the membrane, and of tails of glycoproteins (HA, NA). Lipids are incorporated as a 'random sample' of the bilayer. Other cellular glycoproteins (anchored triangular heads) and embedded membrane proteins (ovals) are excluded. (Courtesy D. C. Wiley.)

acid. In simple enveloped RNA viruses, the protein in association with RNA is usually denoted N. In Sindbis and Semliki Forest viruses and related structures, the functions of M and N are served by a single protein, conventionally called C (for core). Bunya-viruses lack M completely, and the shape of the particle appears to be determined by the surface glycoproteins (von Bonsdorff & Petterson, 1975). More complex internal structures are present in the larger enveloped RNA viruses, such as retroviruses. Glycopro-teins are anchored by hydrophobic tails that span the bilayer, usually terminating in small 'cytoplasmic domains'. It is generally believed that the anchor is α-helical where it passes through the lipid, and indeed the corresponding hydrophobic sequences are almost universally about 25 residues long, bounded by charged (usually basic) residues at both ends (von Heijne, 1981). An α-helical conformation for a membrane-embedded polypeptide allows all main-chain hydrogen-bonding (H-bonding) functions to be satisfied. Side chains containing more than one H-bonding group (either charged or uncharged) are not found in these sequences, and the frequent appearance of hydroxyl-containing side chains may well be due to their capacity to H-bond back to main-chain carbonyl groups. Specificity in budding probably consists of recognition of the small cytoplasmic domain by appropriate sites on subunits of M.

The organization of a viral membrane is important not only for budding but also for attachment to a cell surface and for fusion with a membrane of the new host. These are functions of the external portions of one or more of the glycoproteins.

EXAMPLES OF VIRUS STRUCTURES

Tobacco mosaic virus

The rod-like TMV particle (Fig. 3) consists of about 2130 subunits (158 amino acid residues) arranged in a helix of $16\frac{1}{3}$ units per turn (for general reviews, see Klug, 1979; Holmes, 1980). The axial rise is 23 Å and the rods are therefore about 3000 Å long by 180 Å in diameter. RNA winds coaxially with the protein, with three nuc-leotides bound to each subunit. The virus can be dissociated by several methods into protein and RNA. Various reassociated forms of the purified protein have been characterized as a function of pH, ionic strength and other variables (see below), and crystals of a

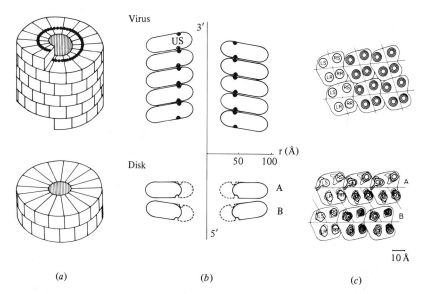

Fig. 6. Packing of subunits in the TMV protein disk (lower part) and in the virus (upper part). (a) The disk has 34 subunits in two rings of 17; the virus, $16\frac{1}{3}$ subunits per helical turn. RNA is sandwiched between helically arranged subunits in the virus, three nucleotides per subunit. (b) Subunits in the A and B layers of the disk have different tilts, both of which are different from the tilt of subunits in the virus. The innermost part of the subunit is disordered on the disk (dashed outline), perhaps to facilitate RNA incorporation. (c) Cross-section through subunits in the disk and in the virus, viewed in a radial direction from outside toward the centre. The disk sections are representations from an electron-density map; the virus sections are schematic. The major part of each subunit consists of four α-helices, (symbolized by LS, RS, LR, RR for left and right 'slewed' and 'radial' helices respectively), running in a roughly radial direction. In the transition from disk to helix, subunits slide over each other by about 10 Å. Because of the change in tilt, lateral contacts are bent somewhat, but not completely shifted like the axial bonds. (From Klug, 1979.)

particularly important form, the so-called disk, have allowed molecular details of the subunit to be visualized (Bloomer *et al*, 1978).

The disk is a 34 subunit, two-layer structure; subunits in each layer have the same axial orientation, and the interactions between them are closely related to the interactions present in the virus itself (Fig. 6). A transition from disk to helix, which can indeed occur under conditions of rapid pH drop (Klug & Durham, 1971), might be pictured as follows. Break the two rings at one point; dislocate to form a 'lockwasher'; tighten up the helix so formed from 17 to $16\frac{1}{3}$ units per turn. The tilt of the subunits also changes, becoming uniformly $-10°$. The important thing to notice is that lateral interactions do not change significantly, while axial interactions change completely. Moreover, a loop of polypeptide chain that

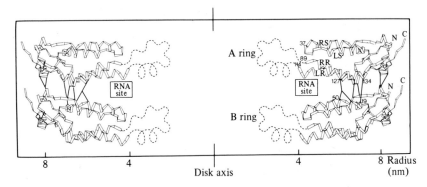

Fig. 7. Section through a TMV disk, as in Fig. 6(*b*), but with a detailed skeletal representation of the polypeptide chain. The abbreviations LS, RS, LR, RR are defined in Fig. 6; N and C represent amino and carboxy termini; some residue numbers are shown on one pair of subunits. (From Klug, 1979.)

forms the innermost part of the subunit is disordered in the disk and ordered in the virus. Since the loop acts as a clamp for RNA in the virus (see Fig. 7), the disorder–order transition is important. A model for the binding of an RNA trinucleotide in TMV has been proposed by Stubbs, Warren & Holmes (1977): the polynucleotide backbone threads between subunits at a radius of about 40 Å, with the bases extended in such a way that they surround the left radial (LR) helix of the upper subunit on three sides. Positively charged residues are in position to neutralize RNA phosphates.

Infectious TMV can be reassembled under physiological conditions from separated protein and RNA (Fraenkel-Conrat & Williams, 1955). This experiment was important historically in showing that the native virus structure is a free energy minimum for the purified components and that assembly can proceed intracellularly or *in vitro* without any additional template. Dissociated TMV protein in fact assembles into a variety of structures, schematically shown in Fig. 8 (Butler & Durham, 1977).

This 'phase diagram' illustrates that there are two classes of structures: helix-related structures at low pH or in the presence of RNA, and disk-related structures at neutral pH or above. The regulation of this polymorphism by pH has been ascribed to a set of carboxylate side chains that titrate anomalously in the virus, due to enforced proximity in the helical assembly (Caspar, 1963). The strain due to the proximity is relieved in the disk, or of course by proton bonding. Threading RNA through the structure compensates for the strain, even above neutral pH where the carboxyl

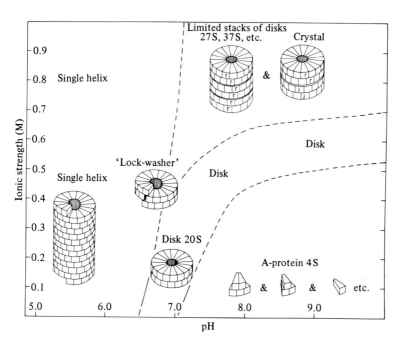

Fig. 8. Forms of association of TMV protein found at various pH and ionic strengths. (Butler & Durham, 1977.)

groups are fully deprotonated. Proton binding is relevant to assembly because the disk is an intermediate. Reassembly *in vitro* proceeds much more rapidly if protein is added as disk rather than as the basic 4S 'A-protein', reflecting a critical function for disk in nucleating assembly as well as a possible role in elongation (Butler & Klug, 1971). The pathway as currently understood is shown in Fig. 9 (Butler, Finch & Zimmern, 1977; Lebeurier, Nicolaeff & Richards, 1977; Klug, 1979).

An initiation loop of RNA, located about 1000 nucleotides from the 3' end of the RNA (just 5' to the coat cistron), inserts into the central hole of a disk (Jonard et al., 1977; Zimmern, 1977). Disorder of the inner loop of polypeptide chain and tilt of the subunit create a pair of loose 'jaws', with the RNA binding site at their interior (Fig. 7). The nucleotide sequence has some suggestive repeats, in particular a G every three bases. Binding of an initial stretch of trinucleotides might be sufficient to shift the axial subunit contacts from disk-like to helix-like. This would dislocate the disk and create a helical growing point. Additions must then occur in two directions, and reassembly experiments clearly show a rapid elongation toward

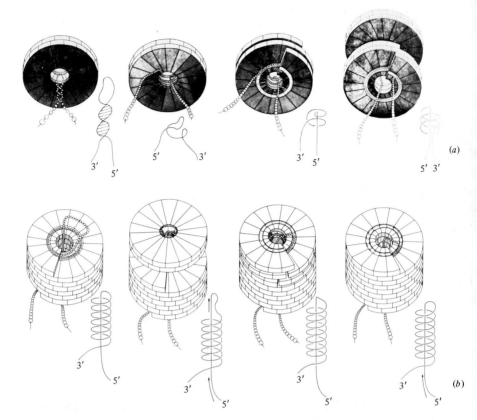

Fig. 9. Schematic views of (a) initiation (b) elongation in TMV assembly. Elongation in the 5' direction occurs by drawing RNA through the central channel in the growing virus particle. (From Klug, 1979.)

the 5' end, drawing RNA up through the hole, and much slower addition toward the 3' end. As shown in Fig. 9, the disk may also be the unit of 5' elongation; 3' elongation must proceed by addition of subunits.

This assembly pathway is strikingly more complicated than simple addition of subunits to the corner of a growing helix, as originally suggested by Watson (1954). It appears to confer significant functional advantages. The assembled viral subunit must be able to accommodate any RNA triplet in its binding site. A small preference for certain sequences (e.g., G in one of the three positions) can give adequate specificity if the initiating region is relatively long (Holmes, 1980). Hence the significance of the disk in nucleation. During elongation, assembling virus is a 'melting' protein that unwinds helical stems in the RNA. Presentation of a loop of RNA at the 5' growing end can permit cooperative addition of subunits, in order to drive unwinding of a stem at the other end of the axial hole (Klug, 1979).

Tomato bushy stunt virus, turnip crinkle virus and related structures

Our best view of an icosahedral viral shell comes from single-crystal X-ray diffraction studies of TBSV (Harrison *et al.* 1978). This is a structure with 180 coat protein subunits (molecular weight *c.* 42 000), probably one chain of a 80 000 molecular weight protein, and a molecule of single-stranded RNA (*c.* 4800 nucleotides). Its organization is summarized in Fig. 10; the related turnip crinkle virus (TCV) has been shown to have an essentially identical structure (Hogle & Harrison, 1984).

The coat subunit, containing 386 amino acid residues, folds into distinct modules: a projecting domain (P), a domain forming a tightly bonded shell (S), a connecting arm (a), and an internal domain (R). The three symmetrically distinct environments for this subunit are denoted A, B and C. The polypeptide accommodates to these three packing positions by flexion at the hinge between S and P and by an ordering or disordering of part of the arm. Units at positions A and B (60 of each) have one hinge configuration, and the entire N-terminal region (arm and R domain) appears to be spatially disordered. Subunits at position C (60 in all) have another hinge configuration, and the connecting arm is folded in an ordered way along the bottom of the S domain. The R domain is not fixed with respect to the rest of the subunit, and so it cannot be seen in a high-resolution electron-density map, but it may well be folded in a regular way. Thus, our best picture of the N-terminal part of the subunit is of a well-structured R domain flexibly tethered to the S domain, which is held firmly in the viral shell. The disordered part of this tether is very long on A and B (the entire connecting arm) and much shorter on C (probably just a few residues). The packing in the interior of the virus is tight enough so that connectors, R domains and RNA are probably not actually moving about (Munowitz *et al.*, 1980).

The most remarkable feature of the TBSV structure is the way in which the connecting arms of C subunits interdigitate to form an internal framework. As shown in Fig. 11, these arms extend along the inner edge of the S domain and loop around icosahedral

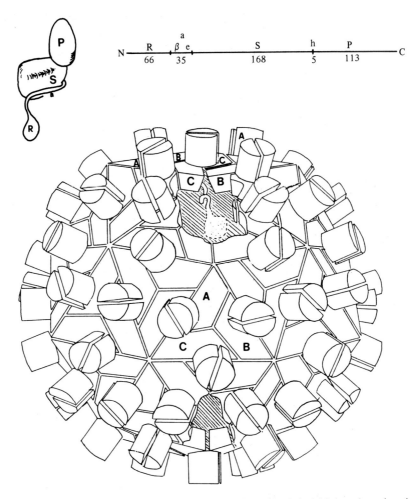

Fig. 10. Architecture of the TBSV particle. The polypeptide chain folds into three domains (R, S, P) with a 35-residue connecting arm (a) between R and S and a hinge (h) between S and P. The number of amino acid residues in each structural module is indicated in the linear diagram. The subunits pack into the virus particle in one of three conformations, denoted A, B and C (see also Fig. 4). S domains of A subunits pack around five-fold axes; S domains of B and C subunits alternate around three-fold axes. The arms of A- and B-position subunits are spatially disordered, extending in various ways toward the particle centre. The arms of C-position subunits fold along the edge of the S domain, as shown in the subunit schematic, and interdigitate around three-fold symmetry axes (see Figs. 11 and 12). The hinge between S and P has one angle on A and B subunits, another angle on C subunits.

three-fold axes. Three such C subunit arms contact each other in this way, and all 60 C subunits form a coherent network (Fig. 12). The function of this framework is to determine the size of the particle – that is, to ensure that the viral shell closes round on itself

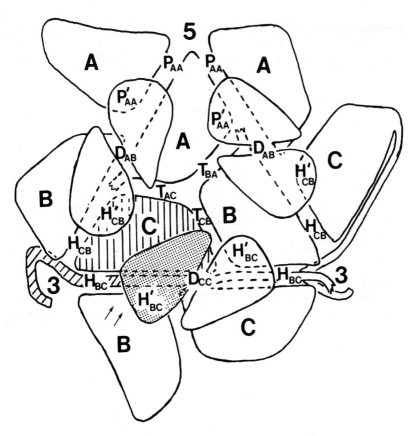

Fig. 11. Subunit contacts in TBSV. Domains are shown in outline. The P domains would protrude from the page, making only two-fold contacts, while the S domains make contacts across all neighbouring symmetry axes. The arms of C subunits are ordered and fold along the base of the S domain, interdigitating with two other such arms around three-fold axes of symmetry. A twofold axis of symmetry is indicated by '5'. The labels H_{BC}, etc., are a nomenclature system for classes of contacts. (Harrison *et al.*, 1978.)

correctly during assembly. This function has been demonstrated by reassembly experiments *in vitro* with TCV (Sorger, Stockley & Harrison, 1984). The virus reassembles from P-domain-linked dimers of the subunit (Golden & Harrison, 1982), and proteolytic cleavage of the connecting arms destroys the ability of the dimer to form 180 subunit shells. Instead, they assemble into 60 subunit 'small particles' (Fig. 13). Trapped, partially completed shells from unmodified protein have the correct curvature. These and other features of TCV reassembly indicate that the conformation of a protein dimer – A/B or C/C – is determined as it enters the growing particle and that the propagating framework of folded, C-position

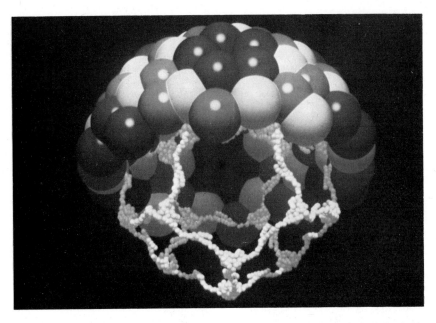

Fig. 12. The arms of all 60 C subunits in TBSV form an internal framework. Subunits in positions A, B and C are black, white and grey respectively. (Courtesy A. J. Olson & N. Max.)

arms is an unambiguous local 'switch' to determine the bonding orientation of the next dimer to join the particle.

The actual interfaces between domains in TBSV show that homologous parts of subunits make similar contacts, as in the original description of 'quasi-equivalence' (Olson et al., 1983). The way in which the C-subunit arms fold into one set of interfaces, however, selects unambiguously which 'quasi-equivalent' choice of side-chain interaction is made. Thus the structure takes advantage of the capacity of the subunit to form specific but flexible bonds without depending on these properties alone to prevent errors.

Packing of RNA in TBSV cannot be visualized directly in electron-density maps derived from X-ray crystallography due to spatial disorder of the nucleic acid and of the protein R domains. The R domains have a large number of positively charged residues, as do the inward-facing S-domain surfaces; together they neutralize about 75% of the RNA phosphates (Hopper, Harrison & Sauer, 1984). The fundamental conclusion appears to be that few (if any) constraints other than compactness need be placed on the configuration of the nucleic acid chain, since R domains are flexibly tethered

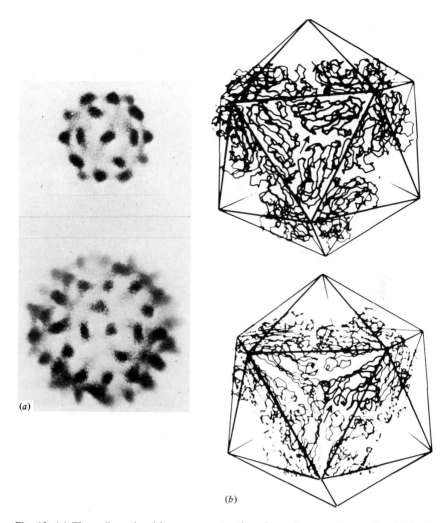

Fig. 13. (a) Three-dimensional image reconstructions from electron micrographs of TBSV (bottom) and TCV 'small particle' (top) (Crowther & Amos, 1971). The relationship is exactly like that of the large and small comma-containing structures in Fig. 4. (b) Packing of S domains in TCV small particles (top) and subunits in STNV (bottom: Liljas *et al.*, 1982). Parts of 15 subunits are shown in each case. By comparing subunits indicated by the arrows, it can be seen that the detailed packing of these similarly folded subunits is rather different in the two cases. (Courtesy A. J. Olson.)

to the rest of the shell. This is probably a general feature of RNA packing in spherical viruses. The absence of strong constraints or detailed spatial arrangement implies that melting out of secondary structure is not required, an important difference from the arrangement in TMV and other helical structures. There is evidence both in TCV and in alfalfa mosaic virus (AMV) for a specific coat protein

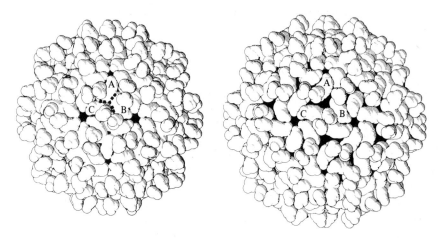

Fig. 14. Compact (left) and expanded (right) TBSV. Pairs of black squares in the compact structure indicate locations of divalent cations that control expansion. Each such interface has two closely spaced sites for binding Ca^{2+}. (From Robinson & Harrison, 1982.)

binding point on the RNA (Zuidema *et al.*, 1983; Sorger *et al.*, 1984). This is plausible both for biological reasons – the virus selectively incorporates its own RNA – and for structural reasons; multiple initiation on a single RNA chain would lead to non-coalescing partial shells and hence to abortive assembly.

TBSV and TCV – and many other spherical RNA plant viruses – undergo a reversible expansion at pH >7 in the absence of Ca^{2+} (Incardona & Kaesberg, 1964; Kruse *et al.*, 1982; Robinson & Harrison, 1982). The expanded particles are about 15% larger in radius than compact ones, but they are by no means 'floppy'. Indeed, expanded TBSV forms crystals that diffract to 8 Å resolution, showing that expansion corresponds to a rearrangement of subunit domains but not to a change in domain conformation (Fig. 14; Robinson & Harrison, 1982). Moreover, many bonding interfaces are conserved, and others containing the cation sites are completely dissociated. One property of the expanded particles not evident from their crystal structure is accessibility of some of the arms and R domains to proteolytic attack (P. Sorger, unpublished). The long flexible link between S and R appears to allow R to extrude through the gaps in the expanded shell. Expanded virus is nonetheless RNase resistant. The highly cooperative transition between expanded and compact particles occurs just at intracellar pH, but its functional significance remains obscure. A reasonable possibility is that expansion represents a first step in disassembly, with subsequent steps requiring one or more factors from the host.

The structure of southern bean mosaic virus (SBMV) has also been determined at high resolution (Abad-Zapatero, et al., 1980). Its subunit has close structural homology to the arm and S domains of TBSV and TCV, but a P domain is totally absent. Moreover, the single major domain of SBMV packs in a manner precisely analogous to the packing of the TBSV S domain, with which it has amino acid sequence similarities (Hermodson et al., 1982; Hopper et al., 1984). The framework of arms is also present. Like TCV, SBMV dissociates into dimers of the coat subunit (B. J. M. Verduin, personal communication); the coherence of these dimers is clearly not dependent on P-domain interactions, suggesting a different or additional function for this domain in TBSV.

A number of other plant viruses have a subunit modularity similar to SBMV and to the R domain-connecting arm-S domain part of TBSV. In brome mosaic virus (BMV) and cowpea chlorotic mottle virus (CCMV), which also have 180 subunits in their icosahedral shells, clusters of positively charged residues are found very near the N terminus (Moosic, 1978). Removal of these residues (1–25) by proteolytic cleavage destroys the ability of viral protein to package RNA (Vriend et al., 1982); further cleavage (to residue 43 or 63) produces subunits that can only form $T = 1$ small particles (Cuillel, Jacrot & Zulauf, 1981). The coat protein of AMV, which makes bacilliform particles of several lengths (corresponding to several packaged RNA molecules), has a similar N-terminal concentration of basic residues, and cleavage at residue 26 restricts reassembly to $T = 1$ structures (Bol, Krall & Brederode, 1974). Satellite tobacco necrosis virus (STNV) is a 60 subunit ($T = 1$) particle (Fig. 13). Its structure is known from high-resolution crystallography: the folded conformation of the protein is strikingly like that of the TBSV S domain or of the major domain in SBMV, although the way in which it packs with its neighbours is rather different (Liljas et al., 1982). There are a number of basic amino acid residues near the inward-projecting N terminus. The first 13 are not clearly seen in the electron-density map, and the next 11 form an α-helical stem. The comparative modularity suggested by these observations is shown diagrammatically in Fig. 15. STNV does not require a connecting arm, since the switch from A/B to C/C interaction is absent in its $T = 1$ shell. Corresponding mechanisms regulating AMV assembly are not yet understood. An exception to the pattern of positively

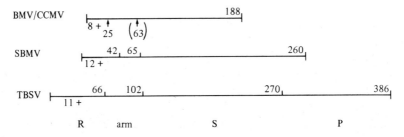

Fig. 15. Modular comparison of primary structures of a number of small plant viruses. The domain assignments for BMV/CCMV are based on cleavage data as described in the text. The structures of TBSV, TCV and SBMV are known. The only significant *sequence* homology seems to be between TBSV (S domain) and SBMV. The numerals above each line indicate residue numbers at boundaries between various structural modules. The indications 25 and 63 for BMV/CCMV show cleavage positions that may be near modular boundaries. The positively charged N-terminal domains have 8, 12, and 11 positive charges as shown.

charged N-terminal modules is the subunit of turnip yellow mosaic virus (TYMV; Peter *et al.*, 1970). In this and related viruses, polyamines neutralize most of the RNA phosphate charges (Cohen & McCormick, 1979). TYMV also lacks an expanded form.

Sindbis and Semliki Forest viruses

These alphaviruses illustrate the simplest organization of a viral membrane. They are closely similar in structure and contain significant sequence homology. Sindbis contains two glycoproteins and a core protein in equimolar proportion (Strauss & Strauss, 1977). Semliki Forest virus (SFV) contains a third glycoprotein, an entirely external segment whose homologue in Sindbis dissociates after processing. All are produced by continuous translation of a single mRNA with appropriate cleavage of the product (Wirth *et al.*, 1977). The complete sequence of the Sindbis message shows that glycoproteins E1 and E2 have polypeptide molecular weights of 49 000 and 47 000 respectively, and that each has two glycosylation sites (Rice & Strauss, 1981). The core protein (C) has a molecular weight of 29 000.

Sindbis particles are isometric and very regular, and small-angle X-ray scattering measurements show that the outer radius is about 350 Å and that there is a lipid bilayer between radii of 210 and 260 Å (Harrison *et al.*, 1971). Electron microscopy reveals an icosahedral surface lattice, apparently T = 4, (Fig. 16; von Bonsdorff & Harrison, 1975, 1978), although recent molecular weight data for SFV are reported to be more consistent with a 180 unit structure (B. Jacrot,

Fig. 16. Freeze-etch electron micrographs of Sindbis virus particles and of hexagonal arrays of Sindbis glycoprotein in lipid/detergent vesicles. The packing in the arrays is essentially isomorphous to the packing in virus particles. Bar = 1000 Å. (From von Bonsdorff & Harrison, 1978.)

personal communication). Microcrystals of SFV (Wiley & von Bonsdorff, 1978) also demonstrate that these viruses have a precise and definite structure. Isolated cores are regular particles of radius 200 Å; they are quite 'smooth', and clear surface features have not been resolved by usual methods of negative contrast (Harrison *et al.*, 1971).

The amino acid sequences of Sindbis (Rice & Strauss, 1981) and SFV (Garoff *et al.*, 1980*b*) core proteins suggest a modular organization similar to that of the plant virus proteins (Fig. 15). The N-terminal regions of about 115 residues have sufficient positive charge to neutralize over half of the approximately 55 RNA phosphates per subunit. These parts show moderate homology between the two viruses. The remainders of the sequences, about 150 residues, do not have strong positive charge, but do show very marked homology.

The equimolar proportionality of C, E1 and E2 (Strauss & Strauss, 1977) and the demonstration that E1 and E2 have anchors that penetrate the bilayer (Gahmberg, Utermann & Simons, 1972; Garoff & Soderlund, 1978) indicate a one-to-one contact between

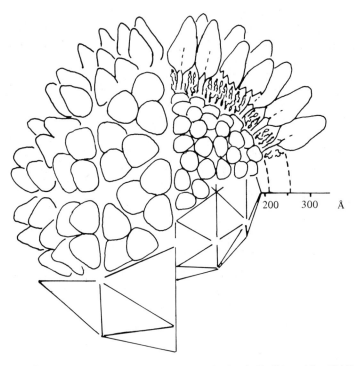

Fig. 17. Schematic representation of the molecular organization in Sindbis and Semliki Forest viruses (SFV). Each unit on the outside represents one E1 and one E2 subunit (E3 in SFV is not shown). Their clustering in trimers gives the grooved patterns seen in Fig. 16. E1 and E2 both have hydrophobic anchors that penetrate the bilayer, which is symbolized in the diagram by the array of lipid molecules between radii of 210 and 255 Å. The T = 4 icosahedral surface lattices characterizing the outer glycoprotein layer and the inner core are shown in the lower part of the drawing.

each core subunit and the 'cytoplasmic' domains of an E1/E2 dimer. E1 has only the dipeptide Arg—Arg on the inside of the bilayer, but E2 has a significantly longer sequence (Garoff *et al.*, 1980 *a*, *b*; Rice & Strauss, 1981). It is therefore reasonable to assign to the core a surface lattice congruent to that of the glycoprotein. Fig. 17 summarizes this view of the organization of the Sindbis membrane. Most of the carbohydrate on Sindbis is inaccessible to glycosidases, unless the particle is disrupted by Triton X-100, suggesting that, as in the case of influenza glycoproteins haemagglutinin and neuraminidase, glycosylated asparagines are found not near the subunit tip, but on lateral surfaces (McCarthy & Harrison, 1977). The tight surface lattice of Sindbis would effectively bury most of the oligosaccharide.

Extensive chymotryptic cleavage of Sindbis and SFV removes the glycoprotein spikes (Compans, 1971; Garoff & Soderlund, 1978), leaving the hydrophobic C-terminal anchors still embedded in the membrane (Utermann & Simons, 1974). These anchors appear to have essentially all the covalently attached palmitic acid, which has been shown to be added post-translationally to Sindbis glycoproteins (Schmidt, Bracha & Schlesinger, 1979; Rice et al., 1982). E1 has 33 uncharged residues between a lysine and a C-terminal Arg—Arg dipeptide; E2 has 27 such residues, followed by a 33-residue cytoplasmic sequence that presumably forms the recognition structure for the core (Rice & Strauss, 1981). This sequence is reasonably well-conserved between Sindbis and SFV, as is the non-basic, C-terminal domain of the core polypeptide.

Treatment of Sindbis with increasing amounts of non-ionic detergent results in a stepwise dissociation of the particle (Helenius & Soderlund, 1973). Moderate ratios of Triton X-100 to virus release nucleocapsid while leaving membrane more or less intact. These membranes, expanded by detergent, tend to fuse in concentrated specimens, and the glycoproteins rearrange to form an extended hexagonal lattice similar in structure to the viral surface (von Bonsdoff & Harrison, 1978). Electron micrographs of freeze-etched, shadowed specimens are shown in Fig. 16 together with comparable freeze-etched views of intact Sindbis. Spontaneous formation of these arrays requires both a high degree of specificity in lateral contacts and sufficient mobility of glycoproteins in the lipid bilayer so that five-fold vertices 'anneal' to six-folds. Formation of extensive arrays of this sort on the surface of an infected cell might be expected to inhibit budding, since formation of five-fold vertices would require removal of a large wedge of protein. It is therefore unlikely that such arrays represent an intermediate stage in assembly. Array formation does indicate precise and reasonably strong glycoprotein–glycoprotein bonds. These non-covalent contacts are presumably part of the driving force for budding, but organization of glycoprotein in an icosahedral lattice is probably determined by regular interactions with C via the anchor structure.

Influenza virus glycoproteins

There are two influenza glycoproteins: the haemagglutinin (HA) and the neuraminidase (NA). HA is the principal antigen, and it contains the functions responsible for binding virus to a cell and for fusing the viral membrane with the endosomal membrane of the

host. The receptor is any sialic-acid-containing surface oligosaccharide, and the NA activity ensures that virus particles do not have
sialic acid residues. The protease-resistant external portions of both
these glycoproteins have been crystallized, and both structures are
now known in detail (Wilson, Wiley & Skehel, 1981; Varghese,
Laver & Coleman, 1983).

HA, like most glycoproteins, is synthesized as a precursor which
loses an N-terminal signal in early processing. It is also later cleaved
at the cell surface to produce a two-chain unit (HA1 and HA2) with
loss of only a single arginine. This interrupted polypeptide is found
as trimers on the viral surface (Wiley, Skehel & Waterfield, 1977).
The protease bromelain separates HA1 and the larger part of HA2
from the C terminus of HA2, which forms the membrane anchor.
The soluble (and still trimeric) fragment, comprising most of the
molecule, is known as BHA: it is this fragment that crystallizes
(Brand & Skehel, 1972; Wiley & Skehel, 1977).

A schematic representation of one of the HA1 and HA2 units in a
BHA trimer is shown in Fig. 18. The trimer axis is vertical and the
membrane would lie in cross section at the bottom of the figure. The
molecule is 135 Å long (from membrane to tip), and the trimer
varies in radius from 15 to 40 Å. There is a long, fibrous region
extending out from the membrane, and a more compact globular
region at the apex. The globular part contains residues entirely in
HA1; the fibrous part has contributions from both HA1 and HA2.
The HA1 chain begins near the membrane and continues in an
almost fully extended conformation for 50 residues before entering
a more compact fold. This first compact module of HA1 includes
residues 50–91 as well as 265–90 on the 'return' of the chain. It
appears to be an 'adaptor' for packing the principal globular domain
against the fibrous stalk. The principal domain, residues 92–264,
contains a binding site for sialic acid, the essential component of any
influenza receptor.

The C-terminal part of HA1 follows a course roughly antiparallel
to much of the N-terminal stretch, and its C terminus (Thr 328) lies
against Pro 21. The N terminus of HA2 is 21 Å from the C terminus
of HA1, indicating that a substantial rearrangement must accom-
pany the cleavage, which deletes only one residue. This initial HA2
sequence is strongly hydrophobic, and a number of correlations link
it with membrane fusion activity (Klenk et al., 1975; Gething, White
& Waterfield, 1978). It lies tucked against the trimer interface,
suggesting that direct participation in fusion must involve significant

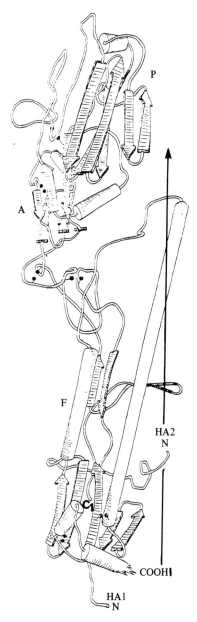

Fig. 18. Representation of the BHA monomer, showing the folding domains described in the text: P, principal domain; A, adaptor; F, fibrous stalk. The three-fold axis relating molecules in a trimer is shown: the trimer contacts are determined principally by packing of the long α-helices. The N terminus of HA2 (the fusion peptide) tucks into the three-fold contact. Note how far it is separated from the C-terminus of HA1 (C1). (Courtesy D. C. Wiley.)

conformational change (see below). The main part of HA2 forms a dramatic antiparallel α-helical hairpin.

The haemagglutinin polypeptide has thus the following approximate modular division: (a) extended terminus of HA1; (b) small compact domain; (c) principal compact domain; (d) extended C terminus of HA1; (e) 'fusion peptide' of HA2; (f) main hairpin of HA2. The membrane anchor, including an 11-residue cytoplasmic domain, would of course follow in the uncleaved molecule. Since the molecule must undergo conformational changes during assembly and during fusion (see below), increased understanding of these transitions may well modify our view of what constitute significant modules.

The gross topology of the HA molecule is thus a loop that begins and ends at the membrane. If the N-terminal signal sequence were to remain uncleaved and anchored in the membrane for any significant period during chain elongation, one could understand this topology in terms of the requirements for folding during biosynthesis with an anchored N terminus (Wilson et al., 1981). Little, if any, of the chain before residue 50 could form a defined structure without the HA2 'stalk'. Thus, a consequence of the loop-like character is that synthesis and folding of the final part of the chain is necessary for folding the initial part. Wilson et al. (1981) have suggested, moreover, that full trimer assembly could be required for establishing a correctly folded HA2. The principal structure that stabilizes the trimer is a three-stranded 'coiled-coil' of the long HA2 α-helices: in fibrous proteins, individual helices of such coiled-coils are not generally stable in solution. Other elements of the HA2 structure might, however, stabilize it even as a monomer: in particular, the shorter helix as well as a significant number of polar and charged residues in the lower portion of the long helix.

The HA chain is glycosylated at seven Asn—x—Thr(Ser) sequences, six of which occur in HA1 (Ward & Dopheide, 1980). They are marked in Fig. 18. Two of the oligosaccharides are 'simple'; the rest contain galactose and fucose in addition to mannose and N-acetyl-glucosamine. Electron density for many of the hexose rings can be identified in the X-ray analysis, but many others are disordered. The most striking conclusion is evident from the positions of the glycosylated asparagines alone: most of the carbohydrate decorates the *lateral* surfaces of the molecule, and few if any of the sugars project outward from the distal part of the

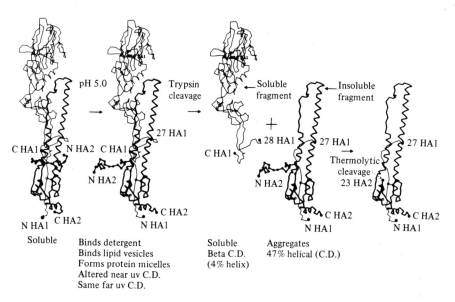

Fig. 19. Schematic representation of the conformational changes in BHA after incubation at pH 5.0. Tryptic cleavage results in two fragments: an aggregating, significantly α-helical piece containing HA2 and part of HA1 and a soluble, significantly β-structured piece containing the adaptor and principal domains of HA1. The structural assignments are based in part on evidence from circular dichroism (CD), as indicated. (Courtesy D. C. Wiley.)

molecule. Accessibility of the carbohydrate (for lectin binding, etc.) therefore depends on packing of HA trimers in the viral surface.

Influenza virus can cause cells to fuse if applied in a medium of low pH (5–5.5) (Helenius et al., 1980; White & Helenius, 1980). This property reflects the response of the virus to the lowered pH of an endosome, the organelle in which virus is found after receptor-mediated uptake. Both viral infectivity and low pH virus-mediated fusion require the post-translational cleavage from HA0 to HA1 and HA2 (Klenk et al., 1975; White, Kartenbeck & Helenius, 1981a), implicating the haemagglutinin molecule. Moreover, the hydrophobic N-terminal sequence of HA2 is homologous to the N terminus of the F1 component of Sendai virus fusion glycoprotein (Gething et al., 1978). Skehel et al. (1982) have demonstrated a conformational change in BHA, triggered at acid pH (between 5.0 and 5.5), that appears to be a good candidate for an effective structure in the fusion process. Acid-treated BHA aggregates in the absence of non-ionic detergent, and peptide bonds at residues 27 and 224 are exposed to the action of trypsin. These bonds are

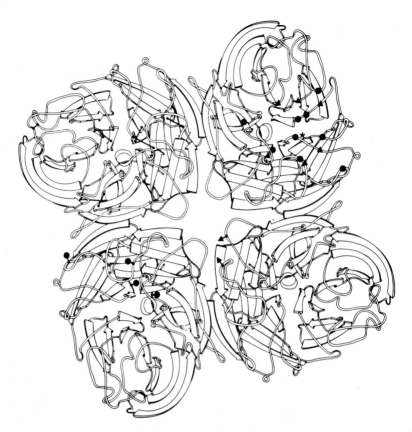

Fig. 20. Neuraminidase tetramer from influenza virus. The diagram shows the four globular heads. Asterisk on upper right-hand subunit indicates one enzymatic active site; disulfides are shown on upper left subunit; carbohydrate attachment sites on the head are shown on lower left subunit. (From Varghese *et al.*, 1983.)

normally buried at subunit interfaces in the trimer, as is the N terminus of HA2 (Fig. 19). The results suggest a major conformational change that could expose the 'fusion peptide'.

Neuraminidase (NA) is a four-fold structure, with globular 'heads' connected to membrane anchors by slender 'stalks' (Colman, Varghese & Laver, 1983; Varghese *et al.*, 1983). The stalks are readily cleaved by pronase, permitting isolation and crystallization of the heads. The membrane anchors are N-terminal – an unusual arrangement – and these sequences also appear to act as signal peptides for secretion (Fields, Winter & Brownlee, 1981). Residues 1–6 are internal and residues 7–35 span the membrane. The highly glycosylated sequence 36–73 (four carbohydrate sites, half of the total) forms the stalk. Electron micrographs suggest that this

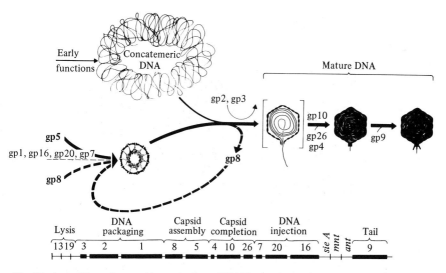

Fig. 21. Assembly pathway of bacteriophage P22. The products of genes 5 and 8 (gp5, gp8) together with several minor species, form proheads. Interaction of a prohead with DNA appears to trigger an expansion that releases the gp8 scaffold (subunits are re-used). DNA coiled in the head is 'sealed in' by addition of the tail. (Courtesy J. King.)

structure may be as long as 100 Å (Laver & Valentine, 1969) implying an extended and perhaps flexible conformation for this part of the protein. The heads have an unusual, propeller-like arrangement of β-sheets (Fig. 20), with an enzymatic active site on each subunit. The four carbohydrate chains on this part of the structure are equally distributed on its outward- and inward-facing surfaces, with one mediating a subunit–subunit contact. Thus, as in HA, oligosaccharides tend to occupy positions where contact with other membrane proteins might occur.

Bacteriophage assembly: P22

Some characteristics of the pathways of bacteriophage assembly have been described in the introduction. Among the simplest of such assemblies is the salmonella phage P22 (Fig. 21; King, Lenk & Botstein, 1973). The particle has an isometric head and a simple tail. The head precursor – the so-called prohead – assembles from the major coat protein (gp5), but only in the presence of an internal scaffold protein (gp8) (King & Casjens, 1974; Earnshaw, Casjens & Harrison, 1976). Several minor proteins are also incorporated. These proteins probably have functions related to DNA packaging,

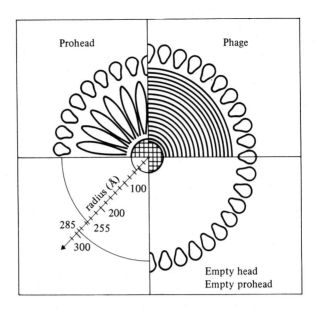

Fig. 22. Diagram of the radial organization of various P22-head-related structures. The protein shell is shown as pear-shaped subunits; the scaffold protein in proheads is represented by elongated structures; the DNA wound into phage is symbolized by concentric rings. The empty heads are found in tail-less mutants: the heads are filled but lose DNA on isolation. The empty proheads are proheads treated with SDS or urea *in vitro* to release the scaffold and expand the gp5 shell. (From Earnshaw *et al.*, 1976.)

since they are not essential for prohead assembly itself. Indeed, proheads can be formed *in vitro* from purified gp5 and gp8 (Fuller & King, 1980). The prohead is about 10% smaller than the head of the mature phage and mild urea or sodium dodecyl sulphate (SDS) treatment of proheads *in vitro* releases gp8 and causes the gp5 shell to expand to its mature size (Fig. 22). A similar expansion occurs *in vivo*, and the released gp8 can participate in further rounds of assembly (King & Casjens, 1974).

The scaffold protein (gp8) does not assemble into larger structures by itself, and in its absence gp5 tends to assemble into aberrant forms (Fuller & King, 1980). Thus, gp8 appears to function in a manner analogous to the framework of arms in TBSV: it co-assembles with gp5, and in so doing it determines the correct size and curvature of the shell. These observations are consistent with the notion that the prohead represents a constrained bonding pattern for gp5, and that conditions promoting release of gp8 also trigger a readjustment to the more stable head size. In this respect, there appears to be some analogy with T4 tail assembly, where the

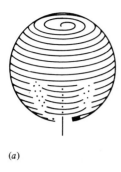 

(a) (b)

Fig. 23. (a) DNA in the head of isometric phages such as P22 is wound in a spool. The spool may not be quite as regular as shown here, and there is probably no fixed sequence for the windings. (Earnshaw & Harrison, 1977). (b) Possible model for insertion of DNA into the phage head to form a smoothly wound structure that can unpackage without entanglement. The leading end of the DNA is attached to the head near the entry point (arrow) and DNA enters axially (Harrison, 1983).

extended sheath is stabilized by interactions with the baseplate and core; release from these 'scaffold' structures leads to a cooperative rearrangement of sheath to the contracted form. A change in subunit packing, producing an expansion of 10%–20% in radius, is a common feature of phage head assembly pathways. In all cases, the expansion occurs when a core structure is released, either by dissociation (as in P22) or by proteolysis (as in λ or T4). Molecular details of the expansion could well be similar to those seen in expansion of TBSV: readjustment of hinges between protein domains, permitting conservation of many inter-subunit contacts and allowing discrete expansion without dissociation.

Mature P22 phage particles have a $T = 7$ icosahedral surface lattice, with projections clustered around hexamer/pentamer and around trimer positions, giving a rather finely divided appearance (Casjens, 1979). The precise ratio of gp8 to gp5 subunits in the prohead is not known (it is approximately $1 : 2$), and a detailed picture of prohead organization is therefore not presently at hand.

DNA is packaged from the concatemeric product of replication into the head precursor, which expands in the process. The packaging requires ATP. Electron microscopy of gently disrupted particles (Richards, Williams & Calendar, 1973) and X-ray scattering from P22 and λ in solution (Earnshaw & Harrison, 1977) indicate that the DNA is wound into a spool-like structure (Fig. 23a). Spatially adjacent segments of the molecule are forced by tight packing to lie parallel to each other, and the relatively regular intersegment spacing is simply governed by the total length of the DNA molecule

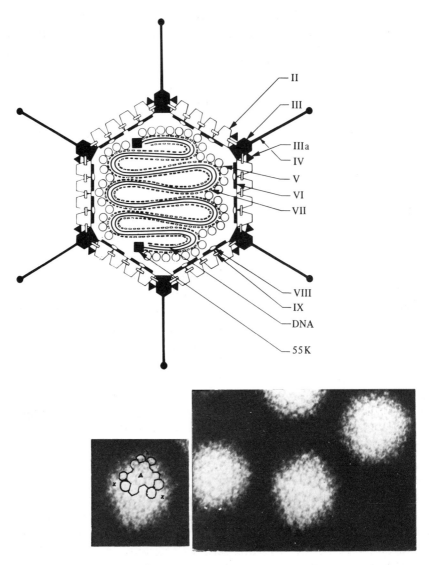

Fig. 24. Diagram showing the location of important substructures in adenovirus (Ginsberg, 1979). The larger inset is an electron micrograph of an adenovirus particle (courtesy R. M. Franklin). The smaller inset is a micrograph with a group of nine outlined, with peripentonal hexons circled, with crosses on pentons, and with a three-fold axis indicated.

and the total volume available to it. A number of experiments with bacteriophage λ indicate that the detailed path of winding of the DNA molecule can vary from particle to particle (Haas, Murphy & Cantor, 1982; Widom & Baldwin, 1983). That is, as DNA is 'pumped' into the head, it probably coils in a manner determined solely by the geometrical constraints of the internal volume and the nature of the packaging machinery. Indeed, as DNA enters, the coil must constantly tighten, requiring that adjacent turns slide along each other and along the inside of the head. The process must ensure that unpackaging can occur without entanglement. One simple picture for packaging is shown in Fig. 23b (see also Harrison, 1983).

Adenoviruses

These relatively complex particles illustrate how larger structures are constructed from subassemblies. The diagram of an adenovirus particle (Fig. 24; Ginsberg, 1979) indicates the location of important substructures (Everitt, Lutter & Philipson, 1975). The strikingly icosahedral outer shell appears at first to have a triangulation number of 25, but in fact the units at the vertices ('pentons') are different from the rest ('hexons'). Moreover, the hexons are trimers, not hexamers, so that the structure is based on a p3 rather than a p6 lattice (Franklin *et al.*, 1971; Crowther & Franklin, 1972). The hexon protein has a molecular weight of 110 000, and a high-resolution X-ray diffraction analysis is nearly complete (Burnett, 1983). Assembly of trimers from newly synthesized protein *in vivo* appears to require another viral encoded protein (the so-called 100 K-protein) that is not part of the final structure (Cepko & Sharp, 1982). In virions, protein VI is also associated with each hexon. Pentons contain two polypeptides – one of 85 000 molecular weight that forms the base, another of 62 000 molecular weight that forms the projecting fibre. A recent report suggests that the penton base may be trimeric (Devaux *et al.*, 1982), which is a curious result, given its five-coordinated position. There are, however, believed to be five copies of protein IIIa at each vertex, presumably to stabilize the presence of the five 'peripentonal' hexons. This sort of symmetry-matching problem occurs in many other structures – e.g., at the attachment of a six-fold symmetry T4 phage tail to a five-fold position on the head.

When adenovirus particles are disrupted by various methods, the outer shell loses first its pentons and peripentonal hexons, and the structure then dissociates into the groups of nine hexons that constitute the rest of a face. The 'groups of nine' are apparently held together by protein IX, which co-purifies with them (Everitt *et al.*, 1973). Analysis of electron micrographs of the groups of nine by rotational filtering shows their three-fold symmetry as well as the three-fold symmetry of the individual hexons (Crowther & Franklin, 1972). The groups of nine are thus significant substructures of the particle, reflecting sets of important protein interactions, but they are not necessarily assembly intermediates.

The core of adenovirus can be isolated as a compact particle, but it does not have a very striking substructure (Brown *et al.*, 1975; Nermut, Harpst & Russell, 1975). It contains a single, linear, double-stranded DNA ($20–25 \times 10^6$ molecular weight) and two basic proteins (V and VII). Protein VII is very arginine-rich, and it can neutralize about half of the DNA phosphates (Everitt, Lutter & Philipson, 1975). How this protein is associated with DNA is not known. Micrographs of gently disrupted cores suggest the DNA might be coiled within them as in a phage head (Brown *et al.*, 1975), rather than being condensed into locally compact structures as in nucleosomes.

STRATEGIES OF ASSEMBLY AND DISASSEMBLY

The variety of virus structures reflects the variety of strategies for assembly, exit, re-entry and disassembly – the 'life history' of a single virion. The second half of this history cannot be just a reversal of the first half, and one or the other must require an irreversible step or some energy-coupled driving mechanism. The classic example is T4 bacteriophage. Assembly is intracellular; it involves irreversible processing steps; it requires ATP to pump DNA into the head; and it holds DNA within the head by addition of a tail structure 'cocked' to contract when triggered by interaction with an appropriate receptor. Disassembly is essentially extracellular and, once triggered, spontaneous. At the other extreme, TMV self assembles *in vitro* and presumably also in the cell. What *disassembles* TMV is not so clear. Some aspects of virus structure relevant to these various strategies are discussed below.

Assembly

The simplest viruses such as TMV, and TCV self assemble *in vitro* and presumably also in the cell. The pathway for TMV involves a defined precursor mode of protein association (the disk), a specific initiation complex of protein and RNA, and a controlled mechanism of propagation by which RNA is drawn into the assembling virus.

Completion occurs when RNA is fully coated. The significance of the disk appears to be two-fold: viral helix must be stable intracellularly only in the presence of RNA, and a collection of subunits is required to recognize the assembly origin. There is also evidence for specific initiation in TCV assembly, which exhibits selectivity for TCV RNA and appears to have a single initiation site per particle (Sorger *et al.*, 1984). Closed-shell structures like TCV, with alternative modes of subunit bonding, must have a mechanism for determining the protein conformation during assembly, so that propagation of the shell results in proper closure. Protein–protein contacts are not flexible enough *in situ* that such a structure can rearrange after assembly – for example, to adjust incorrectly placed five-fold positions. In the case of TBSV and TCV, the conformational switch is simple and straightforward: arms are either ordered or not ordered as a dimer enters the growing shell. It appears likely that size and length in macromolecular assembly will generally be determined by mechanisms as direct as in these viruses and in TMV. For example, there is a protein (the product of gene 54) that might span the length of the T4 tail core (Wagenknecht & Bloomfield, 1978). More elaborate mechanisms, such as the 'cumulative strain' model for determining phage tail length (Kellenberger, 1972), probably suffer from too great a sensitivity to small changes in physico-chemical parameters (temperature, for example, which is not regulated in *E. coli*) and to mutational changes. Direct measuring rods, such as the RNA in TMV and the arms in TBSV, provide a chemically and mutationally robust control.

Assembly of the simpler enveloped viruses, such as Sindbis or vesicular stomatitis virus (VSV), is a two-stage process. Nucleocapsid formation is analogous to formation of non-enveloped structures. There is some evidence for an assembly origin at the 3' end of VSV RNA (Blumberg, Giorgi & Kolakofsky, 1983). The second stage appears also to occur without specific input of energy – that is, once the glycoproteins are correctly translocated and anchored in the plasma membrane, budding occurs spontaneously. In Sindbis, the process is probably driven both by glycoprotein-anchor/core interactions and by the lateral glycoprotein interactions established in the surface. The regular surface structure of influenza C also suggests that glycoprotein contacts can help drive the assembly. It is, however, the anchor/core or anchor/membrane protein contacts that determine the specificity of budding, the shape of the particle, and the overall curvature of the external lattice.

Disassembly

Part of the mechanism by which enveloped viruses enter cells is understood, as a result of observations showing pH-dependent fusion activity in SFV, influenza virus, VSV, and others (White, Kartenbeck & Helenius, 1981*a*; White, Matlin & Helenius, 1981*b*). These viruses are taken up by receptor-mediated endocytosis, and acidification of the endosome induces fusion of viral and endosomal membranes. The nucleocapsid is thereby deposited in the cytoplasm. In influenza virus, two mechanisms are responsible for the reversal of direction from spontaneous budding to spontaneous fusion: a difference in pH between the cell surface during budding and the endosomal interior during infection and proteolytic processing of the haemagglutinin (HA). As described above, the detailed mechanism involves a low-pH-triggered, irreversible conformational change of the HA, exposing a hydrophobic segment that interacts with the target membrane. This segment is found at the N terminus of HA2, a terminus created by a processing step in viral maturation that occurs at the cell surface. This cleavage of HA0 to HA1 plus HA2 primes the virus to undergo fusion when it subsequently encounters acid pH. In Sendai virus, where fusion is pH-independent, apparently only the processing is required.

Disassembly of the nucleocapsid, once deposited in the cytoplasm, is not accounted for by the steps just outlined. In the case of Sindbis and Semliki Forest viruses, the course of protein biosynthesis and the simple icosahedral symmetry of the particles both suggest that cores self assemble from C-protein subunits and RNA. If this is the case, some energy-utilizing disassembly mechanism may be required to reverse the process. VSV nucleocapsid does not need to be disassembled, since plus-strand synthesis occurs with intact nucleocapsids as template.

For non-enveloped viruses, it is probably also useful to distinguish penetration and uncoating. In the case of the plant RNA viruses, it is generally believed that virus particles enter cells of a new host directly, through wounds produced by abrasion or by insects, and that they move from cell to cell through plasmodesmata (Gibbs & Harrison, 1976). Since some viruses also migrate via the plant vascular system, it is reasonable to imagine endocytosis as a route of uptake in some circumstances, but it appears that in many cases uncoating must be a cytoplasmic event, not part of entry itself. In the case of small RNA animal viruses, events are again obscure;

binding of poliovirus to the cell appears to result in release of VP4, but subsequent events have not been traced. Electron micrographs taken shortly after infection show that adenovirus particles arrive in the cytoplasm as intact, organized structures (Dales, 1973; Fitzgerald *et al.*, 1983). These viruses are initially taken up by endocytosis into coated vesicles. It is plausible to suggest that penetration into the cytoplasm might occur by disruption of an endosomal membrane (Fitzgerald *et al.*, 1983). Endosomal lysis might be expected to be less damaging to cellular compartmentalization mechanisms than disruption of lysozomal or Golgi membranes. This sort of 'internal lysis' avoids special mechanisms for getting macromolecular structures across a bilayer, and it results in depositing a virus particle (or a partly disassembled particle) into the cytoplasm. In some cases (such as reovirus) further uncoating does not occur, but uncoating is indeed required for DNA-containing structures such as adenoviruses and papovaviruses. In these cases, we might expect specific cytoplasmic or nuclear disassembly mechanisms, perhaps reflecting more general intracellular processes.

I am grateful to D. C. Wiley for many useful discussions and suggestions and to NIH (Grant No. CA–13202) and NSF (Grant No. PCM–82–02821) for support.

REFERENCES

ABAD-ZAPATERO, C., ABDEL-MIGUID, S. S., JOHNSON, J. E., LESLIE, A. G. W., RAYMENT, I., ROSSMANN, M. G., SUCK, D. & TSUKIHARA, T. (1980). Structure of southern bean mosaic virus at 2.8 Å resolution. *Nature*, **286**, 33–9.

ACHESON, N. H. & TAMM, I. (1967). Replication of Semliki Forest virus: an electron microscopic study. *Virology*, **32**, 128–43.

BAKER, T. S., CASPAR, D. L. D. & MURAKAMI, W. T. (1983). Polyoma virus 'hexamer' tubes consist of paired pentamers. *Nature*, **303**, 446–8.

BALTIMORE, D. (1969). The replication of picornaviruses. In *The Biochemistry of Viruses*, ed. H. B. Levy, pp. 101–76. New York: Marcel Decker.

BANCROFT, J. E. (1970). The self-assembly of spherical plant viruses. *Advances in Virus Research*, **16**, 99–134.

BLOOMER, A., CHAMPNESS, J. N., BRICOGNE, G., STADEN, R. & KLUG, A. (1978). Protein disk of tobacco mosaic virus at 2.8 Å resolution showing the interactions within and between subunits. *Nature*, **276**, 362–8.

BLUMBERG, B. M., GIORGI, C. & KOLAKOFSKY, D. (1983). N protein of vesicular stomatitis virus selectively encapsidates leader RNA *in vitro*. *Cell*, **32**, 559–67.

BOL, J. F., KRALL, B. & BREDERODE, F. TH. (1974). Limited proteolysis of alfalfa mosaic virus: influence on the structural and biological function of the coat protein. *Virology*, **58**, 101–110.

BRAND, C. M. K. & SKEHEL, J. J. (1972). Crystalline antigen from the influenza virus envelope. *Nature New Biology*, **238**, 145–7.

BROWN, D. T., WESTPHAL, M., BURLINGHAM, B. T., WINDERHOFF, U. & DOERFLER, W. (1975). Structure and composition of the adenovirus type 2 core. *Journal of Virology*, **16**, 366–87.

BUTLER, P. J. G. & DURHAM, A. C. H. (1977). Tobacco mosaic virus protein aggregation and the virus assembly. *Advances in Protein Chemistry*, **31**, 187–251.

BUTLER, P. J. G., FINCH, J. T. & ZIMMERN, D. (1977). Configuration of tobacco mosaic virus RNA during virus assembly. *Nature*, **265**, 217–19.

BUTLER, P. J. G. & KLUG, A. (1971). Assembly of the particle of tobacco mosaic virus from RNA and disks of protein. *Nature New Biology*, **229**, 47–50.

BURNETT, R. M. (1983). Structural investigations on hexon, the major coat protein of adenovirus. In *Structural Biology*, ed. A. McPherson & F. A. Jurnak, vol. 2 of *The Viruses*. New York: Wiley.

CASJENS, S. (1979). Molecular Organization of the Bacteriophage P22 Coat Protein Shell. *Journal of Molecular Biology*, **131**, 1–19.

CASJENS, S. & KING, J. (1975). Virus assembly. *Annual Review of Biochemistry*, **44**, 555–611.

CASPAR, D. L. D. (1963). Assembly and stability of the tobacco mosaic virus particle. *Advances in Protein Chemistry*, **18**, 37–121.

CASPAR, D. L. D. & KLUG, A. (1962). Physical principles in the construction of regular viruses. *Cold Spring Harbor Symposia on Quantitative Biology*, **27**, 1–24.

CEPKO, C. L. & SHARP, P. (1982). Assembly of adenovirus major capsid protein is mediated by a nonviron protein. *Cell*, **31**, 407–15.

CHOPPIN, P. W., KLENK, H.-D., COMPANS, R. W. & CALIGUIRI, L. A. (1971). The parainfluenza virus SV5 and its relationship to the cell membrane. In *Perspectives in Virology*, vol. VII, ed. M. Pollard, pp. 127–56. New York: Academic Press.

COHEN, S. S. & McCORMICK, F. P. (1979). Polyamines and virus multiplication. *Advances in Virus Research*, **24**, 331–87.

COLMAN, P. M., VARGHESE, J. N. & LAVER, W. G. (1983). Structure of the catalytic and antigenic sites in influenza virus neuraminidase. *Nature*, **303**, 41–4.

COMPANS, R. W. (1971). Location of the glycoprotein in the membrane of Sindbis virus. *Nature New Biology*, **229**, 114–16.

COMPANS, R. W. & KLENK, H.-D. (1979). Viral Membranes. In *Comprehensive Virology*, vol. 13, ed. H. Fraenkel-Conrat & R. Wagner, pp. 293–407. New York: Plenum Press.

CRICK, F. H. C. & WATSON, J. D. (1956). Structure of small viruses. *Nature*, **177**, 473–5.

CROWTHER, R. A. & AMOS, L. (1971). Three-dimensional image reconstructions of some small spherical viruses. *Cold Spring Harbor Symposia on Quantitative Biology*, **36**, 489–94.

CROWTHER, R. A. & FRANKLIN, R. M. (1972). The structure of the groups of nine hexons from adenovirus. *Journal of Molecular Biology*, **68**, 181–4.

CUILLEL, M., JACROT, B. & ZULAUF, M. (1981). A T = 1 capsid formed by protein of brome mosaic virus in the presence of trypsin. *Virology*, **110**, 63–72.

DALES, S. (1973). Early events in cell–animal-virus interactions. *Bacteriology Review*, **37**, 103–35.

DAY, L. A. & WISEMAN, R. L. (1978). A comparison of DNA packaging in the virions of fd, xf, and Pf1. In *The Single-Stranded DNA Phages*, ed. D. T. Denhardt, D. Dressler & D. S. Ray, pp. 605–26. Cold Spring Harbor: Cold Spring Harbor Laboratory.

DEVAUX, C., ZULAUF, M., BOULANGER, R. & JACROT, B. (1982). Molecular weight of adenovirus serotype 2 capsomers. *Journal of Molecular Biology*, **156**, 927–39.

EARNSHAW, W., CASJENS, S. & HARRISON, S. C. (1976). X-ray diffraction from P22 heads. *Journal of Molecular Biology*, **104**, 387–410.

EARNSHAW, W. & HARRISON, S. C. (1977). DNA arrangement in isometric phage heads. *Nature*, **286**, 598–602.

EVERITT, E., LUTTER, L. & PHILIPSON, L. (1975). Structural proteins of adenovirus. XII. Location and neighbor relationship among proteins of adenovirion type 2 as revealed by enzymatic iodination, immunoprecipitation, and chemical cross-linking. *Virology*, **67**, 197–208.

EVERITT, E., SUNDQUIST, B., PETTERSON, U. & PHILIPSON, L. (1973). Structural proteins of adenoviruses. X. Isolation and topography of low molecular weight antigens from the virion of adenovirus type 2. *Virology*, **52**, 130–47.

FEY, G. & HIRT, B. (1975). Fingerprints of polyoma virus proteins and mouse histone. *Cold Spring Harbor Symposia on Quantitative Biology*, **39**, 235–41.

FIELDS, S., WINTER, G. & BROWNLEE, G. G. (1981). Structure of the neuraminidase gene in human influenza virus A/PR/8/34. *Nature*, **290**, 213–17.

FITZGERALD, D. J. P., PADMANABHAN, R., PASTAN, I. & WILLINGHAM, M. C. (1983). Adenovirus-induced release of epidermal growth factor and pseudomonas toxin into the cytosol of KB cells during receptor-mediated endocytosis. *Cell*, **32**, 607–17.

FRAENKEL-CONRAT, H. & WILLIAMS, R. C. (1955). Reconstitution of active tobacco mosaic virus from its inactive protein and nucleic acid components. *Proceedings of the National Academy of Sciences (USA)*, **41**, 690–8.

FRANKLIN, R. M., HARRISON, S. C., PETTERSON, U., PHILIPSON, L., BRANDEN, C. I. & WERNER, P. E. (1971). Structural studies on the adenovirus hexon. *Cold Spring Harbor Symposia on Quantitative Biology*, **36**, 503–10.

FULLER, M. T. & KING, J. (1980). Regulation of coat protein polymerization by the scaffolding protein of bacteriophage P22. *Biophysical Journal*, **32**, 381–98.

GAHMBERG, C. G., UTERMANN, G. & SIMONS, K. (1972). The membrane proteins of Semliki Forest virus have a hydrophobic part attached to the viral membrane. *FEBS Letters*, **28**, 179–182.

GAROFF, H., FRISCHAUF, A.-M., SIMONS, K., LEHRACH, H. & DELIUS, H. (1980a). The capsid protein of Semliki Forest virus has clusters of basic amino acids and prolines in its amino-terminal region. *Proceedings of the National Academy of Sciences (USA)*, **77**, 6376–80.

GAROFF, H., FRISCHAUF, A.-M., SIMONS, K., LEHRACH, H. & DELIUS, H. (1980b). Nucleotide sequence of cDNA coding for Semliki Forest virus membrane glycoproteins. *Nature*, **288**, 236–41.

GAROFF, H. & SODERLUND, H. (1978). The amphiphilic membrane glycoproteins of Semliki Forest virus are attached to the lipid bilayer by their COOH-terminal ends. *Journal of Molecular Biology*, **124**, 535–49.

GETHING, M. J., WHITE, J. M. & WATERFIELD, M. D. (1978). Purification of the fusion protein of Sendai virus: analysis of the NH_2-terminal sequence generated during precursor activation. *Proceedings of the National Academy of Sciences (USA)*, **75**, 2737–40.

GIBBS, A. & HARRISON, B. D. (1976). *Plant Virology*. London: Arnold.

GINSBERG, H. S. (1979). Adenovirus structural proteins. In *Comprehensive Virology*, vol. 13, ed. H. Fraenkel-Conrat & R. R. Wagner, pp. 409–57. New York: Plenum.

GOLDEN, J. S. & HARRISON, S. C. (1982). Proteolytic dissection of turnip crinkle virus subunit in solution. *Biochemistry*, **21**, 3862–6.

GRIFFITH, J. D. (1975). Chromatin structure: deduced from a minichromosome. *Science*, **187**, 1202–3.

GRIFFITH, J. D., MANNING, M. & DUNN, K. (1981). Filamentous bacteriophage contract into hollow spherical particles upon exposure to a chloroform–water interface. *Cell*, **23**, 747–53.

HAAS, R., MURPHY, R. F. & CANTOR, C. R. (1982). Testing models of the arrangement of DNA inside bacteriophage λ by crosslinking the packaged DNA. *Journal of Molecular Biology*, **159**, 71–92.

HARRISON, S. C. (1983). Packaging of DNA into bacteriophage heads – a model. *Journal of Molecular Biology*, in press.

HARRISON, S. C., DAVID, A., JUMBLATT, J. & DARNELL, J. (1971). Lipid and protein organization in Sindbis virus. *Journal of Molecular Biology*, **60**, 523–8.

HARRISON, S. C., OLSON, A., SCHUTT, C. E., WINKLER, F. K., & BRICOGNE, G. (1978). Tomato bushy stunt virus at 2.9 Å resolution. *Nature*, **276**, 368–73.

HELENIUS, A., KARTENBECK, J., SIMONS, K. & FRIES, E. (1980). On the entry of Semliki Forest virus into BHK-21 cells. *Journal of Cell Biology*, **84**, 404–20.

HELENIUS, A. & SODERLUND, H. (1973). Stepwise dissociation of the Semliki Forest virus membrane with triton X-100. *Biochemica et Biophysica Acta*, **307**, 287–300.

HERMODSON, M., ABAD-ZAPATERO, C., ABDEL-MEGUID, S. S., PUNDAK, S. & ROSSMANN, M. G. (1982). Amino acid sequence of southern bean mosaic virus coat protein and its relation to the three-dimensional structure of the virus. *Virology*, **119**, 133–49.

HOGLE, J. & HARRISON, S. C. (1984). Structure of TCV at 3.2 Å resolution. Submitted to the Journal of Molecular Biology.

HOLMES, K. C. (1980). Protein–RNA interactions during the assembly of tobacco mosaic virus. *Trends in Biochemical Sciences*, **5**, 4–7.

HOPPER, P., HARRISON, S. C. & SAUER, R. (1984). Amino acid sequence of the TBSV coat protein subunit. Submitted to the Journal of Molecular Biology.

INCARDONA, N. L. & KAESBERG, P. (1964). A pH-induced structural change in bromegrass mosaic virus. *Biophysics Journal*, **4**, 11–21.

JASPARS, E. M. J. (1974). Plant viruses with a multipartite genome. *Advances in Virus Research*, **19**, 37–149.

JONARD, G., RICHARDS, K. E., GUILLEY, H. & HIRTH, L. (1977). Sequence from the assembly nucleation region of TMV RNA. *Cell*, **11**, 473–93.

KELLENBERGER, E. (1972). Mechanisms of length determination in protein assemblies. *Ciba Foundation Symposium*, **7**, 295–9.

KIKUCHI, Y. & KING, J. (1975). Genetic control of bacteriophage T4 baseplate morphogenesis. III. Formation of the central plug and overall assembly pathway. *Journal of Molecular Biology*, **99**, 695–716.

KING, J. & CASJENS, S. (1974). Catalytic head assembling protein in virus morphogenesis. *Nature*, **251**, 112–19.

KING, J., LENK, E. & BOTSTEIN, D. (1973). Mechanism of head assembly and DNA encapsulation in *Salmonella* phage P22. II. Morphogenetic pathway. *Journal of Molecular Biology*, **80**, 697–731.

KLENK, H. D., ROTH, R., ORLICK, M. & BLODORN, J. (1975). Activation of influenza A viruses by trypsin treatment. *Journal of Virology*, **68**, 426–39.

KLUG, A. (1979). The assembly of tobacco mosaic virus: structure and specificity. *Harvey Lectures*, **74**, 141–72.

KLUG, A. & CASPAR, D. L. D. (1960). The structure of small viruses. *Advances in Virus Research*, **7**, 225–325.

KLUG, A. & DURHAM, R. A. (1971). The disk of TMV protein and its relation to the helical and other modes of aggregation. *Cold Spring Harbor Symposia on Quantitative Biology*, **36**, 449–60.

KRUSE, J., KRUSE, K. M., WITZ, J., CHAUVIN, C., JACROT, B. & TARDIEU, A. (1982). Divalent ion-dependent reversible swelling of tomato bushy stunt virus and organization of the expanded virion. *Journal of Molecular Biology*, **162**, 393–417.

LAVER, W. G. & VALENTINE, R. C. (1969). Morphology of the isolated hemagglutinin and neuraminidase subunits of influenza virus. *Virology*, **38**, 105–19.

LEBEURIER, G., NICOLAEFF, A. & RICHARDS, K. E. (1977). Inside-out model for self-assembly of tobacco mosaic virus. *Proceedings of the National Academy of Sciences (USA)*, **74**, 149–53.

LILJAS, L., UNGE, T., FRIDBORG, K., JONES, T. A., LOVGREN, S., SKOGLUND, O. & STRANDBERG, B. (1982). Structure of satellite tobacco necrosis virus at 3.0 Å resolution. *Journal of Molecular Biology*, **159**, 93–108.

McCARTHY, M. & HARRISON, S. C. (1977). Glycosidase susceptibility: a probe for the distribution of glycoprotein oligosaccharides in Sindbis virus. *Journal of Virology*, **23**, 61–73.

MOOSIC, J. P. (1978). The primary structure of brome mosaic virus coat protein. PhD thesis, University of Wisconsin, Madison.

MUNOWITZ, M. G., DOBSON, C. M., GRIFFIN, R. G. & HARRISON, S. C., (1980). On the rigidity of RNA in tomato bushy stunt virus. *Journal of Molecular Biology*, **141**, 327–33.

NERMUT, M. V., HARPST, J. A. & RUSSELL, W. C. (1975). Electron microscopy of adenovirus cores. *Journal of General Virology*, **28**, 49–58.

OLSON, A. J., BRICOGNE, G. & HARRISON, S. C. (1983). Structure of tomato bushy stunt virus: the virus particle at 2.9 Å resolution. *Journal of Molecular Biology*, in press.

PETER, R., STEHELM, D., REINBOLT, J., COLLOT, D. & DURANTON, H. (1970). Primary structure of turnip yellow mosaic virus coat protein. *Virology*, **49**, 615–17.

RAYMENT, I., BAKER, T. S., CASPAR, D. L. D. & MURAKAMI, W. T. (1982). Polyoma virus capsid structure at 22.5 Å resolution. *Nature*, **295**, 110–15.

RICE, E. M., BELL, J. R., HUNKAPILLER, M. W., STRAUSS, E. G. & STRAUSS, J. H. (1982). Isolation and characterization of the hydrophobic COOH-terminal domains of the Sindbis virion glycoproteins. *Journal of Molecular Biology*, **154**, 355–78.

RICE, E. M. & STRAUSS, J. H. (1981). Nucleotide sequence of the 26S mRNA of Sindbis virus and deduced sequence of the encoded virus structural proteins. *Proceedings of the National Academy of Sciences (USA)*, **78**, 2062–6.

RICHARDS, K., WILLIAMS, R. & CALENDAR, R. (1973). Mode of DNA packing within bacteriophage heads. *Journal of Molecular Biology*, **78**, 255–9.

ROBINSON, I. K. & HARRISON, S. C. (1982). Structure of the expanded state of tomato bushy stunt virus. *Nature*, **297**, 563–8.

SCHMIDT, M. F. G., BRACHA, M. & SCHLESINGER, M. J. (1979). Evidence for covalent attachment of fatty acids to Sindbis virus glycoproteins. *Proceedings of the National Academy of Sciences (USA)*, **76**, 1682–91.

SKEHEL, J. J., BAYLEY, P. M., BROWN, E. B., MARTIN, S. R., WATERFIELD, M. D., WHITE, J. M., WILSON, I. A. & WILEY, D. C. (1982). Changes in the conformation of influenza virus hemagglutinin at the pH optimum of virus-mediated membrane fusion. *Proceedings of the National Academy of Sciences (USA)*, **79**, 968–72.

SORGER, P. K., STOCKLEY, P. G. & HARRISON, S. C. (1984). *Journal of Molecular Biology*, submitted.

STRAUSS, J. H. & STRAUSS, E. G. (1977). Togaviruses. In *The Molecular Biology of Animal Viruses*, vol. 1, ed. D. Nayak, pp. 111–166. New York: Marcel Dekker.

STUBBS, G., WARREN, S. & HOLMES, K. C. (1977). Structure of RNA and RNA binding site in tobacco mosaic virus from 4 Å map calculated from X-ray fibre diagrams. *Nature*, **267**, 216–21.

UTERMANN, G. & SIMONS, K. (1974). Studies on the amphipathic nature of the membrane proteins in Semliki Forest virus. *Journal of Molecular Biology*, **85**, 569–87.

VARGHESE, J. N., LAVER, W. G. & COLEMAN, P. M. (1983). Structure of the influenza virus glycoprotein antigen neuraminidase at 2.9 Å resolution. *Nature*, **303**, 35–40.

VON BONSDORFF, C. H. & HARRISON, S. C. (1975). Sindbis virus glycoproteins form a regular icosahedral surface lattice. *Journal of Virology*, **16**, 141–5.

VON BONSDORFF, C. H. & HARRISON, S. C. (1978). Hexagonal glycoprotein arrays from Sindbis virus membranes. *Journal of Virology*, **28**, 578–83.

VON BONSDORFF, C. H. & PETTERSSON, R. (1975). Surface structure of Uukuniemi virus. *Journal of Virology*, **16**, 1296–307.

VON HEIJNE, G. (1981). Membrane proteins – the amino acid composition of membrane-penetrating segments. *European Journal of Biochemistry*, **120**, 275–8.

VRIEND, G., HEMMINGA, M. A., VERDUIN, B. J. M., DEWIT, J. L. & SCHAAFSMA, T. J. (1982). Swelling of cowpea chlorotic mottle virus studied by proton nuclear magnetic resonance. *FEBS Letters*, **146–7**, 319–21.

WAGENKNECHT, T. & BLOOMFIELD, V. A. (1978). Length regulation of T4 phage tail tubes. *Biophysics Journal*, **21**, 90A.

WARD, C. W. & DOPHEIDE, T. A. (1980). The Hong Kong (H3) hemagglutinin complete amino acid sequence and oligosaccharide distribution for the heavy chain of A/Memphis/102/72. In *Structure and Variation in Influenza Virus*, ed. W. G. Laver & G. M. Air, pp. 27–38. New York: Elsevier.

WATSON, J. D. (1954). The structure of tobacco mosaic virus. I. X-ray evidence of a helical arrangement of subunits around a longitudinal axis. *Biochimica et Biophysica Acta*, **13**, 10–19.

WHITE, J. & HELENIUS, A. (1980). pH-dependent fusion between the Semliki Forest virus membrane and liposomes. *Proceedings of the National Academy of Sciences (USA)*, **77**, 3273–7.

WHITE, J. KARTENBECK, J. & HELENIUS, A. (1981a). Fusion of Semliki Forest virus with the plasma membrane can be induced by low pH. *Journal of Cell Biology*, **87**, 264–72.

WHITE, J. MATLIN, K. & HELENIUS, A. (1981b). Cell fusion by Semliki Forest influenza, and vesicular stomatitis viruses. *Journal of Cell Biology*, **89**, 674–9.

WIDOM, J. & BALDWIN, R. (1983). Tests of spools models for DNA packaging in phage lambda. *Journal of Molecular Biology*, in press.

WILEY, D. C. & SKEHEL, J. J. (1977). Crystallization and X-ray diffraction studies on the haemagglutinin glycoprotein from the membrane of influenza virus. *Journal of Molecular Biology*, **112**, 343–7.

WILEY, D. C., SKEHEL, J. J. & WATERFIELD, M. D. (1977). Evidence from studies with a cross-linking reagent that the haemagglutinin of influenza virus is a trimer. *Virology*, **79**, 446–8.

WILEY, D. C. & VON BONSDORFF, C. H. (1978). Three-dimensional crystals of the lipid enveloped Semliki Forest virus. *Journal of Molecular Biology*, **120**, 375–9.

WILSON, I. A., WILEY, D. C. & SKEHEL, J. J. (1981). Structural identification of the antibody-binding sites of Hong Kong influenza haemagglutinin and their involvement in antigenic variation. *Nature*, **289**, 373–8.

WIRTH, D., KATZ, F., SMALL, B. & LODISH, H. (1977). How a single Sindbis virus mRNA directs the synthesis of one soluble protein and two integral membrane glycoproteins. *Cell*, **10**, 253–63.

ZIMMERN, D. (1977). The nucleotide sequence at the origin for assembly on tobacco mosaic virus RNA. *Cell*, **11**, 463–82.

Zuidema, D., Bierhuizen, M. F. A., Cornelissen, B. J. C., Bol, J. F. & Jaspars, E. M. J. (1983). Coat protein binding sites on RNA 1 of alfalfa mosaic virus. *Virology*, **125,** 361–9.

THE NATURE OF ANIMAL VIRUS GENETIC MATERIAL

DUNCAN J. McGEOCH

MRC Virology Unit, Institute of Virology, Church Street, Glasgow G11 5JR, UK

INTRODUCTION

Animal virology has experienced a revolution in the last five years. In 1978 the first complete sequence of an animal virus genome, that for SV40 DNA, was published (Fiers *et al.*, 1978; Reddy *et al.*, 1978). Since then published sequence data have increased steadily, so that at the time of writing (June 1983) I have been able to compile a list of some 13 totally sequenced animal virus genomes, as well as a mountain of smaller scale sequences. Over the same period, nucleotide sequencing technology has steadily advanced, to the point where it is now possible, for *any* virus, to determine the complete nucleotide sequence (Sanger *et al.*, 1982; Dunn & Studier, 1983). The amount of work necessary to obtain this varies greatly from case to case. Nonetheless, I assert that in no case does the estimated work input exceed what (by the standards of biological science) would be regarded as a moderate to large, but certainly feasible, experimental project. We thus stand at the edge of an era of knowing about many virus genomes in exhaustive detail. Problems of access to the huge amount of sequence data now published have led to the establishment of national and international sequence data libraries. In April 1983, the library organized by the European Molecular Biology Laboratory contained 391 292 residues of sequences classified as 'viral/phage', and this represented 35% of the total library content (*EMBL Nucleotide Sequence Data Library News*, no. 2).

This paper is concerned with examining the nature of animal virus genomes in terms of their chemical and physical properties and in terms of what we can presently understand about their organization and informational content. It thus primarily considers virus genomes through examples of known nucleotide sequences, attempting to achieve the general through examination of the particular. The discussion falls into three main parts. The first is scene-setting and could be headed 'Theory'. It deals with definitions of what constitutes a virus genome for our present purposes, with statistical properties of sequences, and with relations between sequences. The second part treats two aspects of sequence interpretation which correspond to widespread phenomena: information overlay and consensus sequences. In the third part I examine aspects of sequences of different virus types. Because of space limitations, and also because of our present state of incomplete (although substantial) knowledge and interpretation, this part is rigorously eclectic, and is organized under three headings: RNA viruses; small DNA viruses; and large DNA viruses.

Present day nucleotide sequencing systems are powerful and are undoubtedly capable of analysing virus genome primary structures with great accuracy (Sanger, Nicklen & Coulson, 1977; Maxam & Gilbert, 1980). However, the resulting descriptions of genomes do have limitations, and it is well to register these at the start of this discussion. Many genomes exhibit special features, or must be regarded as possessing the potential for such features (see Matthews, 1982). These include: single strand nicks or gaps in otherwise duplex nucleic acids; modifications, such as methylation; and terminal structures, such as covalently linked proteins (in adenoviruses and picornaviruses), or covalently linked strands of duplex DNA (in poxvirus), or methylated cap structures (in some RNA viruses). Most present sequencing practice, for both DNA and RNA, involves cloning into a prokaryotic vector, or making a copy *in vitro*, or both, and information on any special structures is necessarily lost in such procedures.

Furthermore, it is emerging that some regions of some virus genomes are intrinsically highly variable, for instance, certain variable copy number reiterated sequences in herpesviruses (see section on large DNA viruses). Examination of the sequence of one clone of such a region thus gives an incomplete and static picture. I regard such variability as qualitatively different from variation due to high rates of distributed point mutation (Reanney, this volume).

Finally, it is certainly true that cloning procedures can induce changes. There is a case in the genome of herpes simplex virus, where all attempts so far to clone a certain restriction fragment have yielded partially deleted progeny (Spaete & Frenkel, 1982; B. Matz and N. Stow, personal communications). These classes of complication could thus impose a limit of precision and interpretation. Given a reasonable level of cross-checking in a conscientiously executed sequencing project, they should present very infrequent problems.

THE GENETIC NATURE OF VIRUSES

This chapter is concerned with structure and functionality of virus genomes, with no overt cognizance of viruses as pathogens. I present here a statement of the nature of the virus genome, suitable for these purposes.

Viruses are genetic elements. In common with all other genetic elements, the virus can exist inside a cell, where its genetic functions may be expressed and it may replicate. Unlike all other genetic elements, the virus has the specific capacity also to exist in a functionally inert, extracellular form. The virus genome encodes information to enable its own replication, and information to enable its extracellular existence. In addition this implies that, at some level, encoded information exists which enables the transitions between the two states. Minimally, information in the first category might represent an origin for replication of the virus genome by cell machinery. At increasing levels of complexity of virus genome structure, this category could correspond to encoding of polypeptides, first, to participate directly in replication, and, second, to participate indirectly by redirecting the cell's metabolism. The almost universal form of information to enable extracellular existence has been the encoding of polypeptide to form some version of a protective coat for the genome. However, I note that with viroids (which *are* viruses by these criteria, although not further treated here) the second coding function apparently comprises specification of a secondary structure for the naked genome (see Sänger, this volume).

Some virions contain, apart from the packaged genome, other factors which function inside the newly infected cell to establish infection. The best known examples of this are the carrying of RNA transcriptases by negative strand RNA virions (Baltimore, Huang & Stampfer, 1970), and reverse transcriptase by retroviruses (Baltimore, 1970; Temin & Mizutani, 1970). The most developed case is found with poxviruses; the vaccinia virion contains at least 12 distinct enzyme species associated with cytoplasmic expression of the DNA genome (McFadden & Dales, 1982). The phenomenon may be more widespread than has been suspected: for instance, recent work with herpes simplex virus has indicated that the virus carries a factor necessary for efficient transcription of immediate early genes (Post, Mackem & Roizman, 1981; Mackem & Roizman, 1982; Cordingley, Campbell & Preston, 1983).

Virus genomes have been found to be constructed of either RNA or DNA. In retroviruses, they are RNA at one stage and DNA at another. This fact, that many (or most) virus genomes are RNA, is now such a resounding cliche that we have become blind to its unique nature: viruses are the *only* known class of genetic system to include RNA genomes. There is no clear answer as to why this should be so. It may correlate with the fact that RNA replication is error-prone and can thus only be viable where a large replication burst allows reselection at each generation. However, this argument may be circular: imputing an intrinsically and uniquely error-prone nature to all RNA replication is simplistic, since the existing RNA replication systems are also the product of evolution, and are presumably as precise, or imprecise, as they need to be.

Virus genomes, and in particular DNA virus genomes, cover a great size range. The smallest characterized DNA virus genome (hepatitis B; Tiollais, Charnay & Vyas, 1981) is 3200 residues in length; the largest (cytomegalovirus; Geelen *et al.*, 1978) is around 250 000 residues. In small viruses, and as a broad generalization, roughly half of the genome encodes replication functions, and the other half comprises genes for virion components. As we look at larger genomes we might expect to see some radically new class of function emerging. The only case really fitting this category may be the carrying of transforming genes by retroviruses (Bishop, 1983). In general, however, larger genomes encode more complex virions, and they are replicated in more elaborate ways which are less dependent on host cell functions, but the rough 'half and half' rule still holds, with (for example) genes for enzymes of nucleotide metabolism being assigned for the replication class.

GLOBAL PROPERTIES OF SEQUENCES

This section deals with broad statistical measures characterizing the ordering of residues in a nucleic acid. The crudest of such measures is a listing of base composition. I remark here only that small viruses tend to have base compositions close to their host cell genome values, while values for large genome viruses are less constrained. Much greater insight into the ordering of residues in a sequence is obtainable from a listing of the relative frequencies of the 16 possible dinucleotides.

A sequence of given base composition could be constructed (on paper) in various ways to give quite different results. For example, we could specify the ordering as all the As, followed by the Ts, and so on; or as repeats of some tetranucleotide, or by random selection of each successive residue. The dinucleotide (or doublet) frequencies for each sequence type would be distinct, and evidently would give an indication of the constraints exhibited by the sequence. An experimental scheme for such nearest-neighbour analysis of DNA was described by Josse, Kaiser & Kornberg (1961), long before the advent of modern sequencing techniques. This system was applied to study of virus nucleic acids by Subak-Sharpe and his colleagues, who found that each virus type exhibited a characteristic pattern. For small animal viruses, this pattern resembled the host cell's genome DNA pattern rather closely (in particular with a remarkably low frequency of 5'-CG); large DNA virus patterns did not so resemble host DNA (Subak-Sharpe et al., 1966; Morrison, Keir, Subak-Sharpe & Crawford, 1967; Hay & Subak-Sharpe, 1968).

The doublet frequency set of a virus DNA or RNA genome must reflect the pressures to which the sequence has been subjected in its evolution. These include: randomizing tendency of mutations; constraints resulting from the amino acid compositions and orders, of encoded proteins; constraints from any systematic preference between synonymous codons; constraints on codon ordering; and constraints from special uses of certain sequences, and from conformational requirements of the sequence. The use of doublet frequencies for comparisons between sequences was set on an objective statistical basis by Elton (1975).

Gatlin (1972) in a book entitled *Information Theory and the Living System*, used information theory to derive two functions, computable from doublet frequencies, which measure separately the amount of redundancy in a sequence due to divergence from uniform base composition, and the amount of redundancy due to divergence of successive residues from independence. With the advent of large amounts of data from nucleotide sequencing, these ideas have been revived and developed for statistical descriptions of sequence properties computed from the primary sequence data. Further statistical measures have also been introduced to examine, for instance, special features associated with the use of triplet codons (Lipman & Maizel, 1982; Lipman et al., 1982; Smith, Waterman & Sadler, 1983). Although not emphasized in this account, these measures are of fundamental importance. They represent the conceptual tools appropriate for examination of the

basic nature of a sequence, after the usual primary goal of interpretation in terms of gene layout has been achieved.

RELATIONS BETWEEN SEQUENCES

Comparisons between sequences are an important aspect of
sequence analysis, in several contexts: (*a*) in analysis of functional
sites, as discussed below in the section on consensus sequences (in
this case, similarity between sequences is generally considered as
resulting from functional constraints rather than evolutionary relatedness – that is, the sequences are analogous rather than
homologous); (*b*) in evaluation of clearly related sequences, to
obtain a measure of relatedness and to examine divergences; (*c*) in
examination of sequences which are not closely related, in order to
test the validity of a suspected relationship. It should be noted that
random alignment of any two sequences will give some coincidence
of residue types, which would be scored in a simple homology
count. For random sequences containing equal proportions of four
nucleotides, such coincidence will be distributed about 25%.

It has become clear that, except in closely related cases, the
objective alignment of two sequences to maximize homology is not a
trivial task. Two classes of computerized procedures are now used.
The first is the matrix plotting method, in which homologies are
presented graphically (Maizel & Lenk, 1981; Pustell & Kafatos,
1982). Sequences to be compared are aligned on the *x* and *y* axes of
a graph, and the coordinates of locally homologous regions are
marked. This method is qualitative and requires significant user
interpretation, but can be excellent for presenting the whole field of
comparison between two sequences.

The second method depends on computer implementation of
algorithms which search rigorously and exhaustively for optimal
alignments (including addition/deletion changes) and which, for
each alignment, give a quantitative measure of homology (Sellers,
1980; Smith and Waterman, 1981; Gotoh, 1982). Two aspects
require care with this class of procedure. First, relative penalties
applied to introduction of mismatches and of insertion/deletion
changes are user-specified. 'Optimal alignment' is therefore *not* an
absolute. Second, the significance of low level homology detection
has to be evaluated by using comparisons between shuffled sequences as a baseline, and significant assumptions about the general

constraints operating on real sequences are made in producing such reference sequences. Ideally, all recognized constraints should be preserved (for instance, at least nearest-neighbour frequencies as well as base composition) and this has not generally been done (Fitch, 1983).

Finally, I note that with the development of the theoretical and computer software tools described in this section it is possible to evaluate relatedness of sequences in an hierarchical way, by use of the successively broader criteria of homology, doublet frequency, and information theory measures (Lipman *et al.*, 1982; see section on RNA viruses, below).

INFORMATION OVERLAY

Barrell, Air & Hutchinson (1976) described a situation in the genome of bacteriophage ϕX174 in which the polypeptide-coding DNA sequence of one gene was wholly contained in the coding sequence of another, with the two polypeptides being translated from separate reading frames. This first description of the phenomenon of 'overlapping genes' has since been followed by many others, for instance in RNA phages (Atkins *et al.*, 1979; Beremond & Blumenthal, 1979); in DNA animal viruses (Fiers *et al.*, 1978); and in RNA animal viruses (Lamb & Lai, 1980; Porter, Smith & Emtage, 1980).

Various transcriptional mechanisms are invoked to express such loci, reflecting the transcriptional strategies of the various systems. These phenomena have been widely interpreted as illustrating the efficiency of utilization of small virus genomes. This is probably a valid interpretation, but I consider that it has been over-emphasized and that two qualifications are appropriate. First, the phenomenon is not unique to *small* viruses, since an example is now available in the 160 000 bp genome of herpes simplex virus, which does not in other ways exhibit pronounced 'compression' of genome information (F. J. Rixon & D. J. McGeoch, unpublished). The initial unique association with small virus genomes was evidently due to the relative ease of study and interpretability of their genome structures. My second point concerns the evolutionary value of such systems. The functionality of overlapping genes is evident. However, they must impose additional constraints on genome sequence, and the price for their development could be a loss of evolutionary plasticity. I regard overlapping genes as examples of opportunistic variation, which can occur in any situation where the gene expression machinery allows the possibility and the level of constraints on the genome sequence is favourable.

Polypeptide-coding overlap constitutes only one example (but the most forceful) of what I now term 'information overlay', the encoding of more than one signal in the same DNA or RNA sequence. Some other overt overlays relate to the expression of overlapping genes or transcription units, for instance overlay of RNA splice signals on polypeptide-coding sequences (Lamb & Lai, 1980); overlay of promoter sequences on polypeptide-coding sequences (McLauchlan & Clements, 1983; F. J. Rixon & D. J. McGeoch, unpublished); and overlay of polyadenylation signals on coding sequences. There also exist other well characterized overlay classes: in tobacco mosaic virus RNA, signals for initiation and propagation of genome packaging are overlaid on polypeptide-coding RNA (Goelet et al., 1982).

Possibly the most general and important class of overlay information concerns nucleic acid conformation. A clear example of this is the secondary structure adopted by phage MS2 RNA, which acts in a translational control mechanism (Kastelein et al., 1982). In an analogous way, sequences in duplex DNA could encode functional variations in local conformation independently of their polypeptide-coding or other functions. It has been claimed that repetitive trends are discernible in the sequences of eukaryotic DNAs which form nucleosome structures, and that a resulting conformational variation facilitates DNA bending in the nucleosome (Trifonov, 1980). This general class of phenomena could be very widespread, but it is certainly intrinsically elusive and hard to study systematically.

CONSENSUS SEQUENCES

It is clear that many processes involving specific sites on nucleic acids do not correlate with unique or easily defined nucleotide sequences. In an early example of this phenomenon, comparisons of promoter sites for *Escherichia coli* RNA polymerase showed that many different DNA sequences can act as promoters (for a modern version, see Hawley & McClure, 1983). Alignment of such sequences and abstraction of the common elements produced a *consensus* sequence. Consensus sequences associated with many classes of

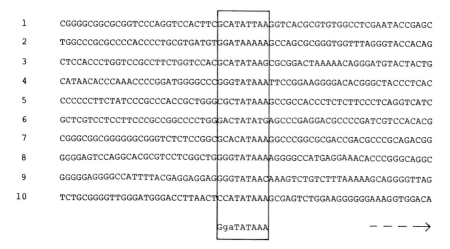

Fig. 1. Sequences near transcription initiation sites for RNA polymerase II in the genome of herpes simplex virus type 1.

Ten examples are listed of DNA sequences adjacent to mRNA start sites. The strand listed is the mRNA sense strand, and mRNA 5'-termini are in the 11 residue region at the right, as indicated by the arrow. The derived 'TATA box' consensus is indicated below. Sources of data are: (1) TK gene (Preston & McGeoch, 1981); (2) and (3) early promoters in U_l (McLauchlan & Clements, 1983); (4) and (5) promoters in the glycoprotein C region of U_l (Frink et al., 1983); (6) and (7) immediate early promoters in R_s (Murchie & McGeoch, 1982); (8) – (10) promoters in U_s (D. J. McGeoch, A. Dolan, S. Donald & F. J. Rixon, unpublished).

event in nucleic acid metabolism of animal viruses have been described, including sites involved in transcription, translation and DNA metabolism. A spectrum of types is discernible. At one end we have sequences involved in more or less well specified ways with characterized functional entities. RNA polymerase promoters in prokaryote and eukaryote systems fall in this class. Next are sequences clearly associated with a specific process, but presently lacking characterized 'hardware', for example, splice donor and acceptor sequences associated with splicing of eukaryote mRNA (Mount, 1982). Finally there are examples lacking defined function. In this category I would place sequences far upstream from eukaryotic mRNA sites. As far as viruses are concerned, depending on the strategy of replication, consensus sites may be virus specific (for instance, transcription signals in poxviruses or in RNA viruses), or more general (for instance, transcription signals for RNA polymerase II, in DNA viruses other than poxviruses).

As an example of consensus, Fig. 1 lists sequences in the genome of herpes simplex virus which are in the neighbourhood of RNA polymerase II sites. Clearly there are some features in common, but very little that is completely constant. There are several reasons why this might be so. First, such promoters differ in their power (the rate of transcript production) and in their regulation. Many promoters, for instance, are not recognized by the cell's unmodified transcription apparatus early in infection. The second reason is more basic: we are forced to concede that a significant component of the variability is a basic consequence of the nature of molecular interactions between polymerase and DNA.

For work on promoter function, this means that a different experimental or interpretative attack must be tried. For work on interpretation of DNA sequence data in terms of transcription patterns and so on, the next question is: can the available consensus be used to deduce correctly the positions of previously uncharacterized promoters in a known DNA sequence? The answer is disappointing: such searches may fail to find real promoters and commonly give false positives. The test is therefore not usefully *predictive*. It is, however, *correlative*, in that, knowing the position in a DNA sequence of the 5'-terminus of an mRNA, the appropriate consensus elements can most often be identified. Consensus sequences associated with RNA polymerase splice sites are similarly limited in that they generally give too many false positives to be useful in prediction. The so-called 'polyadenylation' consensus (Fitzgerald & Shenk, 1981) $A_T^A TAAA$ is more powerful: it can give a few false positives, but probably does not miss real sites, and it may be possible to improve its discrimination by extending the consensus to include conserved neighbouring sequence elements (J. McLauchlan & J. B. Clements, personal communication).

The conclusion from such inspections, that many site-specific processes are not associated with unique sequences, prompts examination of processes which do involve unique sequences. The best studied examples are the triplet code for amino acids, and the recognition sites for many type II restriction enzymes. The genetic code contains examples of unique association of structure and function (UGG, and only UGG, encodes tryptophan; AUG, and only AUG, encodes methionine internally in polypeptides). However, if its discovery had proceeded historically by different routes, many aspects of the code might today be regarded as consensus sequences; for instance, glycine is encoded by the consensus GGN. Instead, these aspects are (correctly) regarded as exemplifying the code's redundancy, and can be understood in light of theories on codon–anticodon interactions and multiple tRNAs (Crick, 1966).

However, there are two areas where the operation of the genetic code is now assuming a significant aspect of consensus. This excludes the case of mitochondrial genetic codes, which is clearly not applicable to this discussion. First, until recently it was considered that in eukaryotic mRNAs the AUG sequence found nearest to the 5′-terminus constituted the initiation codon for translation (Kozak, 1980). Recent sequence data have now shown this not to be invariably the case. With poliovirus RNA, for instance, it is considered that translation starts at the ninth AUG (Kitamura *et al.*, 1981; Racaniello & Baltimore, 1981). In the thymidine kinase gene of herpes simplex virus, translation is initiated independently at the first, the second and the third AUG (Preston & McGeoch, 1981; Marsden, Haarr & Preston, 1983). Since these are in the same reading frame, this results in a protein product variable at its N-terminus. Kozak (1981) has examined the sequence environments of initiator codons and has produced a consensus of surrounding sequence which correlates with translational initiator activity. The mechanistic basis of this effect remains obscure.

The second instance of the genetic code apparently involving a consensus element concerns translation termination. In *E. coli* it has been shown that efficiency of suppression of a terminator codon depends on adjacent sequences (Bossi, 1983; Miller & Albertini, 1983). This, of course, represents an experimentally constructed situation, but the notion that termination efficiency can depend on sequence context has relevance to virus gene expression: it is known that terminator codon readthrough proteins are produced with retroviruses (Philipson *et al.*, 1978) and with tobacco mosaic virus (Pelham, 1978; Goelet *et al.*, 1982) and it seems that in these cases the sequence adjacent to the terminator may have evolved so as to allow some suppression by a member of the normal tRNA population.

Turning to consideration of restriction nuclease sites, it is clear that many are uniquely defined, while others are defined by a precisely delineated consensus (Modrich & Roberts, 1982; Roberts, 1982). The perceived role of a true restriction endonuclease demands selection for an all or none effect, rather than some graded response to similar sequences. Even in this case, however, experimental manipulation of conditions *in vitro* can change many enzymes' specificities and broaden an unique recognition site to a consensus (Modrich & Roberts, 1982).

Before discussing the nature of variable sequence sites, I consider it appropriate to express my own misgivings about the widespread use of the 'consensus' technique. Except in the best cases, and where the relevant machinery is well defined (as with *E. coli* RNA polymerase), the construction of a consensus can only be a crude *ad hoc* presentation of common features, with inadequate definition of variability or range of application. Within these limitations, such a device is reasonable and could be valuable; however, it is prone to being invested with a specious validity as representing the 'essence' of a functional site.

The physicochemical properties of DNA and RNA are very considerably more complex than representation as a string of Roman characters on paper allows for. The local thermodynamic stability of base-paired nucleic acids is known to be sequence dependent as well as composition dependent (Gotoh & Tagashira, 1981; Patel *et al.*, 1983). Certain functional groups of different nucleotide residues occupy equivalent positions in a DNA double helix, and could thus make the same possible contacts in DNA–protein binding (Sadler, Waterman & Smith, 1983). Finally, evidence is accumulating that local DNA sequence can have pronounced effect on the conformation of a DNA duplex; such conformational variety could be involved in specific DNA–protein interaction (Dickerson, 1983; Nordheim & Rich, 1983).

One can thus discern mechanisms by which different sequences could behave similarly. A sequence involved in a protein–nucleic-acid interaction must be subject to a different set of constraints from a nucleic acid sequence functional by virtue of some special intrinsic property (such as, ease of local strand separation). In the latter case one could imagine the sequence being of variable length, for instance, whereas in the former case any multiple contacts should preclude such variability. All this underlines the desirability of direct biochemical characterization of processes involving such variable sequence regions.

RNA VIRUSES

This section addresses characteristics of the genomes of RNA animal viruses. Discussing this diverse set of genomes in one undivided category carries some imputation of hubris. In defence, I note that aspects of RNA virus genomes are discussed at length elsewhere in this volume. I deal here only with two related topics: the possibilities for structural variation of RNA genomes, and an example of hierarchic analysis of sequences.

RNA animal virus genomes may be positive sense, single strand RNA, or negative sense, single strand RNA or double strand RNA (see Matthews, 1982). Of the positive strand viruses, complete genome sequences are known for strains of the picornavirus, polio virus (Kitamura *et al.*, 1981; Racaniello & Baltimore, 1981; Nomoto *et al.*, 1982), and substantial data for the alphaviruses, Semliki Forest and Sindbis (Garoff *et al.*, 1980; Rice & Strauss, 1981), and for a coronavirus (Armstrong, Smeekens & Rottier, 1983).

Negative strand genomes may be either segmented or mono-molecular. In the former class are orthomyxoviruses (influenza viruses) and bunyaviruses. Influenza virus RNA sequences have probably been more extensively characterized than any other virus RNAs. All eight segments of one strain of influenza A virus (PR/8/34) have been completely sequenced (see Winter & Fields, 1982), and there are extensive data on other influenza A strains and some influenza B data (reviewed by McCauley & Mahy, 1983). Substantial but not complete sequences are available for several bunyaviruses (Bishop *et al.*, 1982).

In the group of monomolecular negative strand genomes, no complete sequences have been determined. However, there are extensive data for the rhabdovirus, vesicular stomatitis virus, and some for rabies virus (McGeoch, 1979; Rose, 1980; Schubert *et al.*, 1980; Anilionis, Wunner & Curtis, 1981; Gallione *et al.*, 1981; Rose & Gallione, 1981).

This discussion deals nowhere with retroviruses, but I note here that three complete genome sequences have been published (Reddy, Smith & Aaronson, 1981; Shinnick, Lerner & Sutcliffe, 1981; van Beveren *et al.*, 1981; Schwartz, Tizard & Gilbert, 1983). From the viewpoint of this paper, retrovirus genomes are neither fish nor fowl: by the criteria developed above on the essence of 'virus', they are as much DNA genomes as RNA genomes, and many of their properties are better understood as such.

All classes of RNA virus genomes for which we have comparative sequence data show variation in sequence between strains of serotypes, especially in genes encoding external structural proteins, which are primary targets for selection by antibody response in the host population. For some viruses which are clearly related by criteria of virion structure and genome organization, such genes have diverged to the point where the sequence relations are only just detectable. An example of this is the comparison of external glycoprotein genes of the two rhabdoviruses (vesicular stomatitis virus and rabies virus) where low level homology between the amino acid sequences can be seen, but little homology between the RNA sequence is detectable (Rose *et al.*, 1982).

Such variation has evidently proceeded by point mutation and by small addition/deletion events. However, large scale structural alterations or rearrangements appear to be rare. Apart from the posssible absence of appropriate molecular mechanisms to enable such changes, this may reflect the highly specialized forms into which these genomes have evolved, in which only the most restrained gestures of structural variations are permissible. In picornaviruses, the only large scale variation discernible is the existence of a variable length poly(C) tract near the 5'-terminus of the genome, in some types but not others (Black *et al.*, 1979).

In orthomyxoviruses, one large scale genome variation has been described. This is the absence in influenza C virus of a genome RNA segment corresponding to the neuraminidase-encoding segment of influenza A and B (Palese *et al.*, 1980). This correlates with the distinct cell surface structures, not involving neuraminic acid, which are recognized by influenza C viruses (Kendal, 1975), and is readily rationalized as the loss or gain of a complete viral 'chromosome', as opposed to a covalent rearrangement. The status of neuraminidase in influenza A and B viruses has elements of an accessory element rather than an ultimately indispensible function: mutants temperature sensitive in neuraminidase can be released from the host cell surface by the action of exogenously supplied neuraminidase, and are then infectious (Palese, Tobita, Ueda & Compans, 1974). Thus, it is not unreasonable to consider construction of a conditionally lethal mutant of influenza A or B, lacking any neuraminidase gene and dependent for viability on addition of neuraminidase enzyme. However, this is the only case in which such a possibility seems even conceivable.

I turn now to a brief account of the hierarchical analysis carried out by Lipman *et al.* (1982) on the sequence of three influenza A haemagglutinin (HA) genes of different subtypes. The most specific comparisons were carried out by homology analysis between each pair of HA sequences on successive overlapping 240-base subsec-

HA1 →) (← HA2

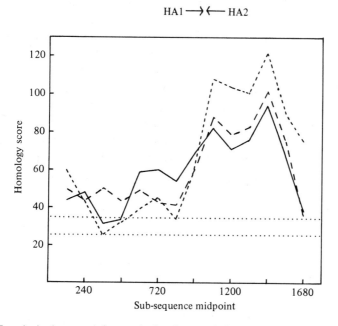

Fig. 2. Homologies between influenza A virus haemagglutinin genes.
  Homologies were measured between three pairs of HA genes of differing subtypes using an algorithm of Smith & Waterman (1981) on overlapping 240 residue regions. The graph plots these values against the midpoints of sequences examined. The dotted lines indicate background homology as determined by comparisons between randomized sequences. Redrawn from the data of Lipman *et al.* (1982). -------, H3/Havl comparison; ———, H2/Havl comparison; ———, H3/H2 comparison.

tions of the genes. The results of this are illustrated in Fig. 2. The downstream portions of the genes show high homologies, but the upstream portions show low homologies, descending to background. These two sections correspond, respectively, to the HA2 and HA1 segments of the mature processed protein. At this level of analysis, the results correlate well with the fact that the HA1 segment contains the variable, antigenic sites of the molecule (Wiley, Wilson & Skehel, 1981).

  The next more general level of analysis examines sequences by statistical comparisons of doublet frequencies. By such tests, the HA1 sequences are indistinguishable from each other, and the HA2 sequences are also indistinguishable. Thus, the HA1 sequences show similar constraints, not apparent in the first-level homology analysis. These correlate well with the observation that, between types, HA1 *amino acid* sequences show great variation, but that this is within a *conserved* framework, visible in the constancy of cysteine

residues and hydrophobic regions (Min Jou *et al.*, 1980). On the
other hand, comparison of the total doublet frequencies summed for
the three HA1 subtypes with the corresponding data for HA2
sequences shows that the HA1 and HA2 *are* distinguishable by
doublet analysis. When the sequences are each dissected into three
subsequences corresponding separately to the first, second and third
codon positions, then only the third codon position sequences of
HA1 and HA2 are statistically distinguishable. Considering the
nature of the genetic code, whereby the third codon position has
least correlation with amino acid specification, this result can be
taken as showing that the two sections of the gene set show
differences due to different constraints on amino acid sequences
encoded, but similarities below the level of amino acid specification
– as, for example, in requirements of RNA secondary structure or
translation efficiency.

The most general comparison level uses the measures of information theory. These give the result that all the HA1 sequences are
nearer random in terms both of extent of divergence from non-
uniform base composition and in terms of divergence from independence of occurrence of successive residues. This relative tendency to randomness of the HA1 encoding sequences correlates with
HA1s nature as a selected target for mutation to new antigenic
forms.

In summary, this section contends that the main mechanism of
genome variation open to RNA viruses is cumulative micro-change.
Hierarchical analysis of influenza HA genes shows that such changes
are subject to distinct classes of constraint. The possibilities for large
scale genome changes and rearrangements seem few. This is in
contrast to the great variety of RNA genome types in existence; it is
held that evolution into these forms has imposed severe constraints
on further variation.

SMALL DNA VIRUSES

The viruses discussed in this section are hepatitis B virus, parvoviruses, and papovaviruses, whose genomes fall in the range of
3200–7800 residues.

Hepatitis B virus has the distinction of possessing the shortest
characterized genome (DNA or RNA) of any animal virus. The
genome consists of a DNA circle of 3200 residues. One DNA strand

contains a nick, the other a gap of variable size (Summers, O'Connell & Millman, 1975). These two interruptions are offset so that the whole retains a circular form. Total sequencing of the genome has shown that all the virus genes are probably transcribed from one strand, although definition of transcriptional products is still incomplete (Tiollais *et al.*, 1981). Apart from any other interest, the organization of this extraordinarily small genome will have lessons for our perception of relative compactness in other systems.

The parvovirus genome consists of a single stranded DNA of some 5100 residues (see Ward & Tattersall, 1978; Berns & Hauswirth, 1982). There are two groups of parvoviruses: helper dependent and autonomous. The former package both plus and minus strands of DNA separately into virions. The autonomous parvoviruses package minus strand almost exclusively, although there are exceptions. Unlike their bacteriophage counterparts, ϕX174 and M13, parvovirus DNAs are linear. Nucleotide sequence studies have shown that they possess the potential to form stable hairpin structures at each end; this explains the early observations that the virion DNAs could serve as efficient template/primers for *E. coli* DNA polymerase (McGeoch, Crawford & Follett, 1970). The total sequence for one autonomous parvovirus, minute virus of mice, has been published (Astell *et al.*, 1983). Interpretation in terms of gene coding shows a moderately simple transcription pattern, with two promoters and several splicing events (Pintel *et al.*, 1983).

Papovaviruses are regarded as comprising two genera, the polyomaviruses and the papillomaviruses (see Matthews, 1982). Molecular biology of the first group, which contains among others SV40 and polyoma, has been intensively studied for many years. The 5200 bp circular double stranded DNA of SV40 was the first animal virus genome to be completely sequenced (Fiers *et al.*, 1978; Reddy *et al.*, 1978; for sequence see Tooze, 1980), and since then two other members of the group have been sequenced (Yang & Wu, 1979; Deininger *et al.*, 1980; Soeda *et al.*, 1980).

It is instructive to look at the large effort being expended on SV40 some five years after publication of the sequence. Polyomaviruses encode two classes of genes, early and late, on opposite strands of the genome, and their expression strategy utilizes a moderately complex mRNA splicing system. Today it is rather clear that 'pre-sequence' biochemistry and genetics could never have satisfactorily resolved the nature of the virus genes and events in their

expression. The genomes of SV40 and polyoma, although detect-
ably related by homology analysis of the sequence data (Pustell &
Kafatos, 1982), exhibit an interesting difference: polyoma DNA
encodes three early proteins through variable splicing of a single
primary transcript, while the SV40 sequence allows only two
species. For our present purpose, the lesson from these well studied
genomes is that knowledge of the sequence was essential to achieving a good understanding of their gene organization, but that such
understanding only came after substantial further work with the
sequence data available as a 'groundplan'.

Study of papilloma viruses has suffered from the lack of *in vitro*
culture systems. Work on these viruses, like that with hepatitis B
virus, has been changed totally by the advent of cloning and
sequencing techniques. Two papilloma virus DNAs, a human and a
bovine strain, have now been cloned and sequenced (Chen *et al.*,
1982; Danos, Katinka & Yaniv, 1982). Papillomavirus DNAs, like
those of polyomaviruses, are closed double strand circles, but, at
7800 bp, they are around 50% larger than SV40 or polyoma DNA.
From the sequences it is deduced that, unlike polyomaviruses, their
genes are all encoded in one orientation. Splicing of mRNAs has to
be invoked to interpret the sequence data. No homology can be
detected with SV40 sequences. The distinct gene organization,
together with genome size difference, lack of homology, and
differences in mode of genome existence in transformed cells, opens
the question as to whether papillomaviruses and polyomaviruses
should not now be assigned to separate families.

LARGE DNA VIRUSES

This section describes genome properties of large DNA animal
viruses. I include in this category herpesviruses and poxviruses, both
of which have double stranded DNA genomes of the order of
1×10^8 daltons. The location of adenoviruses (genome size 0.3×10^8
daltons) in the abrupt classification used in this article is left
undefined. The discussion is almost completely directed to herpesviruses and, within that family, mainly to herpes simplex virus.

The classification and genome properties of the family herpesviridae have been recently discussed (Matthews, 1982; Roizman,
1982). Genomes of this family are diverse in a number of basic
attributes. The base composition of genome DNAs range from 32%

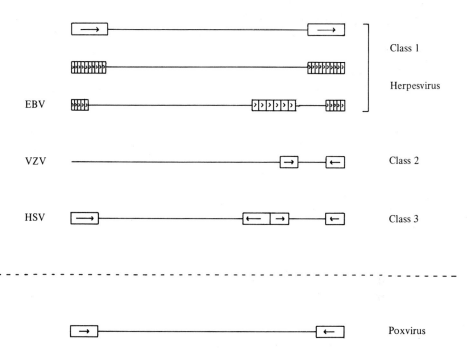

Fig. 3. Genome arrangements of large DNA viruses.

The structures of herpesvirus DNAs are presented, classified according to the arrangement of repeat sequences, as proposed by Honess & Watson (1978). Unique regions are indicated by solid lines and repeated elements by boxes, with arrows indicating orientation. Viruses discussed in the text are indicated: herpes simplex virus (HSV), Epstein-Barr virus (EBV) and varicella zoster virus (VZV). Also shown is a representation of vaccinia virus DNA (Wittek *et al.*, 1978). The diagrams do not preserve scale.

to 75% G + C; that is, over the whole range of values considered accessible for polypeptide coding purposes (Woese, 1967). Genome molecular weights range from 0.8×10^8 to 1.5×10^8. The genomes typically, but not without exception, contain large scale repeated elements; however, these vary widely in arrangement, as depicted in Fig. 3. Finally, the biology of this virus family varies considerably: three groupings (designated alpha-, beta- and gammaherpesviruses) can be delineated according to biological characteristics, but groupings by any of the genomic structural criteria outlined above do not correlate with biological groupings.

The genome of herpes simplex virus (HSV) consists of a duplex DNA molecule of 100×10^6 daltons, about 160 000 bp (Becker, Dym & Sarov, 1968; Kieff, Bachenheimer & Roizman, 1971). The molecule consists of two sequence components, a long region (L) of about 135 000 bp and a short region (S) of 25 000 bp. Each of these comprises a unique region bounded by a pair of repeat sequences in opposite orientations. A sequence of some 400 bp, designated the 'a' sequence, is present as a direct repeat at each terminus, with one or more copies present internally in opposite orientation at the 'joint' between the L and S regions (Sheldrick & Berthelot, 1974; Wadsworth, Jacob & Roizman, 1975; Delius & Clements, 1976; Wilkie, 1976). Any preparation of HSV virion DNA consists of four isomers, in which the L and S components are each independently oriented in one sense or the other about the central 'joint'.

The HSV genome is some 30 times larger than the DNAs of polyoma or SV40 discussed in the last section. Thus, we are now considering sequence arrangements on an altogether different scale from that applicable to small DNA viruses and RNA viruses. This discussion cannot be a detailed description of all aspects of the HSV genome, and it concentrates on what I regard as characteristic structural features, at the expense of aspects dealing with the genome's functioning. The features described are based largely on DNA sequence studies. We are engaged in several DNA sequencing projects on the genome of HSV type 1 and this discussion is illustrated by data from the short unique (U_s) region of 13 000 bp and the short repeats (IR_s/TR_s) of 6000 bp. The former has been totally sequenced and the latter is near completion at the time of writing.

A very notable aspect of herpesvirus DNAs is the wide range of $G + C$ contents exhibited. HSV-1 DNA is towards the high $G + C$ limit of the set, with an overall $G + C$ content of 67% (Kieff et al., 1971). This value is not uniform over the whole genome. In particular, the short repeat region has a very high $G + C$ content: the sequenced region of this element, about 70% of the 6000 bp total, has a $G + C$ value of 78% (Davison & Wilkie, 1981; Murchie & McGeoch, 1982; D. J. McGeoch, S. Donald, A. Dolan & D. Brauer, unpublished). R_s contains polypeptide-coding DNA for only one gene, the 4200 bp gene specifying the prominent immediate early protein, Vmw 175 (Rixon, Campbell & Clements, 1982). The coding sequences for this polypeptide are a remarkable 80% $G + C$. The base distribution varies for the three codon positions; the third position is 95% $G + C$. Thus, the overall base composition of the DNA is such as to restrict the amino acid composition of encoded polypeptide. However, the maximum occurrence of G and C in third position of codons allows greatest amino acid coding freedom for the existing total base composition. This pronounced

		Copy number variability	Polypeptide-coding
1	GGAGCGG GGGGA	Probable	No
2	GCCGGGGAGGG CTGGG	Probable	No
3	CGAGGGGCGG GAGGGGG	Probable	No
4	GGGCGGAGGAGGGGG GACGCGG	Yes	No
5	CCTCCACCCCC TCGACCACCA	Not known	Yes
6	TGG GTGGGTGGGGAG	Yes	No
7	CCCCGG TCTCCCCGGGAG	Yes	Yes

Fig. 4. Reiterated sequences in the short region of herpes simplex virus (HSV) type 1.
Sequence data are listed for each of the families of tandemly reiterated sequences found in the short region of HSV-1 strain 17 DNA. Each sequence is listed in the orientation of the nearest transcription unit, and is given as the 'xy' permutation (see text), with a gap between the x and y components. Knowledge of copy number variability, and polypeptide-coding status are also indicated. The partial genome map indicates locations and orientations of the sequences.

triplet periodicity of G + C content is not observed in non-coding regions of R_s. This region of the genome thus appears to have developed to an extreme of base composition. However, the nature of the evolutionary events involved remains obscure. Of particular note is the fact that the mRNA for Vmw 175 is expressed immediately on virus infection and is apparently translated by the unmodified machinery of the host cell.

Another feature of herpesvirus genomes, which is emerging as a widespread phenomenon, is the existence of tandemly reiterated DNA sequences. In HSV-1, such reiterations, varying from 12 bp to 54 bp, have been detected (Fig. 4). They are always present in a non-integral number of copies – that is, with a partial copy at one end of the set. There is thus an alternative way of viewing such repeats: if the partial end sequence is called 'x' and the remainder of the repeat unit 'y', then the whole set is $---xyxyxyx---$, or $---(xy)_nX$.

This structure presumably relates to the mechanisms of generation and maintenance of the reiterations.

In well characterized examples, the reiterations have been observed to vary in copy number in individual clones. In any given example, the repeats are exact, with any divergent form generally being found only at an extremity of the set. These characteristics of copy number variation and sequence homogeneity are probably mediated by some form of unequal crossing over. The possibility of recombinational events generating such forms was discussed by Smith (1976) prior to their discovery. In HSV, this class of reiterations was first detected in non-polypeptide-coding locations (Davison & Wilkie, 1981). These early examples had extreme base compositions and simple sequence characteristics, and gave every impression of being weeds growing in regions not subject to rigorous selection. More recently, examples have been found in polypeptide-coding DNA. These latter are multiples of three in length, that is, they encode reiterated amino acid sequences (D. J. McGeoch, A. Dolan, S. Donald & F. J. Rixon, unpublished).

Three sets of direct repeats of similar general aspect have been described in the genome of the gammaherpesvirus Epstein-Barr virus (EBV). Two of these, with reiteration lengths of 103 and 124 bp, and G + C contents of 84%, are thought not to be polypeptide-coding (Dambaugh & Kieff, 1982). The third comprises a 708 bp array, composed of the three related elements GCAGGA, GCAGGAGGA and GGGGCAGGA, and is transcribed and translated (Heller, van Santen & Kieff, 1982). The EBV genome also contains two other direct repeat elements which are of larger size and are not, for the purpose of this discussion, considered in the same class as those above (though this may be erroneous). These are the major repeat features of the genome (Fig. 3): the variable number direct 500 bp repeats at the termini of the genome, and the 3071 bp direct internal repeats. The sequence of the latter has been published (Cheung & Kieff, 1982). Interestingly, they conform to the $(xy)_n x$ format mentioned above, where x is an 1850 bp, G + C rich sequence and y is a 1221 bp sequence.

Returning to the HSV genome, the final sequence feature to be described is the nature of the junctions between the two copies of the short repeat element, and the short unique sequence. Two immediate early genes lie in the vicinity of these junctions: the gene for immediate early protein Vmw 68 is situated across the IR_s/U_s junction and the gene for Vmw 12 occupies the same position with respect to TR_s/U_s (Clements, McLauchlan & McGeoch, 1979; Watson, Preston & Clements, 1979; Marsden *et al.*, 1982). DNA sequencing and mRNA mapping have shown that the promoters and the bulk of each mRNAs 5'-non-coding region lie in the repeat elements, and are thus identical for the two genes, while the whole of each polypeptide-coding region lies in U_s, and so the encoded proteins are completely distinct (Murchie & McGeoch, 1982; Rixon & Clements, 1982; Watson & vande Woude, 1982). This presents a clearly delimited and, it might be said, aesthetically rather satisfying picture of these regions of the genome. Recently, however, the general significance of this view has been abruptly upset by a similar analysis of the junctions between short repeat and short unique regions in the genome of varicella zoster virus (VZV), performed by A. J. Davison (unpublished).

VZV is, like HSV, an alphaherpesvirus. However, its evolutionary path evidently diverged from that of HSV 'in the deeps of time': the G + C content of VZV DNA is 46% (21 percentage points lower than that of HSV); VZV DNA has a molecular weight of only 80×10^6; and it has no long repeat regions (Dumas *et al.*, 1981; Ecker & Hyman, 1982; Straus *et al.*, 1982; Davison & Scott, 1983). Cross-hybridization experiments have shown that, on a gross scale, the arrangement of genes in the two genomes is colinear (Davison & Wilkie, 1983). In VZV there are only 4 genes in U_s, as opposed to probably 12 in HSV U_s. Three of the VZV genes can be aligned with three in HSV (Fig. 5) by homology analysis of predicted polypeptides using a matrix plotting program (Pustell & Kafatos, 1982). This region of the two genomes is then seen to be related, as follows.

As an *ad hoc* convention, all divergent changes are viewed as having occurred in VZV. First, U_s of VZV contains one major deletion relative to HSV. Second, the repeat regions have encroached on U_s of VZV: the gE gene of HSV is wholly within U_s, whereas in VZV the homologous 70 K gene runs into TR_s. In VZV, the arrangement of genes around the R_s/U_s junctions is the reverse of that found with HSV: a gene runs from U_s into each repeat. The coding sequences of one of these (the 11 K protein gene) are totally in U_s and terminate precisely at the U_s/TR_s boundary, while the coding sequences of the other gene (for the 70 K protein) run from U_s into the body of TR_s. While the VZV U_s region is 6000 bp, as compared with 13 000 in HSV-1, R_s in VZV is 7500 bp; that is, some 1500 bp longer than in HSV. Thus in VZV the homologue of either

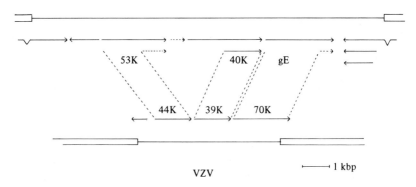

Fig. 5. Comparisons of genes in the short unique regions of herpes simplex virus (HSV) and varicella zoster virus (VZV) DNAs.

The HSV map indicates the positions of mRNAs as deduced from DNA sequencing and mRNA mapping. The species indicated by dotted lines are predicted from sequence data but corresponding RNA mapping data is not yet available (D. J. McGeoch, A. Dolan, S. Donald & F. Rixon, unpublished). The VZV map is based on DNA sequencing only (A. J. Davison, unpublished). Three homologous pairs of genes are marked.

the Vmw 68 gene or the Vmw 12 gene of HSV probably lies inside the repeat. Sequence analysis of the whole R_s regions of the two genomes will resolve this possibility.

Thus, in comparing these two systems, we see that the large scale form of repeats and unique sequences is the same, while the detailed arrangement of genes around the R_s/U_s junction is quite different. In both instances the junctions lie close to the extremities of polypeptide-coding regions in such a way as to preserve the coding uniqueness of the pair of trans-junction genes. These data are consistent with the notions that, in herpesvirus genomes of this form, the repeats appear to be dynamic entities (in an evolutionary sense), capable of expanding until limited by polypeptide-coding requirements, and that the nature of genes encoded by the repeat is not of primary importance.

As mentioned above, the genomes of HSV and VZV have a broadly similar arrangement of sequences, in common with several other alphaherpesviruses, as detected by hybridization (Davison & Wilkie, 1983). The resolution of such experiments is of the order of 10^4 bp. At a finer level of discrimination, of the order of 10^3 bp, changes become evident: addition/deletion of genes and varying lengths of inverted repeats. At the level of 10^1–10^2 bp we observe the phenomenon of variable copy number tandem reiteration within one genome species. Finally, all these changes are superposed on sequences which seem prone to evolution to extremes of base composition.

Features associated with sequences of 10^3 bp or greater are clearly not relevant in comparing the situation in herpesviruses with that in small DNA viruses or RNA viruses. However, the classes of variation which are found in herpesvirus DNA at finer resolution (high variability of copy number in tandem reiterations, base composition changes) do not appear to occur with smaller genomes. This is probably because in the small virus genomes, constraint on genome size (mediated at least in part through packaging requirements) and associated, extensive informational overlay act to maintain genome form. In addition, small genome size implies large dependence on host genome functions, involving additional constraints.

I suggest that at some point in the genome size spectrum, these several constraining factors lose their overriding power and that large scale variation of genome structure then becomes an accessible option. On this view, the kinds of repeats and other features exhibited by herpesvirus genomes correlate with 'being a large DNA virus genome' rather than with 'being a herpesvirus genome'. It is interesting to note that the comparably sized vaccinia virus genome contains large scale inverted repeats (Fig. 3) and that these each contain two regions of 70-bp tandem reiterations (Wittek & Moss, 1980). Such arrangements are distinctly reminiscent of herpesvirus structures, although the biology of poxviruses, with their cytoplasmic replication mode, is totally distinct. Finally, in this vein, arguments could be made that the various large scale repeated elements in the genomes of herpesviruses are of less fundamental import than has been accorded them since their discovery nine years ago (Sheldrick & Berthelot, 1974), being primarily a consequence of variation in a relatively unconstrained environment.

CONCLUSION

Extrapolating from current rates of progress, we should know complete genome sequences for all major classes of animal viruses within three years. Again judging from current activities, after determination of a genome sequence there will follow a period in which the theme is elaborated, with analysis of mutants, variants and related species. In less than ten years, however, one might expect to see the end of the main age of virus genome sequencing. In this paper I have tried to indicate the ways in which systematic analysis of sequence data might develop. One direction involves statistical and informational analysis of sequences; conceptual tools for such activities now exist. Another road involves facing up to the rigours of three-dimensional structural chemistry to interpret subtleties of conformation inherent in nucleic acid sequences, and to attempt to extract information from the masses of predicted polypeptide sequences generated from nucleic acid sequencing.

These are inherently difficult topics and the work input presently required for such research directions is enormous, so it is possible that many genome sequences will never be 'fully' interpreted in this sense. This view, however, ignores advances in techniques. For instance, the application of computer technology to biology is in many ways still quite limited. Increases in computing power and advances in artificial intelligence approaches may bring remarkable changes in sequence data interpretation. Finally, I still consider that the main importance of genome sequence determination is that it supplies a detailed (if not fully interpretable) blueprint, a reference for virologists in development of new ideas and experiments. Thus, determination of a genome sequence marks the end of the beginning of investigation of the virus.

I acknowledge the help given by Drs Craig Pringle, Nigel Stow and Russell Thompson in critically reviewing the manuscript. I thank Dr Andrew Davison for discussion and access to unpublished data. I thank Jennifer McGeoch, Paul McGeoch and Adam McGeoch for their forbearance and help.

REFERENCES

ANILIONIS, A., WUNNER, W. H. & CURTIS, P. (1981). Structure of the glycoprotein gene in rabies virus. *Nature*, **294**, 275–8.

ARMSTRONG, J., SMEEKENS, S. & ROTTIER, P. (1983). Sequence of the nucleocapsid gene from murine coronavirus MHV-A59. *Nucleic Acids Research*, **11**, 883–91.

ASTELL, C. R., THOMSON, M., MERCHLINSKY, M. & WARD, D. C. (1983). The complete DNA sequence of minute virus of mice, an autonomous parvovirus. *Nucleic Acids Research*, **11**, 999–1018.

ATKINS, J. F., STEITZ, J. A., ANDERSON, C. W. & MODEL, P. (1979). Binding of mammalian ribosomes to MS2 phage RNA reveals an overlapping gene encoding a lysis function. *Cell*, **18**, 247–56.

BALTIMORE, D. (1970). RNA-dependent DNA polymerase in virions of RNA tumour viruses. *Nature*, **226**, 1209–11.

BALTIMORE, D., HUANG, A. S. & STAMPFER, M. (1970). Ribonucleic acid synthesis of vesicular stomatitis virus. II. An RNA polymerase in the virion. *Proceedings of the National Academy of Sciences (USA)*, **66**, 572–6.

BARRELL, B. G., AIR, G. M. & HUTCHINSON, C. A. (1976). Overlapping genes in bacteriophage φX174. *Nature*, **264**, 34–41.

BECKER, Y., DYM, H. & SAROV, I. (1968). Herpes simplex virus DNA. *Virology*, **36**, 184–92.

BEREMOND, M. N. & BLUMENTHAL, T. (1979). Overlapping genes in RNA phage: a new protein implicated in lysis. *Cell*, **18**, 257–66.

BERNS, K. I. & HAUSWIRTH, U. W. (1982). Organization and replication of parvovirus DNA. In *Organization and Replication of Viral DNA*, ed. A. S. Kaplan, pp. 147–72. Boca Raton: The Chemical Rubber Company Press.

BISHOP, D. H. L., GOULD, K. G., AKASHI, H. & CLERX-VAN HAASTER, C. M. (1982). The complete sequence and coding content of snowshow hare bunyavirus small (S) viral RNA species. *Nucleic Acids Research*, **10**, 3703–13.

BISHOP, J. M. (1983). Cancer genes come of age. *Cell*, **32**, 1018–20.

BLACK, D. N., STEPHENSON, P., ROWLANDS, D. J. & BROWN, F. (1979). Sequence and location of the poly(C) tract in aphtho- and cardiovirus RNA. *Nucleic Acids Research*, **6**, 2381–90.

BOSSI, L. (1983). Context effects: translation of UAG codon by suppressor tRNA is affected by the sequence following UAG in the message. *Journal of Molecular Biology*, **164**, 73–87.

CHEN, E. Y., HOWLEY, P. M., LEVINSON, A. D. & SEEBURG, P. H. (1982). The primary structure and genetic organization of the bovine papillomavirus type 1 genome. *Nature*, **299**, 529–34.

CHEUNG, A. & KIEFF, E. (1982). Long internal direct repeat in Epstein-Barr virus DNA. *Journal of Virology*, **44**, 286–94.

CLEMENTS, J. B., MCLAUCHLAN, J. & MCGEOCH, D. J. (1979). Orientation of herpes simplex type 1 immediate early mRNAs. *Nucleic Acids Research*, **7**, 77–92.

CORDINGLEY, M. G., CAMPBELL, M. E. M. & PRESTON, C. M. (1983). Functional analysis of a herpes simplex virus type 1 promoter: identification of far-upstream regulatory sequences. *Nucleic Acids Research*, **11**, 2347–65.

CRICK, F. H. C. (1966). Codon–anticodon pairing: the wobble hypothesis. *Journal of Molecular Biology*, **19**, 548–55.

DAMBAUGH, T. R. & KIEFF, E. (1982). Identification and nucleotide sequences of two similar tandem direct repeats in Epstein-Barr virus DNA. *Journal of Virology*, **44**, 823–33.

DANOS, O., KATINKA, M. & YANIV, M. (1982). Human papillomavirus 1a complete DNA sequence: a novel type of genome organization among papovaviridae. *The EMBO Journal* **1**, 231–6.

DAVISON, A. J. & SCOTT, J. E. (1983). Molecular cloning of the varicella-zoster virus genome and derivation of six restriction endonuclease maps. *Journal of General Virology*, in press.

DAVISON, A. J. & WILKIE, N. M. (1981). Nucleotide sequences of the joint between the L and S segments of herpes simplex virus types 1 and 2. *Journal of General Virology*, **55**, 315–31.

DAVISON, A. J. & WILKIE, N. M. (1983). Location and orientation of homologous sequences in the genomes of five herpesviruses. *Journal of General Virology*, in press.

DEININGER, P. L., ESTY, A., LAPORTE, P., HSU, H. & FRIEDMANN, T. (1980). The nucleotide sequence and restriction enzyme sites of the polyoma genome. *Nucleic Acids Research*, **8**, 855–60.

DELIUS, H. & CLEMENTS, J. B. (1976). A partial denaturation map of herpes simplex type 1 DNA: evidence for inversions of the unique DNA regions. *Journal of General Virology*, **33**, 125–33.

DICKERSON, R. E. (1983). Base sequence and helix structure variation in *B* and *A* DNA. *Journal of Molecular Biology*, **166**, 419–41.

DUMAS, A. M., GEELEN, J. L. M. C., WESTSTRATE, M. W., WERTHEIM, P. & VAN DER NOORDAA, J. (1981). *Xba*I, *Pst*I and *Bgl*II restriction maps of the two orientations of the varicella-zoster genome. *Journal of Virology*, **39**, 390–400.

DUNN, J. J. & STUDIER, F. W. (1983). Complete nucleotide sequence of bacteriophage T7 DNA and the locations of T7 genetic elements. *Journal of Molecular Biology*, **166**, 477–535.

ECKER, J. R. & HYMAN, R. W. (1982). Varicella zoster DNA exists as two isomers. *Proceedings of the National Academy of Sciences (USA)*, **79**, 156–60.

ELTON, R. A. (1975). Doublet frequencies in sequenced nucleic acids. *Journal of Molecular Evolution*, **4**, 323–46.

FIERS, W., CONTRERAS, R., HAEGEMAN, G., ROGIERS, R., VAN DE VOORDE, A., VAN HEUVERSWYN, H., VAN HERREWEGHE, J., VOLKAERT, G. & YSEBAERT, M. (1978). Complete nucleotide sequence of SV40 DNA. *Nature*, **273**, 113–20.

FITCH, W. F. (1983). Random sequences. *Journal of Molecular Biology*, **163**, 171–6.

FITZGERALD, M. & SHENK, T. (1981). The sequence 5'-AAUAAA-3' forms part of the recognition site for polyadenylation of late SV40 mRNAs. *Cell*, **24**, 251–60.

FRINK, R. J., EISENBERG, R., COHEN, G. & WAGNER, E. K. (1983). Detailed analysis of the portion of the herpes simplex virus type 1 genome encoding glycoprotein C. *Journal of Virology*, **45**, 634–47.

GALLIONE, C. J., GREEN, J. R., IVERSON, L. E. & ROSE, J. K. (1981). Nucleotide sequences of the mRNAs encoding the vesicular stomatitis virus N and NS proteins. *Journal of Virology*, **39**, 529–35.

GAROFF, H., FRISCHAUF, A. M., SIMONS, K., LEHRACH, H. & DELIUS, H. (1980). Nucleotide sequence of cDNA coding for Semliki Forest virus membrane glycoproteins. *Nature*, **288**, 236–41.

GATLIN, L. (1972). *Information Theory and the Living System*. New York & London: Columbia University Press.

GEELEN, J. L. M. C., WALIG, C., WERTHEIM, P. & VAN DER NOORDAA, J. (1978). Human cytomegalovirus DNA. I. Molecular weight and infectivity. *Journal of Virology*, **26**, 813–16.

GOELET, P., LOMONOSOFF, G. P. BUTLER, P. J. G., AKAM, M. E., GAIT, M. J. & KARN, J. (1982). Nucleotide sequence of tobacco mosaic virus RNA. *Proceedings of the National Academy of Sciences (USA)*, **79**, 5818–22.

GOTOH, O. (1982). An improved algorithm for matching biological sequences. *Journal of Molecular Biology*, **162**, 705–8.

GOTOH, O. & TAGASHIRA, Y. (1981). Stabilities of nearest-neighbour doublets in double-helical DNA determined by fitting calculated melting profiles to observed profiles. *Biopolymers*, **20**, 1033–42.

HAWLEY, D. K. & McCLURE, W. R. (1983). Compilation and analysis of *Escherichia coli* promoter DNA sequences. *Nucleic Acids Research*, **11**, 2237–55.

HAY, J. & SUBAK-SHARPE, H. (1968). Analysis of nearest neighbour frequencies in the RNA of a mammalian virus: encephalomyocarditis virus. *Journal of General Virology*, **2**, 469–72.

HELLER, M., VAN SANTEN, V. & KIEFF, E. (1982). Simple repeat sequence in Epstein-Barr virus DNA is transcribed in latent and productive infections. *Journal of Virology*, **44**, 311–20.

HONESS, R. W. & WATSON, D. H. (1977). Unity and diversity in the herpesviruses. *Journal of General Virology*, **37**, 15–37.

JOSSE, J., KAISER, A. D. & KORNBERG, A. (1961). Enzymatic synthesis of deoxyribonucleic acid. VIII. Frequencies of nearest-neighbour base sequences in deoxyribonucleic acid. *Journal of Biological Chemistry*, **236**, 864–75.

KASTELEIN, R. A., REMAT, E., FIERS, W. & VAN DUIN, J. (1982). Lysis gene expression of RNA phage MS2 depends on a frameshift during translation of the overlapping coat protein gene. *Nature*, **295**, 35–41.

KENDAL, A. P. (1975). A comparison of 'influenza C' with prototype myxoviruses: receptor-destroying activity (neuraminidase) and structural polypeptides. *Virology*, **65**, 87–99.

KIEFF, E. D., BACHENHEIMER, S. L. & ROIZMAN, B. (1971). Size, composition and structure of the deoxyribonucleic acid of herpes simplex virus subtypes 1 and 2. *Journal of Virology*, **8**, 125–32.

KITAMURA, M., SEMLER, B. L., ROTHBERG, P. G., LARSEN, G. R., ADLER, C. H., DORNER, A. J., EMINI, E. A., HANECAK, R., LEE, J. J., VAN DER WERF, S., ANDERSON, C. W., & WIMMER E. (1981). Primary structure, gene organization and polypeptide expression of poliovirus RNA. *Nature*, **291**, 547–53.

KOZAK, M. (1980). Evaluation of the 'scanning model' for initiation of protein synthesis in eucaryotes. *Cell*, **22**, 7–8.

KOZAK, M. (1981). Possible role of flanking nucleotides in recognition of the AUG initiator codon by eukaryotic ribosomes. *Nucleic Acids Research*, **9**, 5233–52.

LAMB, R. A. & LAI, C. J. (1980). Sequence of interrupted and uninterrupted mRNAs and cloned DNA coding for the two overlapping non structural proteins of influenza virus. *Cell*, **21**, 475–85.

LIPMAN, D. J. & MAIZEL, J. (1982). Comparative analyses of nucleic acid sequences by their general constraints. *Nucleic Acids Research*, **10**, 2723–39.

LIPMAN, D. J., SMITH, T. F., BECKMAN, R. J. & WATERMAN, M. S. (1982). Hierarchical analysis of influenza A haemagglutinin gene sequences. *Nucleic Acids Research*, **10**, 6375–89.

McCAULEY, J. W. & MAHY, B. W. J. (1983). Structure and function of the influenza virus genome. *Biochemical Journal*, **211**, 281–94.

McFADDEN, G. & DALES, S. (1982). Organization and replication of poxvirus DNA. In *Organization and Replication of Viral DNA*, ed. A. S. Kaplan, pp. 173–90. Boca Raton: CRC Press.

McGEOCH, D. J. (1979). Structure of the gene N: gene NS intercistronic junction in the genome of vesicular stomatitis virus. *Cell*, **17**, 673–81.

McGEOCH, D. J., CRAWFORD, L. V., & FOLLETT, E. A. C. (1970). The DNAs of three parvoviruses. *Journal of General Virology*, **6**, 33–40.

MACKEM, S., & ROIZMAN, B. (1982). Regulation of α genes of herpes simplex virus: the α 27 gene promoter – thymidine kinase chimera is positively regulated in converted L cells. *Journal of Virology*, **43**, 1015–23.

McLAUCHLAN, J. & CLEMENTS, J. B. (1983). Organization of the herpes simplex virus type 1 transcription unit encoding two early proteins with molecular weights of 140 000 and 40 000. *Journal of General Virology*, **64**, 997–1006.

MAIZEL, J. V. & LENK, R. P. (1981). Enhanced graphic matrix analysis of nucleic acid and protein sequences. *Proceedings of the National Academy of Sciences (USA)*, **78**, 7665–9.

MARSDEN, H. S., HAARR, L. & PRESTON, C. M. (1983). Processing of herpes simplex virus proteins and evidence that translation of thymidine kinase mRNA is initiated at three separate AUG codons. *Journal of Virology*, **46**, 434–45.

MARSDEN, H. S., LANG, J., DAVISON, A. J., HOPE, R. G. & MACDONALD, D. M. (1982). Genomic location and lack of phosphorylation of the HSV immediate-early polypeptide IE 12. *Journal of General Virology*, **62**, 17–27.

MATTHEWS, R. E. F. (1982). Classification and nomenclature of viruses: fourth report of the international committee on taxonomy of viruses. *Intervirology*, **17**, 1–199.

MAXAM, A. M. & GILBERT, W. (1980). Sequencing end-labelled DNA with base-specific chemical cleavages. *Methods in Enzymology*, **65**, 499–560.

MILLER, J. H. & ALBERTINI, A. M. (1983). Effects of surrounding sequence on the suppression of nonsense codons. *Journal of Molecular Biology*, **164**, 59–71.

MIN JOU, W., VERHOEYEN, M., DEVOS, R., SAMAN, E., FANG, R., HUYLEBOV., FIERS, W., THRELFALL, G., BARBER, C., CAREY, N., & EMTAGE, S. (1980). Complete structure of the haemagglutinin gene from the human influenza A/Victoria/3/75 (H3N2) strain as determined from cloned DNA. *Cell*, **19**, 683–96.

MODRICH, P. & ROBERTS, R. J. (1982). Type-II restriction and modification enzymes. In *Nucleases*, ed. S. M. Linn & R. J. Roberts, pp. 109–54. Cold Spring Harbor: Cold Spring Harbor Press.

MORRISON, J. M., KEIR, H. M., SUBAK-SHARPE, H. & CRAWFORD, L. V., (1967). Nearest neighbour base sequence analysis of the deoxribonucleic acids of a further three mammalian viruses: simian virus 40, human papilloma virus and adenovirus type 2. *Journal of General Virology*, **1**, 101–8.

MOUNT, S. M., (1982). A catalogue of splice junction sequences. *Nucleic Acids Research*, **10**, 459–72.

MURCHIE, M.-J. & McGEOCH, D. J. (1982). DNA sequence analysis of an immediate-early gene region of the herpes simplex virus type 1 genome (map co-ordinates 0.950 to 0.978). *Journal of General Virology*, **62**, 1–15.

NOMOTO, A., OMATA, T., TOYODA, H., KUGE, S., HORIE, H., KATASOKA, Y., GENBA, Y., NAKANO, Y. & IMURA, N. (1982). Complete nucleotide sequence of the attenuated poliovirus Sabin 1 strain genome. *Proceedings of the National Academy of Sciences (USA)*, **79**, 5793–7.

NORDHEIM, A. & RICH, A. (1983). Negatively supercoiled simian virus 40 DNA contains Z-DNA segments within transcriptional enhancer sequences. *Nature*, **303**, 674–9.

PALESE, P., RACANIELLO, V. R., DESSELBERGER, U., YOUNG, J. & BAEZ, M. (1980). Genetic structure and genetic variation of influenza viruses. *Philosophical Transactions of the Royal Society of London Series B*, **288**, 299–305.

PALESE, P., TOBITA, K., UEDA, M. & COMPANS, R. W., (1974). Characterization of temperature sensitive influenza virus mutants defective in neuraminidase. *Virology*, **61**, 397–410.

PATEL, D. J., IKUTA, S., KOZLOWSKI, S., & ITAKURA, K. (1983). Sequence dependence of hydrogen exchange kinetics in DNA duplexes at the individual base pair level in solution. *Proceedings of the National Academy of Sciences (USA)*, **80**, 2184–8.

PELHAM, H. R. B., (1978). Leaky UAG termination codon in tobacco mosaic virus RNA. *Nature*, **272**, 469–71.

PHILIPSON, L., ANDERSSON, P., OLSHEVSKY, U., WEINBERG, R., BALTIMORE, D. & GESTELAND, R. (1978). Translation of MuLV and MSV RNAs in nuclease-treated reticulocyte extracts: enhancement of the *gag-pol* polypeptide with yeast suppressor tRNA. *Cell*, **13**, 189–99.

PINTEL, D., DADACHANJI, D., ASTELL, C. R. & WARD, D. C. (1983). The genome of minute virus of mice, an autonomous parvovirus, encodes two overlapping transcription units. *Nucleic Acids Research*, **11**, 1019–38.

PORTER, A. G., SMITH, J. C. & EMTAGE, J. S. (1980). Nucleotide sequence of influenza virus RNA segment 8 indicates that coding regions for NS1 and NS2 proteins overlap. *Proceedings of the National Academy of Sciences (USA)*, **77**, 5074–8.

POST, L. E., MACKEM, S. & ROIZMAN, B. (1981). Regulation of genes of herpes simplex virus: expression of chimeric genes produced by fusion of thymidine kinase with α gene promoters. *Cell*, **24**, 555–65.

PRESTON, C. M., & McGEOCH, D. J. (1981). Identification and mapping of two polypeptides encoded within the herpes simplex virus type 1 thymidine kinase gene sequences. *Journal of Virology*, **38**, 593–605.

PUSTELL, J. & KAFATOS, F. C. (1982). A high speed, high capacity homology matrix: zooming through SV40 and polyoma. *Nucleic Acids Research*, **10**, 4765–82.

RACANIELLO, V. R., & BALTIMORE, D. (1981). Molecular cloning of poliovirus cDNA and determination of the complete nucleotide sequence of the viral genome. *Proceedings of the National Academy of Sciences (USA)*, **78**, 4887–91.

REDDY, E. P., SMITH, M. J. & AARONSON, S. A. (1981). Complete nucleotide sequence and organisation of the Moloney murine sarcoma virus genome. *Science*, **214**, 445–50.

REDDY, V. B., THIMMAPPAYA, B., DHAR, R., SUBRAMANIAN, K. N., ZAIN, S., PAN, J., GHOSH, P. K., CELMA, M. L. & WEISSMAN, S. M. (1978). The genome of simian virus 40. *Science*, **200**, 494–502.

RICE, C. M. & STRAUSS, J. H. (1981). Nucleotide sequence of the 26S mRNA of Sindbis virus and deduced sequence of the encoded virus structural proteins. *Proceedings of the National Academy of Sciences (USA)*, **78**, 2062–6.

RIXON, F. J., CAMPBELL, M. E. & CLEMENTS, J. B. (1982). The immediate early mRNA that encodes the regulatory polypeptide Vmw 175 of herpes simplex virus type 1 is unspliced. *The EMBO Journal*, **1**, 1273–7.

RIXON, F. J. & CLEMENTS, J. B. (1982). Detailed structural analysis of two spliced HSV-1 immediate-early mRNAs. *Nucleic Acids Research*, **10**, 2241–56.

ROBERTS, R. J. (1982). Restriction and modification enzymes and their recognition sequences. *Nucleic Acids Research*, **10**, r117–r140.

ROIZMAN, B. (1982). The family Herpesviridae; general description, taxonomy and classification. In *The Herpesviruses*, vol. 1. pp. 1–23. New York & London: Plenum Press.

ROSE, J. K. (1980). Complete intergenic and flanking gene sequences from the genome of vesicular stomatitis virus. *Cell*, **19**, 415–21.

ROSE, J. K., DOOLITTLE, R. F., ANILIONIS, A., CURTIS, P. J. & WUNNER, W. H. (1982). Homology between the glycoproteins of vesicular stomatitis virus and rabies virus. *Journal of Virology*, **43**, 361–4.

ROSE, J. K. & GALLIONE, C. J. (1981). Nucleotide sequences of the mRNAs encoding the vesicular stomatitis virus G and M proteins determined from cDNA clones containing the complete coding regions. *Journal of Virology*, **39**, 519–28.

SADLER, J. R., WATERMAN, M. S. & SMITH, T. F. (1983). Regulatory pattern identification in nucleotide sequences. *Nucleic Acids Research*, **11**, 2221–31.

SANGER, F., COULSON, A. R., HONG, G. F., HILL, D. F. & PETERSEN, G. B. (1982). Nucleotide sequence of bacteriophage λ DNA. *Journal of Molecular Biology*, **162**, 729–73.

SANGER, F., NICKLEN, S. & COULSON, A. R. (1977). DNA sequencing with chain terminating inhibitors. *Proceedings of the National Academy of Sciences (USA)*, **74**, 5463–7.

SCHUBERT, M., KEENE, J. D., HERMAN, R. C. & LAZZARINI, R. A. (1980). Site on the vesicular stomatitis virus genome specifying polyadenylation and the end of the L gene mRNA. *Journal of Virology*, **34**, 550–9.

SCHWARTZ, D. E., TIZARD, R. & GILBERT, W. (1983). Nucleotide sequence of Rous sarcoma virus. *Cell*, **32**, 853–69.

SELLERS, P. H. (1980). The theory and computation of evolutionary distances: pattern recognition. *Journal of Algorithms*, **1**, 359–73.

SHELDRICK, P. & BERTHELOT, N. (1974). Inverted repetitions in the chromosome of herpes simplex virus. *Cold Spring Harbor Symposia on Quantitative Biology*, **39**, 667–79.

SHINNICK, T. M., LERNER, R. A. & SUTCLIFFE, J. G. (1981). Nucleotide sequence of Moloney murine leukaemia virus. *Nature*, **293**, 543–8.

SMITH, G. P. (1976). Evolution of repeated DNA sequences by unequal crossover. *Science*, **191**, 528–35.

SMITH, T. F. & WATERMAN, M. S. (1981). Identification of common molecular subsequences. *Journal of Molecular Biology*, **147**, 195–7.

SMITH, T. F., WATERMAN, M. S. & SADLER, J. R. (1983). Statistical characterization of nucleic acid sequence functional domains. *Nucleic Acids Research*, **11**, 2205–20.

SOEDA, E., ARRAND, J. R., SMOLAR, N., WALSH, J. E. & GRIFFIN, B. E. (1980). Coding potential and regulatory signals of the polyoma virus genome. *Nature*, **283**, 445–53.

SPAETE, R. R. & FRENKEL, N. (1982). The herpes simplex virus amplicon: a new eukaryotic defective-virus cloning-amplifying vector. *Cell*, **30**, 295–304.

STRAUS, S. E., OWENS, J., RUYECHAN, W. T., TAKIFF, H. E., CASEY, T. A., VANDE WOUDE, G. F. & HAY, J. (1982). Molecular cloning and physical mapping of varicella-zoster virus DNA. *Proceedings of the National Academy of Sciences (USA)*, **79**, 993–7.

SUBAK-SHARPE, H., BURK, R. R., CRAWFORD, L. V., MORRISON, J. M., HAY, J. & KEIR, H. M. (1966). An approach to evolutionary relationship of mammalian DNA viruses through analysis of the pattern of nearest neighbour base sequence. *Cold Spring Harbor Symposia on Quantitative Biology*, **31**, 737–48.

SUMMERS, J., O'CONNELL, A. & MILLMAN, I. (1975). Genome of the hepatitis B virus: restriction enzyme cleavage and structure of DNA extracted from Dane particles. *Proceedings of the National Academy of Sciences (USA)*, **72**, 4597–601.

TEMIN, H. & MIZUTANI, S. (1970). RNA-dependent DNA polymerase in virions of Rous sarcoma virus. *Nature*, **226**, 1211–13.

TIOLLAIS, P., CHARNAY, P. & VYAS, G. N. (1981). Biology of hepatitis B virus. *Science*, **213**, 406–11.

TOOZE, J. (1980). *DNA tumor viruses*. Cold Spring Harbor: Cold Spring Harbor Press.

TRIFONOV, E. N. (1980). Sequence-dependent deformational anisotropy of chromatin DNA. *Nucleic Acids Research*, **8**, 4041–53.

VAN BEVEREN, C., VAN STRAATEN, F., GALLESHAW, J. A. & VERMA, I. M. (1981). Nucleotide sequence of the genome of a murine sarcoma virus. *Cell*, **27**, 97–108.

WADSWORTH, S., JACOB, R. J. & ROIZMAN, B. (1975). Anatomy of herpes simplex virus DNA. II. Size, composition, and arrangement of inverted terminal repetitions. *Journal of Virology*, **15**, 1487–97.

WARD, D. C. & TATTERSALL, P. (1978). *Replication of Mammalian Parvoviruses*. Cold Spring Harbor: Cold Spring Harbor Press.

WATSON, R. J., PRESTON, C. M. & CLEMENTS, J. B. (1979). Separation and characterisation of herpes simplex virus type 1 immediate-early mRNAs. *Journal of Virology*, **31**, 42–52.

WATSON, R. J. & VANDE WOUDE, G. F. (1982). DNA sequence of an immediate-early gene (IE mRNA-5) of herpes simplex virus type 1. *Nucleic Acids Research*, **10**, 979–91.

WILEY, D. C., WILSON, I. A. & SKEHEL, J. J. (1981). Structural identification of the antibody-binding sites of Hong Kong influenza haemagglutinin and their involvement in antigenic variation. *Nature*, **289**, 373–8.

WILKIE, N. M. (1976). Physical maps for herpes simplex type 1 DNA for restriction endonucleases *Hind*III, *Hpa*I and *X.bad*. *Journal of Virology*, **20**, 222–33.

WINTER, G. & FIELDS, S. (1982). Nucleotide sequence of human influenza A/PR/8/34 segment 2. *Nucleic Acids Research*, **10**, 2135–43.

WITTEK, R., MENNA, A., MUELLER, H. K., SCHUEMPERLI, D., BOSELEY, G. & WYLER, R. (1978). Inverted terminal repeats in rabbitpox virus and vaccinia virus DNA. *Journal of Virology*, **28**, 171–81.

WITTEK, R. & MOSS, B. (1980). Tandem repeats within the inverted terminal repetition of vaccinia virus DNA. *Cell*, **21**, 277–84.

WOESE, C. R. (1967). *The Genetic Code: the Molecular Basis for Genetic Expression*. New York: Harper & Row.

YANG, R. C. A. & WU, R. (1979). BK virus DNA: complete nucleotide sequence of a human tumor virus. *Science*, **206**, 456–62.

THE VAGARIES OF VIRAL EVOLUTION: THE EXAMPLE OF POLIOVIRUS REPLICATION INITIATION

DAVID BALTIMORE

Whitehead Institute for Biomedical Research, Cambridge, Massachusetts 02139, USA; Department of Biology and Center for Cancer Research, Massachusetts Institute of Technology, Cambridge, Massachusetts 02139, USA

Viruses were discovered at the end of the nineteenth century but for many years all that was known about them was that they were very, very small (Hughes, 1977; Waterson & Wilkinson, 1978). How a living organism could be so tiny was puzzling and a few investigators realized that anything so small must have distilled out the essence of biologic organization (Muller, 1922). Learning the truth of that perception awaited the discovery of the coding potential in nucleic acids. Only then did it become clear that a virus is a set of genes protected from environmental degradation, provided with a mechanism for entering living cells and sometimes endowed with a few enzymes for initiating the infection process. This realization could only have come about once the fundaments of molecular biology had been established. In fact, viruses have often played a central role in the discovery of molecular biological mechanisms because they demonstrate such mechanisms with special clarity.

As their life cycles have been revealed, one overpowering realization has been the diversity of the mechanisms evolved by viruses. Leaving aside the question of how viruses did evolve (Baltimore, 1980), a virologist cannot help but marvel at the ingenuity and intricacy of virologic mechanisms. One might have thought that viruses, being so small, would have used only essential mechanisms for nucleic acid replication and transcription. We find, however, an extraordinary diversity of detailed mechanisms many of which have no obvious counterpart in host cells and therefore seem to be processes unique to viruses.

Among the consistencies of virology is the compactness of viral genomes: coding sequences often overlap as do controlling elements. We have come to expect that almost every nucleotide in a viral genome will have at least one function, if not a number of functions in different contexts. The compact nature of viral genomes

probably constrains their evolution both because there is no substra-
tum of excess DNA or free space where new capacities can be tested
and because it is difficult to mutate one function of an overlapping
and tightly integrated set. In spite of this potential rigidity, each
class of viruses is unique, suggesting that each evolved quite
separately. This economy contrasts with the contemporary view of
mammalian genomes as sparsely inhabited structures, peppered
with non-coding intervening sequences, long spaces between genes
and a tremendous burden of parasitic DNA.

The diversity of viruses is evident both in their structure and in
the mechanisms by which they carry through their life cycle. Not
only are there DNA and RNA viruses, but many subtypes of each
with extensive variety in their mechanisms of replication and
transcription. In special cases, replication and transcription are the
same event – as in picornaviruses – but generally they are separate
processes because virion nucleic acid and mRNA are not congruent.
I suggested that the overall strategies of viral transcription lent
themselves to grouping in six categories and thus proposed that six
classes of virus be distinguished (Baltimore, 1971). Investigations
during the last 12 years have confirmed the basic differences
between the six classes but have shown the inadequacy of the system
in categorizing the extensive diversity of mechanisms found within
many of the classes. For instance, among the negative strand RNA
viruses, influenza-type viruses show a large number of differences
from those viruses with a single, negative stranded RNA genome.
Lumping together all double-strand DNA viruses in a single class
misses the fundamental distinction between the cytoplasmic DNA
viruses which encode their own gene expression system and the
nuclear DNA viruses which involve themselves more insidiously
with pre-existing cellular machinery. A final example is the hepatitis
B virus which, although a DNA virus, appears to use a reverse
transcription mechanism during its intracellular growth (Summers &
Mason, 1982).

To review the remarkable diversity found among animal viruses
would require writing a textbook of molecular virology. Rather than
attempting to provide such a general summary, I shall describe only
recent results on one well-defined system: the replication of
poliovirus RNA. Poliovirus-RNA-dependent RNA synthesis illustrates a uniquely virologic mechanism because only viruses appear
to use RNA as a genetic material. By concentrating attention on the
initiation of RNA synthesis, we focus on the most important and
ingenuous event of the replication process; progression of an RNA

polymerase down a template is a repetitive process which we would assume is carried out basically in the same way for all nucleic acid replications.

HISTORY OF THE PROBLEM

Investigations in the 1960s of poliovirus replication showed that three types of nucleic acid are involved in the poliovirus replication process (Baltimore, 1969; Levintow, 1974). The first was intra-cellular single-stranded RNA, then thought to be homogeneous and indistinguishable from virion RNA. We now know that the intracellular and virion forms are identical through their 7440 nucleotides and 3'-poly(A); they differ in that the virion RNA has a protein, VPg, linked to its 5'-terminus, while much of the intracellular RNA lacks the protein (Flanegan et al., 1977; Lee et al., 1977). Both RNAs are 'plus' strands, RNA of the same sequence as messenger RNA.

The second type of RNA found in infected cells is a double-stranded RNA molecule which we believe to be a by-product of the replication process. It is historically and practically a very important molecule because it contains both the plus-strand of RNA and a complete, complementary minus-strand of RNA.

The third kind of RNA in infected cells is the replicative intermediate RNA, which also contains a minus-strand. It, how-ever, is a metabolically active structure in the cell. The replicative intermediate represents molecules caught in the process of replication; its minus-strand serves as a template for synthesizing plus-strands. There is presumably a fourth kind of nucleic acid in infected cells, the minus-strand replicative intermediate, but it has never been detected.

Because there are two different strands of RNA found in infected cells, the plus- and minus-strands, poliovirus replication must involve two initiation steps, one for the synthesis of plus-strands and one for the synthesis of minus-strands. I shall review here evidence showing that both initiation processes involve a protein primer.

THE MODEL

The observation that poliovirus virion RNA has a 22-amino acid protein, VPg, attached to its 5'-most phosphate was sufficient

Model of poliovirus RNA replication

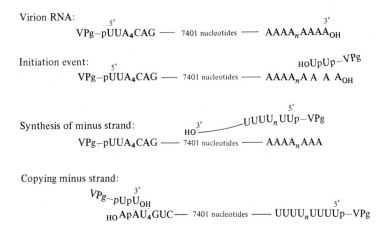

Fig. 1. Proposed model of poliovirus RNA replication. The model suggests that initiation takes place using a preformed molecule of VPg-pUpU. It hybridizes to the 3'-poly(A) of poliovirus RNA and then is elongated by the viral replicase to form the minus strand of poliovirus RNA. The minus strand is copied following hybridization of VPg-pUpU to the two 5'-As.

evidence for both groups who initially studied it to suggest that poliovirus replication might be initiated by the protein (Flanegan *et al.*, 1977; Lee *et al.*, 1977). Further investigations showed that nascent plus-strand molecules in the replicative intermediate, as well as minus-strands, contain VPg at their 5'-end (Nomoto *et al.*, 1977; Pettersson, Ambros & Baltimore, 1978). Those two facts were consistent with the notion that both plus- and minus-strand RNA are initiated by VPg. It was difficult to obtain any further evidence for such a model from the study of intact infected cells, necessitating the development of a system for poliovirus RNA replication *in vitro*.

An explicit model of initiation of RNA synthesis by VPg is shown in Fig. 1; it postulates formation of VPg-pUpU for reasons that will become evident. The model highlights synthesis of minus-strands on a plus-strand template because that reaction can be studied *in vitro*. Synthesis of plus-strand on a minus-strand template has been difficult to examine because of the non-availability of a ready source of minus-strand templates.

DEVELOPMENT OF AN *IN VITRO* REPLICATION SYSTEM

RNA-dependent RNA synthesis was first accomplished *in vitro* with a picornavirus in 1962 (Baltimore & Franklin, 1962) and was very soon extended to poliovirus (Baltimore *et al.*, 1963). This reaction, however, was studied in whole cytoplasmic extracts of infected cells and we could adduce no evidence of *de novo* initiation of RNA chains. Further advances had to await purification of the replication enzymes in the absence of template to produce a usable reconstituted system.

The clue to purifying the poliovirus replicase came with the demonstration in 1974 that the 3'-poly(A) on poliovirus RNA plays a necessary role in the infective cycle (Spector & Baltimore, 1974). That was shown by removing the poly(A) from viral RNA and thereby virtually abolishing its infectivity. Because removal of poly(A) had no effect on translation of viral RNA (Spector, Villa-Komaroff & Baltimore, 1975), it seemed that replication might involve the poly(A). Furthermore, the location of poly(A) on the 3'-end of the plus-strand suggested that its function in replication might be to initiate minus-strand synthesis. The initial event in minus-strand synthesis would therefore be poly(A)-dependent poly(U) synthesis and thus the poliovirus replicase might be specially designed to use poly(A) as a template. With this rationale in mind, we utilized poly(A) in our search in infected cells for a poly(U) polymerase. To avoid having to assay both initiation and elongation, we provided an oligo(U) primer onto which we hoped poly(U) synthesis would occur.

The strategy worked: infected cells could be shown to contain a poly(U) polymerase that would use poly(A) as a template (Flanegan & Baltimore, 1977). We called the enzyme that was purified p63 because of its apparent molecular weight; it is the C-terminal segment of the viral polyprotein encoded at the 3'-end of viral RNA (Fig. 2). p63 can unambiguously be identified as the 3'-most viral protein because antibodies made to the amino acids encoded at the 3'-terminus of poliovirus RNA inhibit poly(A)-dependent poly(U) synthesis by this enzyme (Baron & Baltimore, 1982a). The real molecular weight of p63 is 53 000 (Kitamura *et al.*, 1981; Semler *et al.*, 1981), but we continue to designate it p63 for convenience.

Once the poly(U) polymerase was discovered, it involved only a small change of magnesium concentration to demonstrate that the

Gene organization of poliovirus

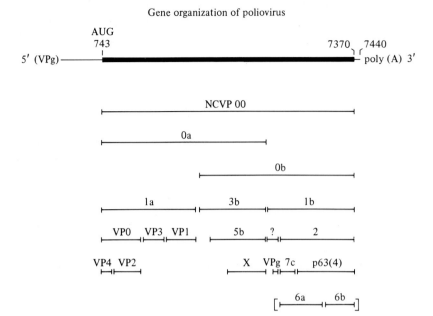

Fig. 2. Gene organization of poliovirus. The positions of the various poliovirus proteins are indicated.

enzyme could copy viral RNA (Dasgupta, Baron & Baltimore, 1979). Remarkably, this reaction did not require an oligo(U) primer and thus appeared to be true initiation of viral RNA synthesis.

Attempts to purify the poliovirus replicase were initially stymied by loss of activity. This set-back became an advance when it was discovered that two proteins were necessary for the initiation of replication, one being p63 and the other being a host cell protein of 67 000 molecular weight (Dasgupta, Zabel & Baltimore, 1980; Baron & Baltimore, 1982b). Although the host protein has been purified, its activity has yet to be characterized (Baron & Baltimore, 1982b). It is, however, absolutely required for any poliovirus replication.

Using its poly(U) polymerase activity, the poliovirus replicase has been purified to apparent homogeneity (Flanegan & Baltimore, 1979; van Dyke & Flanegan, 1980; Baron & Baltimore, 1982b). As will become evident, however, it apparently contains a very important contaminant. Purified host factor and purified replicase together are able to produce full-length copies of poliovirus RNA (Baron & Baltimore, 1982c). If presented with a preparation enriched in minus-strand RNA, the replicase host factor preparations will make a plus-strand copy (Baron, 1981) and therefore this combination of enzymes appears to be totally sufficient for both steps in the poliovirus RNA replication cycle.

INVOLVEMENT OF VPg

The ability of the host factor/replicase combination to replicate poliovirus RNA would seem to imply that replication is independent of VPg. Not being prepared to accept this conclusion, we have investigated whether antibodies to VPg would inhibit the replicase/ host factor activity, implying that a source of VPg would be present in the reaction mixture. To do these experiments it was necessary to generate anti-VPg antibodies. These could not be produced by injecting VPg into animals because insufficient amounts of VPg were available either from virions or from infected cells. We therefore undertook the chemical synthesis of VPg.

VPg was produced chemically using standard methods of peptide synthesis (Baron & Baltimore, 1982d). Two preparations were made: one was the 14 C-terminal residues of VPg and the other was the complete 22 amino acid structure. They are designated VPg(14/22) and VPg(22/22). Both of these, when coupled to bovine serum albumin and injected into rabbits, induced the synthesis of antibody which would immunoprecipitate either of the artificial VPgs (Baron & Baltimore, 1982d). Furthermore, both would immunoprecipitate a series of polypeptides from infected cells which included all of those known to contain the VPg sequence. The antibodies also immunoprecipitated molecules containing the VPg sequence which had not previously been identified and are called pre-VPgs. In our initial experiments, no detectable free VPg was found in infected cells but, as indicated below, that observation has turned out to be false.

To examine the effect of anti-VPg antibodies on the replicase reactions, various amounts of antibody were added to three reaction systems. In one reaction, replicase, poly(A) and oligo(U) were added and the effect of the antibody on poly(U) synthesis was examined. In the second reaction, replicase, host factor, poliovirus RNA and oligo(U) were added to examine whether oligo(U)-primed poliovirus RNA replication would be sensitive to the antibody. In the third reaction no primer was added; it consisted of

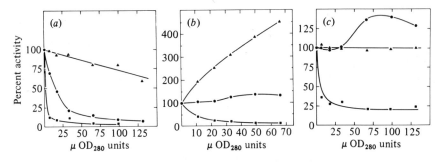

Fig. 3. Effect of anti-peptide antibodies on poliovirus polymerase activities. Three preparations of affinity-purified antibodies were tested: anti-VPg (●—●), anti-p63 (■—■) and
control anti-pp60^{src} (▲—▲) (Baron & Baltimore, 1982a, e). Three reactions were studied: (a)
poliovirus RNA replication with replicase and host factor and without a primer; (b) poliovirus
RNA replication primed with oligo(U) and lacking host factor; (c) replicase-directed
synthesis of poly(U) on a poly(A) template. The selective effect of anti-VPg on the first
reaction is evident.

host factor, replicase and poliovirus RNA. Three antibody preparations were used in these experiments: a control anti-pp60^{src} serum,
an anti-replicase serum and the anti-VPg serum.

The results of the experiment clearly showed that anti-VPg
inhibits the initiation of poliovirus replication *in vitro* (Fig. 3; Baron
& Baltimore, 1982e). Whereas the anti-replicase antibodies inhibited all three of the reaction systems, anti-VPg inhibited only the
primer-dependent copying of poliovirus RNA. This demonstrates
that VPg is involved in that reaction but not in the pre-primed
elongation reaction. Thus, although no intentional source of VPg is
added to the reactions *in vitro*, VPg must be present and must be
involved in the initiation reaction.

There are only two possible sources of the VPg in the replicase
reaction. One is on the poliovirus RNA itself, where covalently
bound VPg could, in principle, be acting as an initiator for replication. This was easily ruled out by treatment of the RNA with
protease before addition to the reaction. Such protease treatment
had no effect on the template activity of the RNA (Crawford &
Baltimore, unpublished). The other possible source of VPg is in the
replicase preparations. The host factor could not contain VPg
because it is isolated from uninfected cells and VPg is a viral
protein. We are presently examining whether VPg, or a derivative
of it, is present in the replicase preparations. Before undertaking
those experiments, we examined what forms of VPg might be
expected to be present.

VPg IN INFECTED CELLS

Although our initial experiments (Baron & Baltimore, 1982*d*), and those of others (Semler *et al.*, 1982), indicated that no VPg was present in infected cells, more recent experiments using anti-VPg antibodies have shown the presence of a significant pool of VPg (Crawford & Baltimore, unpublished). Apparently, in previous work, the VPg was eluted from polyacrylamide gels after electrophoresis, during the staining reaction. Using newly available enhancing agents for fluorography, it has been possible to maintain VPg in the gel and demonstrate its presence. Thus either free VPg or one of its precursors could, in principle, be the initiating molecule.

The initiation of RNA synthesis by VPg could either be a concerted reaction, in which initiation is rapidly followed by elongation, or a two-step reaction in which VPg is first initiated with a small number of nucleotides and then is elongated separately (as in Fig. 1). To examine the possibility of a two-step reaction, we have sought an initiated form of VPg in infected cells. Such an initiated molecule might be expected to have one or a few uridylate residues attached to the tyrosine moiety on VPg. We approached this problem by ^{32}P-labelling of infected cells, immunoprecipitating the labelled proteins and examining them by polyacrylamide gel electrophoresis. When this was done, a ^{32}P-containing low molecular weight protein was identified. It migrated somewhat more slowly than VPg-pUp which was prepared by digestion of viral RNA with pancreatic ribonuclease. Its migration rate, as well as a number of other properties recently ascertained, suggests that this molecule is VPg-pUpU (Crawford & Baltimore, unpublished).

PRESENT MODEL

The existence of VPg-pUpU in infected cells has now changed our concept somewhat of how poliovirus RNA synthesis is initiated. We believe that either VPg or a precursor of VPg is a substrate for an enzyme which can add two uridylate residues to the tyrosine of the protein. This molecule binds tightly to replicase and the complex then binds to RNA in a specific manner so that it is capable of initiating synthesis of either plus or minus strand (Fig. 1). The plus-strand 5′ sequence is VPg-UUAAAACAG and therefore we would expect that the first two Us come from the addition to VPg. It is possible that this model is somewhat incorrect in that the Us might be coded by the template, but investigations currently under way should decide that issue.

THE FUTURE

There are a number of quite obvious experiments which will tell us whether the new model of initiation of poliovirus replication is correct. A more important issue is whether this initiation mechanism is unique to poliovirus or has a counterpart in uninfected cells. The only other system where initiation of replication by a protein has been clearly demonstrated is in the replication of adenovirus DNA. There, a precursor to the 5' terminal protein on the DNA has been found to be deoxycytidylated in a reaction which is dependent on viral DNA (Lichy, Horowitz & Hurwitz, 1981). It would be very surprising if protein-initiated nucleic acid synthesis were peculiar to viruses.

More generally, this particular initiating system illustrates the variety of mechanisms evolved by viruses. It is difficult at this juncture to understand why picornavirus RNA initiation should involve a protein primer while other viral and non-viral systems can use a ribonucleoside triphosphate for initiation. Perhaps we are seeing the results of a complex evolution of a messenger RNA towards a self-replicating structure. In that case, it could be that picornaviruses evolved quite separately from minus-strand RNA viruses or double-stranded RNA viruses (Baltimore, 1980).

Margaret Baron and Nigel Crawford provided most of the experimental results described here and Margaret's many suggestions improved the text. Their work was supported by grants from the National Institute of Allergy and Infectious Diseases and The National Cancer Institute.

REFERENCES

BALTIMORE, D. (1969). The replication of picornaviruses. In *The Biochemistry of Viruses*, ed. H. B. Levy, pp. 101–76. New York: Marcel Dekker.

BALTIMORE, D. (1971). Expression of animal virus genomes. *Bacteriological Reviews*, **35**, 235–41.

BALTIMORE, D. (1980). Evolution of RNA viruses. In *Genetic Variation of Viruses*, ed. P. Palese & B. Roizman. *Annals of the New York Academy of Sciences*, **354**, 492–8.

BALTIMORE, D. & FRANKLIN, R. M. (1962). Preliminary data on a virus-specific enzyme system responsible for the synthesis of viral RNA. *Biochemical and Biophysical Research Communications*, **9**, 388–92.

BALTIMORE, D., FRANKLIN, R. M., EGGERS H. J. & TAMM, I. (1963). Poliovirus-induced RNA polymerase and the effects of virus-specific inhibitors on its production. *Proceedings of the National Academy of Sciences (USA)*, **49**, 843–9.

BARON, M. H. (1981). Initiation of poliovirus replication *in vitro*. PhD thesis, Massachussetts Institute of Technology.

BARON, M. H. & BALTIMORE, D. (1982*a*). Antibodies against a synthetic peptide of the poliovirus replicase protein: reaction with native, virus-encoded proteins and inhibition of virus-specific polymerase activities *in vitro*. *Journal of Virology*, **43**, 969–78.

BARON, M. H. & BALTIMORE, D. (1982*b*). Purification and properties of a host cell protein required for poliovirus replication *in vitro*. *Journal of Biological Chemistry*, **257**, 12 351–8.

BARON, M. H. & BALTIMORE, D. (1982*c*). *In vitro* copying of viral positive strand RNA by poliovirus replicase: characterization of the reaction and its products. *Journal of Biological Chemistry*, **257**, 12 359–66.

BARON, M. H. & BALTIMORE, D. (1982*d*). Antibodies against the chemically synthesized genome-linked protein of poliovirus react with native virus-specific proteins. *Cell*, **28**, 395–404.

BARON, M. H. & BALTIMORE, D. (1982*e*). Anti-VPg antibody inhibition of the poliovirus replicase reaction and production of covalent complexes of VPg-related polypeptides and newly made RNA. *Cell*, **30**, 745–52.

DASGUPTA, A., BARON, M. H. & BALTIMORE, D. (1979). Poliovirus replicase: a soluble enzyme able to initiate copying of poliovirus RNA. *Proceedings of the National Academy of Sciences (USA)*, **76**, 2679–83.

DASGUPTA, A., ZABEL, P. & BALTIMORE, D. (1980). Dependence of the activity of the poliovirus replicase on a host cell protein. *Cell*, **19**, 423–9.

FLANEGAN, J. B. & BALTIMORE, D. (1977). A poliovirus-specific primer-dependent RNA polymerase able to copy poly(A). *Proceedings of the National Academy of Sciences (USA)*, **74**, 3677–80.

FLANEGAN, J. B. & BALTIMORE, D. (1979). Poliovirus poly(U) polymerase and RNA replicase have the same viral polypeptide. *Journal of Virology*, **29**, 352–60.

FLANEGAN, J. B., PETTERSSON, R. F., AMBROS, V., HEWLETT, M. J. & BALTIMORE, D. (1977). Covalent linkage of a protein to a defined nucleotide sequence at the 5'-terminus of the virion and replicative intermediate RNAs of poliovirus. *Proceedings of the National Academy of Sciences (USA)*, **74**, 961–5.

HUGHES, S. S. (1977). *The Virus: A History of the Concept*. New York: Science History Publications.

KITAMURA, N., SEMLER, B. L., ROTHBERG, P. G., LARSEN, G. R., ADLER, C. J., DORNER, A. M., EMINI, E. A., HANECAK, R., LEE, J. J., VAN DER WERF, S., ANDERSON, C. W. & WIMMER, E. (1981). Primary structure, gene organization and polypeptide expression of poliovirus RNA. *Nature*, **291**, 547–53.

LEE, Y. F., NOMOTO, A., DETJEN, B. M. & WIMMER, E. (1977). A protein covalently linked to poliovirus genome RNA. *Proceedings of the National Academy of Sciences (USA)*, **74**, 59–63.

LEVINTOW, L. (1974). The reproduction of picornaviruses. In *Comprehensive Virology*, vol. 2, ed. H. Fraenkel-Conrat & R. R. Wagner, pp. 109–69. New York: Plenum Press.

LICHY, J. H., HOROWITZ, M. S. & HURWITZ, J. (1981). Formation of a covalent complex between the 80 000-dalton adenovirus terminal protein and 5'-dCMP *in vitro*. *Proceedings of the National Academy of Sciences (USA)*, **78**, 2678–82.

MULLER, H. J. (1922). Variations due to change in the individual gene. *American Naturalist*, **56**, 32–50.

NOMOTO, A., DETJEN, D., POZZATTI, R. & WIMMER, E. (1977). The location of the poliogenome protein in viral RNAs and its implication for RNA synthesis. *Nature*, **268**, 208–13.

PETTERSSON, R. F., AMBROS, V. & BALTIMORE, D. (1978). Identification of a protein linked to nascent poliovirus RNA and to the polyuridylic acid of negative strand RNA. *Journal of Virology*, **27**, 357–65.

SEMLER, B. B., ANDERSON, C. W., HANECAK, R., DORNER, L. F. & WIMMER, E. (1982). A membrane-associated precursor to poliovirus VPg identified by immunoprecipitation with antibodies directed against a synthetic heptapeptide. *Cell*, **28**, 405–12.

SEMLER, B. L., ANDERSON, C. W., KITAMURA, N., ROTHBERG, P. G., WISHART, W. L. & WIMMER, E. (1981). Poliovirus replication proteins: RNA sequence encoding P3-1b and the sites of proteolytic processing. *Proceedings of the National Academy of Sciences (USA)*, **78**, 3464–8.

SPECTOR, D. H. & BALTIMORE, D. (1974). Requirement of 3′-terminal polyadenylic acid for the infectivity of poliovirus RNA. *Proceedings of the National Academy of Sciences (USA)*, **71**, 2983–7.

SPECTOR, D. H., VILLA-KOMAROFF, L. & BALTIMORE, D. (1975). Studies on the function of polyadenylic acid on poliovirus RNA. *Cell*, **6**, 41–4.

SUMMERS, J. & MASON, W. (1982). Replication of the genome of a hepatitis B-like virus by reverse transcription of an RNA intermediate. *Cell*, **29**, 403–15.

VAN DYKE, T. A. & FLANEGAN, J. B. (1980). Identification of poliovirus polypeptide p63 as a soluble RNA-dependent RNA polymerase. *Journal of Virology*, **35**, 732–40.

WATERSON, A. P. & WILKINSON, L. (1978). *An introduction to the History of Virology*. Cambridge University Press.

EXPLORING CARCINOGENESIS WITH RETROVIRUSES

J. MICHAEL BISHOP

George W. Hooper Foundation and Department of Microbiology and Immunology, University of California, San Francisco, California 94143, USA

Malignancy is adequately described as a breakdown of one or several growth controlling systems, and the genetic origin of this breakdown can hardly be doubted.

(Jacob & Monod, 1961)

INTRODUCTION

Two schools of thought have long battled over tumour viruses. One school has argued that we should search for viruses in human cancer, that viruses must cause the disease. The other school has held that since there may be many causes of cancer, we would be better off seeking the central molecular mechanisms by which the disease arises: tumour viruses should be used to ferret out the genetic and chemical processes that cause the cancer cell to run amok. But there is no longer reason for battle; both views have been vindicated. Viruses have been found in human cancer and it is possible that they contribute to the genesis of the disease. And tumour viruses have revealed to us a set of human genes whose actions may lie at the heart of every cancer, no matter what its cause, and whose functions are accessible to biochemical dissection. An enemy has been found – it is part of us – and we have begun to understand the lines of its attack.

The role of viruses in the etiology of human cancer is considered elsewhere (Weiss, this volume). Here I show how one family of tumour viruses – the retroviruses – has been used to unveil what may be the final common pathway to tumorigenesis. The value of retroviruses in cancer research can be traced to two distinctive properties. (*a*) Retroviruses are mutagens (Varmus, 1983*a*), and the mutations they cause in host cells are on occasion tumorigenic (Varmus 1983*b*). Since the locations of these mutations within cellular DNA can be traced, we can be led to cellular genes that contribute to tumorigenesis. (*b*) Retroviruses recombine with great facility, both among themselves (Vogt, 1977) and with the genome

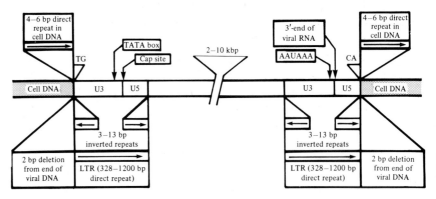

Fig. 1. The canonical provirus of retroviruses. The drawing illustrates common features of retroviral DNA that has been integrated into cellular DNA (see Varmus, 1983a, for details).

of the host cell (Bishop, 1983a). As a consequence, retroviruses are transducing agents – a property that has brought to light cellular genes which are potentially tumorigenic (Bishop, 1983b). As these principles unfolded, a marvellous and unanticipated symmetry also emerged. The cellular genes whose mutation by retroviruses may initiate tumorigenesis, and the tumorigenic cellular genes transduced by retroviruses, belong to overlapping – perhaps consubstantial – families. In studying the one, we also study the other. This symmetry will pervade much of what follows.

FIRST PRINCIPLES

The ability of retroviruses to mutate and transduce genes can be laid to distinctive features of viral structure and replication. The genes of retroviruses are carried in single-stranded RNA within virions but are copied into double-stranded DNA by reverse transcriptase early in the viral life cycle (Varmus & Swanstrom, 1982). Viral DNA is then integrated into chromosomal DNA of the host cell, where it displays several diagnostic features (Fig. 1): internal elements that encode viral proteins; large terminal repeats (known as LTRs) that are in turn bounded by short inverted repeats; characteristic dinucleotides at each terminus; and short direct repeats of cellular DNA joined to the ends of viral DNA. The provirus is a self-contained transcriptional template, equipped with all the signals necessary to direct the enzymatic machinery of the host cell in the production of viral RNA. Most of these signals reside within the LTR; nucleotide sequences to direct the initiation of transcription (a 'promoter'); an independent domain that enhances expression of viral and other genes, perhaps by facilitating or strengthening the action of the promoter (an 'enhancer'); and signals for the termination of transcription and the polyadenylation of the completed transcript.

The structure of retroviral proviruses evokes the image of transposable genetic elements found in both bacteria and eukaryotic cells (Varmus, 1983a). The behaviour of retroviruses displays features that are also reminiscent of transposons. First, retroviruses transpose genes, although they do so through the agency of RNA within viral particles rather than by direct transposition of DNA from one position within cellular DNA to another. If proviruses can transpose directly without the intervention of an RNA intermediate, the capability has yet to be recognized. Second, expression of cellular genes can be influenced in either a positive or a negative manner by insertion of a provirus within or near the genes (Varmus, 1983a, b). Third, proviruses can suffer deletions that extend substantial distances into adjacent cellular DNA (Varmus, 1983a). Fourth, proviruses can excise from cellular DNA by homologous recombination between the 5' and 3' LTRs, leaving behind a solitary LTR (Varmus, 1983a).

The genomes of retroviruses are diploid (Varmus & Swanstrom, 1982). As a result, heterozygous particles arise with ease, and these predispose to genetic recombination (Vogt, 1977). Usually, it is viral genes that recombine, but if other RNAs are encapsidated as part of a heterozygous genome, these too can combine with viral RNA (Goldfarb & Weinberg, 1981). The genetic promiscuity of retroviruses has been attributed to the frequency with which reverse transcriptase may cross over from one unit of the diploid genome to the other while in the process of copying viral RNA into DNA, in the manner of classical 'copy-choice' mechanisms for recombination (Coffin, 1979). Whatever its mechanism, the promiscuity may play a large part in the transduction of cellular genes by retroviruses (see below).

Integration of retroviral DNA can elicit mutations in much the same manner as does the transposition of other mobile genetic elements: it can damage and thus inactivate cellular genes by disrupting their structure (Varmus, 1983a); and it can influence the expression of cellular genes, either by damaging their own controlling elements, or by bringing the genes under the sway of powerful viral regulatory signals (Varmus, 1983b). The rubric for these events is 'insertional mutagenesis'. Since the newly integrated viral DNA is easily found by molecular procedures, cellular genes afflicted by insertional mutagenesis can likewise be unearthed (Varmus, 1983b).

Inactivation of cellular genes by integration is a blatant affair: insertion of foreign DNA into the midst of a gene can destroy or obscure the signals required for transcription and splicing, or wreck the coding domain of the gene. By contrast, activation of cellular genes by integration is a more subtle event whose mechanism remains uncertain, but which for the moment we attribute to the ability of proviral LTRs to enhance the expression of nearby genes (for example, see Luciw et al., 1983). The property of enhancement can be localized to a specific domain almost entirely within the LTR but separate from the components of the transcriptional promoter (Luciw et al., 1983). Enhancement occurs irrespective of whether the LTR is located upstream or downstream of the harried gene, and with the LTR pointed in either direction (Payne, Bishop & Varmus, 1982).

There are at least three ways in which the enhancer within LTRs might act: by facilitating integration of viral DNA; by augmenting the intrinsic performance of transcriptional promoters; or by allowing transcription to occur at positions in chromatin that might otherwise be silent. The first of these has been eliminated from contention by experiment (Luciw et al., 1983) and, in any event, would not account for enhancement of genes whose residence in the cellular genome predated integration of viral DNA (Varmus, 1983b). The second proposal has not been subjected to a decisive test, but provisional evidence indicates that it is not correct. For example, an LTR acting in cis to enhance the expression of a thymidine kinase gene appears to have no effect on the constitutive activity of the promoter for the gene (P. Luciw, unpublished). It therefore seems likely that enhancement by LTRs is best described as the activation of what would otherwise be an inert gene.

Retroviruses clearly benefit from their enhancers. Since the choice of the cellular sites for integration appears to be made virtually at random (Varmus & Swanstrom, 1982; Varmus, 1983a), there is no guarantee that proviruses will be situated at positions in chromatin where transcription was active prior to integration. The enhancing element in the LTR may serve to counter this uncertainty by greatly increasing the likelihood of transcription from the provirus, irrespective of its locus within chromatin. It is notable that there are retroviruses whose LTRs are devoid of measurable enhancing activity; as a rule, these viruses establish productive infection less readily than do retroviruses that are blessed with enhancement. Whatever its benefit might be to the virus, enhancement by the LTR is no gift to the cell. Inappropriate activation of cellular genes in the vicinity of proviruses represents idle meddling at the least, lethal intervention at the worst (see below).

Transduction of cellular genes by retroviruses came to light because at least some of the transduced genes confer the dramatic property of rapid tumorigenesis on the transductants (Bishop, 1983a, b). There is no reason to believe that transduction need be limited to genes that are potentially oncogenic. Transduction by retroviruses appears, however, to be a rare event and assays with adequate selective power for other forms of transduced genes have yet to be deployed.

A retrovirus intent upon the seizure of a cellular gene must solve the following problems. First, the transduced gene must be made an intrinsic part of the replicating viral genome: two recombinations occur, so that the cellular gene is joined on both the right and the left to viral RNA and is thus contained entirely within the viral genome; and the resident signals for initiation and termination of transcription from the cellular gene should be discarded, or else these might release the transduced gene from the jurisdiction of viral signals and thwart the need to replicate viral and cellular nucleic acids as a unit. Second, if the transduced gene is to be expressed (as it is in all extant examples), its configuration must allow translation under the direction of either its own signals or those provided by the viral genome. Third, with subtle exceptions (Bishop, 1983a), introns in the transduced portion of the gene must be eliminated. This may not be an *a priori* prerequisite for transduction, but it is dictated by fact: the transduced genes found within retroviruses are spliced versions of their cellular progenitors (Bishop, 1983a). Fourth, transduction by retroviruses represents duplication rather than piracy: the cell retains the object of transduction unmarred; the transducing virus gains only a facsimile, generally or perhaps inevitably incomplete.

Transduction by retroviruses amounts to recombination between virus and cell. The mechanism by which this might occur could call on two features of viral replication. First, integration is a form of recombination between viral and cellular genomes and presents

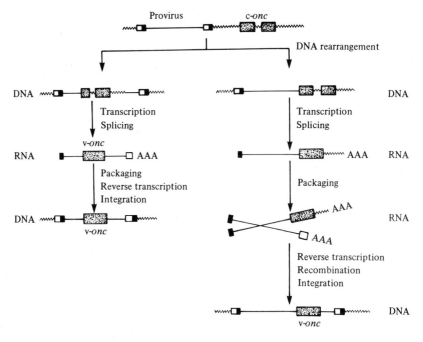

Fig. 2. Transduction of cellular oncogenes by retroviruses. The drawing illustrates how cellular proto-oncogenes might be transduced by pre-existent retroviruses. The scheme begins with an intact retroviral provirus fortuitously integrated upstream from a cellular proto-oncogene. The characteristic terminal redundancies of the provirus are illustrated by black and white boxes, exons of the cellular proto-oncogene by stippled boxes, viral nucleic acid by straight lines, and cellular nucleic acid by jagged lines. A postulated rearrangement of DNA begins transduction and could take either of two forms, as illustrated. Further details are given in Bishop, 1983*a*.

itself as a strong (but not exclusive) candidate for the first step in transduction. Second, generation of heterozygous virions sets the stage for recombination between the components of the diploid retroviral genome. If cellular RNA bearing the gene to be trans-duced can be brought within virions, opportunity for recombination between cellular and viral genes arises. These considerations have engendered three sorts of models for transduction: encapsidation of cellular RNA bearing the object of transduction, followed by at least two recombinations during reverse transcription to complete transduction; integration followed by rearrangement of DNA that immediately incorporates part or all of a cellular gene within a functional provirus (as illustrated on the left in Fig. 2); and a multi-step choreography that combines several of the capabilities of retroviruses (integration, evocation of deletions, encapsidation of RNA that is in part cellular, and recombination as a consequence of heterozygosis) to accomplish transduction (as illustrated on the right in Figure 2).

The first of these models arose from the finding that retroviruses can indeed encapsidate ample quantities of cellular RNA, particularly if the encapsidation of the viral genome itself fails for want of an adequate supply of viral RNA (Levin & Seidman, 1979), or because of a genetic defect in the viral genome (Linial, Medeiros & Hayward, 1978). There is no adequate precedent for the second model, although it remains alive because of recent findings in one instance of transduction (Takeya & Hanafusa, 1983). The third model conforms to a variety of experimental facts (for example, see Swanstrom *et al.*, 1983) and remains the prevailing favourite.

No matter which model (or models) may eventually prove correct, transduction by retroviruses can damage the transduced gene in at least two ways. First, exclusion of portions of the cellular gene from the transductant could drastically change (truncate) the coding unit of the gene. Second, two error-prone polymerases participate in replication of the transduced gene: reverse transcriptase and RNA polymerase II. We have it on good authority that reverse transcriptase is prodigiously mutagenic (Gopinathan *et al.*, 1979), and I presume that the same may be true of RNA polymerase II (to which no editing function has been attributed) when it shares the responsibility for replicating genes – as it does in the life cycle of retroviruses.

TUMORIGENESIS BY RETROVIRUSES

The idiosyncrasies of retroviruses presage two forms of tumorigenesis. First, by analogy with other forms of mutation, insertional mutagenesis is potentially oncogenic. Although inactivation of genes by integration has been demonstrated experimentally (Varmus, 1983*a*), this mode of mutagenesis has yet to be implicated in tumorigenesis by retroviruses – perhaps because mutations induced in this manner are likely to be recessive. By contrast, activation of cellular genes by insertional mutagenesis has been firmly implicated in oncogenesis (Varmus, 1983*b*). Second, numerous strains of retroviruses carry transduced genetic loci ('oncogenes') that are tumorigenic (Bishop & Varmus, 1982; Bishop, 1983*a*). Genetical analyses indicate that the oncogenes of retroviruses are responsible for both the initiation and maintenance of neoplastic growth, and it

Table 1. *Oncogenic avian retroviruses*

Virus	Oncogene	Principal tumors	Cells transformed in culture
Rous sarcoma virus	v-*src*	Sarcomas	Fibroblasts
Y73 sarcoma virus	v-*yes*	Sarcomas	Fibroblasts
Fujinami sarcoma virus	v-*fps*	Sarcomas	Fibroblasts
UR2 sarcoma virus	v-*ros*	Sarcomas	Fibroblasts
Myelocytomatosis virus	v-*myc*	Carcinomas, sarcomas, myelocytic leukaemia	Epithelial and myelomonocytic cells; fibroblasts
Avian erythroblastosis virus	v-*erb*-A/B	Erythroleukemia, sarcomas	Erythroblasts, fibroblasts
Avian myeloblastosis virus	v-*myb*	Myeloblastic leukaemia	Myelomonocytic cells
Avian leukosis virus	None	B-cell lymphomas	None
Avian leukosis virus	None	Erythroleukaemia	None
Chicken syncytial virus	None	B-cell lymphomas	None
Myeloblastosis-associated virus	None	Nephroblastomas	None

is therefore possible that the action of these genes suffices to elicit the complete malignant phenotype (but see below).

Both modes of tumorigenesis are exemplified in a suite of retroviruses isolated from chickens (Table 1). A number of these viruses possess oncogenes that induce characteristic forms of malignancy in birds. Tumorigenesis is rapid (birds are often dead of malignancy within one to two weeks of infection), and cells that are analogous to the tissues attacked *in vivo* can be transformed to neoplastic growth in culture. The remainder of the viruses have no oncogenes and instead wreak their havoc by activating cellular genes (Varmus, 1983*b*). The tumorigenesis that follows infection is slow and progressive, in contrast to tumorigenesis by transduced oncogenes, and transformation of cells in culture has not been achieved. Three features unite these otherwise disparate forms of viral oncogenesis. First, the tumorigenicity of each virus is restricted to one or a few tissues. Second, the specificity of oncogenesis is of a genetic nature, since it is determined by the transduced oncogenes in one instance and by the cellular genes activated in the other. Third, the two forms of tumorigenesis probably have the same provenance: the transduced oncogenes and the cellular genes whose activation can initiate tumorigenesis often represent similar if not identical functions and are drawn from a very few gene families, perhaps from a Superfamily with a single distant ancestor (Bishop, 1983a).

THE NATURE AND MECHANISMS OF TUMORIGENESIS BY RETROVIRAL ONCOGENES

Spontaneous tumorigenesis in living organisms is characteristically a protracted and multi-step affair. The immediacy of neoplastic transformation by retroviruses carrying transduced oncogenes stands in sharp contrast to progressive tumorigenesis and suggests that the oncogenes are dominant effectors of neoplastic growth which can act alone to induce malignant growth. This view has rarely been carefully scrutinized, however, and there is now reason to suspect that it might be incorrect – at least in some instances. The most explicit challenge comes from recent findings with lymphoid tumours induced by the oncogene v-abl of the Abelson Murine Leukaemia Virus. Following prolonged propagation of these tumours in animals, some have lost v-abl (along with the remainder of the leukaemia virus provirus; see Grunwald et al., 1982) and, in its place, have acquired either of two genetic lesions that may now be responsible for sustaining neoplastic growth: appearance of a cellular gene with the ability to transform cells in culture (Lane, Neary & Cooper, 1982); or chromosomal rearrangements that appear to activate a cellular gene (c-myb) known to be the progenitor of a retroviral oncogene (Muschinski et al., 1983a). It is not yet known whether one or the other of these two lesions must come into play before v-abl can be fully tumorigenic, but the emergence of these lesions in concert with the action of v-abl suggests that they have a necessary role in tumorigenesis.

If the action of individual retroviral oncogenes does not suffice to elicit malignancy, what might the deficiency be? Establishment of indefinite growth – the circumvention of senescence – is a prominent feature of the malignant phenotype, yet there are instances where oncogenes fail to elicit this feature. In the parlance, the genes fail to display an 'establishment function'. (a) Transformation of chicken cells by several oncogenes of avian retroviruses (v-src, v-myc, and v-erb-B) fails to establish indefinite growth. Tumours expand, metastasize and eventually kill infected birds because of virus spread that continually recruits new cells into tumour masses (Teich et al., 1982). (b) On at least some occasions, chicken erythroid cells

Table 2. *Possible functions of retroviral oncogene products*

Tumorigenicity	Oncogene	Properties of product
Sarcomas B-Lymphoma	*src, yes, fps/fes, ros* *abl*	Tyr-specific protein kinase on/in plasma membrane
Carcinomas, sarcomas, and myelocytic leukaemia Myeloblastic leukaemia Osteosarcomas	*myc* *myb* *fos*	Nuclear proteins
Erythroleukaemias Sarcomas	*erb*-B, SFFV[a]-*env* *fms*	Membrane glycoproteins
Erythroleukaemias and sarcomas	*ras*	GTP-binding protein on/in plasma membrane
Sarcomas	*mos*	Phosphoprotein in cytosol
Sarcomas	*sis*	Cytoplasmic homologue of PDGF[b]

[a] SFFV, spleen focus forming unit.
[b] PDGF, platelet-derived growth factor.

transformed by the oncogene v-*erb*-B develop the property of indefinite growth when an intercurrent and presently obscure event renders their growth and tumorigenicity independent of the oncogene's activity (H. Beug & T. Graf, personal communications). (*c*) A mutant gene isolated from the cells of a human bladder carcinoma can transform previously established lines of cells to neoplastic growth (Reddy *et al.*, 1982; Tabin *et al.*, 1982; Taparowsky *et al.*, 1982), but it elicits established growth in primary cultures from mammalian embryos only if supplemented with any of several known genes which by themselves are also incapable of fully transforming the cells (see below). (*d*) Adenoviruses and papovaviruses have made these issues most explicit by relegating the establishment function and the other events of neoplastic transformation to separate genetic domains within their genomes (Flint, 1981). Taken at face value, the available data now suggest that some of the known oncogenes serve primarily an establishment function, some provide other aspects of neoplastic transformation, and few or none may be capable of unassisted tumorigenesis.

Even though we do not yet know where the transduced oncogenes of retroviruses stand in the hierarchy of tumorigenic events, it is from these genes that we have gained our most promising glimpse of biochemical mechanisms leading to neoplastic growth. Table 2 summarizes our present knowledge of how these genes may act. The information is fragmentary and generally superficial, but one message seems clear. Oncogenes display a provocative diversity: some encode tyrosine-specific protein kinases, some surely do not; at least one is a close homologue of a normal polypeptide growth factor; some attack in the nucleus of the cell, some at intracellular membranes, some within the cytoplasm, some at the plasma membrane. There is little correlation between what we now know of how oncogenes act and their tumorigenicities (Table 2). Four tyrosine-specific protein kinases induce sarcomas, but a fifth does not. Two glycoproteins which congregate in what may be juxtanuclear membranes induce erythroleukaemias, but a third causes sarcomas instead. And three nuclear proteins cause a bewildering array of tumours that presently defy all efforts at synthesis (although two of these proteins – encoded by v-*myc* and v-*myb* – cause leukaemias in the myelomonocytic lineage).

What does this diversity of function signify? That there is more than one way to create a cancer cell, of course. But a greater truth may lie beyond. It seems likely that the growth of cells is regulated by an interdigitating network extending throughout the cell. If that network were touched at any point by an adverse influence and tilted out of balance, cancerous growth might arise. Perhaps the several forms of oncogenes mirror different components of this normal regulatory network, revealing to us how the network performs its vital function. One such revelation is already in hand: the discovery that the products of several retroviral oncogenes are tyrosine-specific protein kinases (see below) inspired the successful search for evidence that phosphorylation of tyrosine might mediate the response of cells to normal growth factors as well (Kolata, 1983). Thus, by studying oncogenes, we are likely to be learning of both cancerous and normal growth at one and the same time. It is an old adage of medical science that study of the abnormal can reveal the normal.

Pleiotropism is the preeminent quality of the malignant phenotype. It was therefore with great satisfaction that biochemists greeted the discovery of the protein kinases encoded by several retroviral oncogenes (Bishop & Varmus, 1982). No biochemical activity could be better designed to elicit a pleiotropic response; each facet of neoplastic transformation might eventually be traced to the phosphorylation of a particular cellular protein. Hopes still run high, but in the face of great frustration. Only a handful of

potential substrates for the viral kinases have been found (Cooper &
Hunter, 1981) and to date, none of these substrates has been
persuasively implicated in the evocation of specific components of
the malignant phenotype, and none offers any indication as to why
the growth of the neoplastic cell is so profoundly disturbed.
Indeed, there is as yet no decisive evidence that phosphorylation of
proteins at tyrosine residues can account for neoplastic transformation, and no indication that tyrosine phosphorylation might mediate
tumorigenesis arising from causes other than viruses (Doolittle *et
al.*, 1983).

We have only frail purchase on other mechanisms by which
oncogenes may transact their business. The large and important
family of *ras* oncogenes encode proteins that bind guanosine
triphosphate (Bishop & Varmus, 1982), but the suggestions as to
how this property might reflect the function of the *ras* genes have
rested entirely on tenuous structural analogies and have been
contradictory (Gay & Walker, 1983; Wierenga & Hol, 1983). The
proteins encoded by several oncogenes (v-*myc*, v-*myb* and v-*fos*)
reside in the nucleus of transformed cells (Donner, Greiser-Wilke &
Moelling, 1982; Abrams, Rohrschneider & Eisenman, 1982; K.-H.
Klempnauer and I. Verma, personal communications), and the *myc*
proteins are reputed to bind to DNA (Donner *et al.*, 1982) and to be
components of chromatin (Bunte *et al.*, 1982). There is analogy
between these findings and the properties of the large T antigens of
papovaviruses (Topp, Lane & Pollack, 1981), but the ability of large
T proteins to bind to DNA has not been persuasively implicated in
neoplastic transformation, so the analogy is for the moment of little
use. The amino acid sequence of the platelet-derived growth factor
(a polypeptide with mitogenic properties) displays extensive homology with the protein encoded by the retroviral oncogene v-*sis*
(Doolittle *et al.*, 1983; Waterfield *et al.*, 1983). There are, however,
striking contrasts between the growth factor and the viral protein:
the former acts by binding to receptors on the surface of cells (Ross
& Vogel, 1978); the latter may never reach the surface of the cells in
which it is synthesized (S. Aaronson, personal communication).

Perhaps the most urgent need in the pursuit of these problems is
the creation of mutations in cellular genes that impart resistance to
the effects of oncogenes. With such mutations, it might be possible
to identify the biochemical functions whose alterations are crucial to
neoplastic transformation, and to describe the means by which
oncogenes affect these functions.

The logic of evolution is making itself felt in the study of oncogenes. Careful comparisons among the amino acid sequences encoded by different oncogenes have revealed homologies that connote ancestral relationships and preservation of function. (*a*) Several of the genes that specify tyrosine-specific protein kinases (v-*src*, v-*fps/fes*, and v-*abl*), and even some that allegedly or indisputably do not (v-*mos*, v-*erb*-B and a serine protein kinase of mammals) possess large catalytic domains (c. 30 kilodaltons in size) that give evidence of having been constructed from common building blocks (Barker & Dayhoff, 1982; and M. Privalsky & R. Ralston, personal communications). These findings will force a reconsideration of whether, contrary to most previous indications, v-*mos* and v-*erb*-B might also encode protein kinases. (*b*) Efforts to trace the genealogy of the retroviral oncogenes v-*myc* and v-*myb* have revealed unexpected kinships: the proteins encoded by these two retroviral oncogenes are related not only to each other but to the E1a domain of the oncogenic region in the genomes of adenoviruses (R. Ralston, unpublished).

It appears that the oncogenes of RNA and DNA tumour viruses may in some instances share evolutionary origins and functions. The relationships appear as identities or resemblances between amino acid residues, and equally important, as preservation of topography in the distribution of structurally variant and invariant regions within the proteins. The quality of the homologies suggests that they reflect responses to selective pressures serving the conservation of function. It is therefore provocative that E1a, v-*myc* and v-*myb* all encode nuclear proteins, and that E1a and v-*myc* can both supply an establishment function to complement the action of a mutant *ras* gene during transformation of cells in culture (see below). Parallel studies of the mechanisms by which E1a, v-*myc* and v-*myb* contribute to neoplastic transformation should prove mutually reinforcing.

CELLULAR ONCOGENES

When the exploration of viral oncogenes began, it was only a dim hope that their mechanisms of action might prefigure abnormalities in cancers of many origins. Hope approached reality with the discovery that the oncogenes of retroviruses are miscreant copies of what we now call proto-oncogenes or, more explicity, cellular oncogenes, found in normal vertebrates and, perhaps, all metazoan organisms (Bishop, 1983a). Isolation and detailed analyses of proto-oncogenes through the agency of molecular cloning have dismissed any residual possibility that the apparent kinships between cellular and viral oncogenes are distant and trivial. The viral genes are faithful images of their cellular homologues, except for changes expected if transduction has occurred along the lines outlined above: transcriptional signals for the cellular gene, and often portions of the coding domain, have been discarded; introns are missing; and occasional nucleotide substitutions have occurred. What remains of the progenitor has become an integral portion of the viral genome, yet gives rise to a protein that in some instances is virtually indistinguishable from its cellular cousin. It is also clear that proto-oncogenes are native to the cell rather than viral genes in disguise: most if not all have introns – in contrast to the genes of retroviruses; in their natural habitat, proto-oncogenes are not joined to viral genetic elements; they are situated at constant chromosomal locations throughout a species and are apparently present in every member of a species; their situation and structure display the stability of cellular genes rather than the tendencies to peregrination that characterize retroviral genes and other mobile elements; and they segregate in the manner of classical Mendelian loci.

Proto-oncogenes are ancient components of the vertebrate root stock – at least several of their members have turned up in insects, echinoderms, annelida and possibly in yeast (Bishop, 1983a; and E. Scolnick & C. Hammond, personal communications). In view of their great age, it comes as no surprise that proto-oncogenes have been conserved with remarkable fidelity across vast reaches of evolutionary time. For example, the amino acid sequences encoding the catalytic domain of the human and *Drosophila src* genes are related to the extent of almost 95% if compared according to the rules enunciated by McLaughlin (M. Simon & R. Ralston, personal communications); only a handful of nucleotide substitutions distinguish the mouse and human versions of *mos* (Watson, Oskarsson & Vande Woude, 1982); and even the topography of the genes in the form of placement of introns can be preserved, a point best illustrated at present by *ras* genes in rats and humans (Chang *et al.*, 1982).

Proto-oncogenes have survived the rigours of natural selection with spectacular success, and it therefore seems likely that they have vital roles in the economy of cell or organism. Accordingly, expression of all but one of these genes (cellular *mos*) has been detected in normal cells in culture (Bishop, 1983*a*), in normal tissues from vertebrate (Bishop, 1983*a*) and invertebrate (Simon, Kornberg & Bishop, 1983) animals, and in yeast (E. Scolnick, personal communication).

THE ONCOGENE HYPOTHESIS REVISITED

Prompted by the conviction that retroviruses might play an important role in spontaneous tumorigenesis, and by the fact that retroviral genomes had been found within the germ lines of vertebrate species, Huebner & Todaro proposed that many if not all forms of neoplasia arise because carcinogenic agents induce the unwanted expression of retroviral oncogenes resident as unwelcome guests in the cell (Todaro & Huebner, 1972). Subsequent findings turned this hypothesis on its head: the cell does harbour latent oncogenes, but these are cellular rather than viral genes, and they engender the oncogenes found in retroviruses. Elements of the original hypothesis survive, however, in the current theories that cast proto-oncogenes as Jekyll and Hyde in DNA. On the one hand, these genes are thought to be essential to the species in which they occur. On the other hand, many different causes of cancer may wreak their havoc by acting on proto-oncogenes to either augment their activity or change their function in some way. How can this contemporary view of cellular oncogenes be tested, and how have its elements fared to date?

THE NORMAL FUNCTIONS OF PROTO-ONCOGENES

What roles do proto-oncogenes play in the affairs of normal cells? We do not know, but we suspect that they are part of the network that regulates cell growth and division, and that at least some of them may help to direct the course of differentiation. These suspicions have their origins in the properties of retroviral oncogenes, which we take to be caricatures of proto-oncogenes. First, it is in the nature of all viral oncogenes to affect the control of cell division. The effects are pathological, of course, but they may nevertheless mirror the capabilities of proto-oncogenes. Second, viral oncogenes behave as if they were designed to work only in certain types of cells. Expression of v-*src* transforms fibroblasts, but has no tangible effect on the phenotype of macrophages (A. Betkowski & H. Oppermann, personal communications). The converse is true for v-*myb*, which can also be expressed in both fibroblasts and macrophages (K.-H. Klempnauer, unpublished), but which transforms only the latter (Beug, Hayman & Graf, 1982). Third, most retroviral oncogenes meddle with differentiation – either to distort an accomplished phenotype (Boettiger & Durban, 1980) or to halt the progress of cells through a developmental lineage (Beug *et al.*, 1982).

There are now several means by which the roles of proto-oncogenes can be sought. The first of these is argument by analogy. Polypeptide growth factors loom ever larger in the control of cell division and embryological development. Recent findings suggest that proto-oncogenes may represent both key and lock in the actions of growth factors. (*a*) The protein encoded by v-*sis* (and presumably the cellular homologue, c-*sis*, as well) may itself be a growth factor, since its amino acid sequence is remarkably similar to that of platelet-derived growth factor (Doolittle *et al.*, 1983; Waterfield *et al.*, 1983). (*b*) Phosphorylation of tryosine (an enzymatic activity first attributed to several viral oncogenes and their cellular progenitors) figures prominently in the response to growth factors. Binding of epidermal growth factor, platelet-derived growth factor and insulin to receptors on the surface of susceptible cells elicits immediate phosphorylation of tyrosine, in the receptors for the growth factors and in other cellular proteins as well (Kolata, 1983). Protein kinases encoded by proto-oncogenes may be part of the enzymatic chain that carries out these phosphorylations. It is assumed but unproven that the phosphorylations help mediate physiological responses to the growth factors.

The role of proto-oncogenes may also be investigated by correlative studies of expression. Expression of at least some proto-oncogenes is restricted to certain developmental lineages (Gonda, Sheiness & Bishop, 1982; G. Ramsay, unpublished) and has been shown to fluctuate during the course of embryogenesis in mice (Muller *et al.*, 1982) and *Drosophila* (M. Simon, unpublished). Correlations of this sort suggest that the functions of proto-oncogenes are tied to differentiation in some way, but whether as cause or effect we could not yet say.

Finally, genetic manipulation may be used in the study of the roles

Table 3. *Modifications of cellular oncogenes that might contribute to tumorigenesis*

	Activate/augment gene expression	Alter gene function
Amplification	+	−
Mutations	+	+
Chromosomal translocations	+	+
Transduction into virus	+	+

of proto-oncogenes. Several proto-oncogenes have now been found in two settings that permit incisive mutational analysis: *Drosophila melanogaster* (Shilo & Weinberg, 1981) and *Saccharomyces cerevisiae* (E. Scolnick & C. Hammond, personal communications). The proto-oncogenes of *Drosophila* are closely akin to those of mammals (for example, see above) and their roles in the development of the insect seem guaranteed to be informative. But we can only hope that there will be physiological validity in comparisons between the proto-oncogenes of metazoan organisms and their very distant cousins in yeast.

PROTO-ONCOGENES AND CANCER

What evidence can we offer that proto-oncogenes might be the common keyboard for all the players in carcinogenesis, and by what means might a proto-oncogene become tumorigenic? We can seek answers to both questions by turning to the genetic anomalies now known to afflict proto-oncogenes (Table 3).

Transduction

Transduction of proto-oncogenes into retroviruses exemplifies two mechanisms by which proto-oncogenes may become oncogenes. First, the transduced gene has been incorporated into a genetic vehicle that assures brisk expression of the gene. No more may be required: experimental manipulation has shown that several proto-oncogenes (c-*mos*, c-*ras*, c-*myc* and c-*fos* at last count) can transform cells in culture if expression of the genes is enhanced extensively (reviewed in Bishop, 1983a; I. Verma & B. Vennstrom, personal communications). Second, transduction by retroviruses can damage genes by deletion, rearrangement and point mutation (reviewed in Bishop, 1983a). Each form of damage might render a previously harmless gene oncogenic. For example, transduction of c-*src* into Rous sarcoma virus appears to have imposed a genetic substitution on the gene (Takeya & Hanafusa, 1983). As a result, the proteins encoded by c-*src* and its progeny, v-*src*, differ substantially at their carboxy-termini, and it is possible that this difference is responsible for the ability of v-*src* to transform cells to neoplastic growth (R. Parker, unpublished).

Gene amplification

Amplification of genes can augment their expresssion by increasing the amount of DNA template available for the production of mRNA (Schimke, 1982). Karyotypic evidence of gene amplification (double-minute chromosomes and homogeneously staining regions within chromosomes) has been recognized in increasing numbers and varieties of malignancies (Schimke, 1982). Explorations of tumour cells bearing these chromosomal abnormalities have revealed amplification of several proto-oncogenes: c-*myc* in a human promyelocytic leukaemia (Collins & Groudine, 1982; Favera, Wong-Staal & Gallo, 1982); c-*myc* in a neuroendocrine tumour that arose as a carcinoma of the human colon (Alitalo et al., 1983); c-Ki-*ras* in a mouse adrenocortical tumour (Schwab et al., 1983); c-*myb* in an adenocarcinoma of the human colon (K. Alitalo, personal communication); and a distant relative of c-*myc* (designated N-*myc*), whose amplification may be a common or even inevitable feature of human neuroblastomas (M. Schwab, unpublished). In each instance, the amplified proto-oncogene is expressed to levels far above those usually encountered in normal cells.

Can gene amplification contribute to the genesis or maintenance of tumours, and if so, does the enhanced expression of one or another proto-oncogene figure in the contribution? Since double-minute chromosomes are readily lost from cells by unequal segregation at mitosis, the survival of amplified DNA in the form of double-minute chromosomes implies that the amplification or its consequences (such as enhanced gene expression) has conferred a selective advantage on the cells bearing the double-minute chromosomes. It therefore seems reasonable to suspect that gene amplification can be an essential event in the genesis of malignant tumours.

The argument does not mandate a role for amplified proto-oncogenes, however, because the domain of amplified DNA is inevitably much larger than a single genetic locus (Schimke, 1982). Nevertheless, proto-oncogenes have appeared with notable consistency amongst the amplified DNA of tumour cells; given the limited number of identified proto-oncogenes (c. 20 at last count) and the vast complexity of mammalian DNA, this consistency is provocative.

Although genetic rearrangements have been reported within amplified DNA (Tyler-Smith & Alderson, 1981), we cannot presently view mutation as a necessary companion of amplification. Rather, it seems likely that the principle contribution of gene amplification to tumorigenesis would be enhancement of gene expression.

DNA rearrangements: chromosomal translocation and insertional mutagenesis

Characteristic translocations between chromosomes have now been recognized as a consistent feature in a wide variety of human malignancies (Rowley, 1983). It has long been postulated – and until recently, generally denied – that these abnormalities might have a causative role in producing tumours. Now the studies of proto-oncogenes and chromosomal translocations have conjoined in a remarkable way to sustain the view that rearrangement of DNA can be a central malady in the cancer cell. Translocations that move a known proto-oncogene from one chromosome to another have been found in Burkitt's lymphoma (c-*myc*) and chronic myelocytic leukemia (c-*abl*) of humans, and mouse plasmacytomas (c-*myc*) (for recent summaries, see Rowley, 1983; Klein, 1983). Similarly, the c-*myb* gene has been rearranged in mouse lymphoid tumours originally induced by v-*abl*, but now growing independently of the viral gene (Muschinski *et al.*, 1983a). Some translocations and rearrangements greatly enhance the expression of the relocated proto-oncogenes (for example, see Muschinski *et al.*, 1983a, b), perhaps by releasing the genes from a *cis*-active repressor, by attaching them to an active transcriptional promoter, or by bringing them under the influence of a transcriptional enhancer. Others have as yet undetermined effects on the proto-oncogenes. It is also possible that some translocations damage the coding unit of proto-oncogenes, evoking a tumorigenic change in the function of the

Table 4. *Insertional mutagenesis of proto-oncogenes*[a]

Virus	Tumour	Proto-oncogene(s)
Avian leukosis virus	B-cell lymphoma	c-*myc*
Avian leukosis virus	Erythroleukaemia	c-*erb*-B
Myeloblastosis-associated virus	Nephroblastoma	?
Mouse mammary tumour virus	Mammary carcinoma	*int*-1/*int*-2
Murine leukaemia virus	T-cell lymphomas	MLVi-1/MLVi-2
Human T-cell leukaemia virus	T-cell leukaemia	?

[a] The table lists potential examples of tumorigenesis by insertional mutagenesis. The affected proto-oncogenes represent (a) previously known loci (c-*myc* or c-*erb*-B), (b) loci identified for the first time by virtue of one mutation (*int*-1, *int*-2, MLVi-1, MLVi-2), and (c) loci whose existence is suspected but not proven (?).

encoded protein. For the moment, there is no guarantee that the inclusion of proto-oncogenes in rearranged DNA is anything other than coincidence. But the consistent positioning of c-*myc* at the break-points of translocations in Burkitt's lymphoma and mouse plasmacytomas, and the enhanced expression of rearranged c-*myb* in mouse lymphoid tumours, cannot be dismissed easily.

Insertional mutagenesis by retroviruses has directly implicated proto-oncogenes in tumorigenesis (Varmus, 1983*a*, *b*) and, in the offing, has provided a reasonable fascimile of what may happen when proto-oncogenes are translocated from one chromosome to another (Klein, 1983). Tumours induced by retroviruses without oncogenes of their own apparently arise because proviral DNA can activate the transcription of neighbouring cellular genes (Varmus, 1983*b*). The possible role of insertional mutagenesis in oncogenesis first emerged from studies of B-cell lymphomas induced in chickens by avian leukosis viruses (Neel *et al.*, 1981; Payne *et al.*, 1981). By a stroke of great good fortune, the gene activated in these tumours was already known to us as the proto-oncogene c-*myc* (Hayward, Neel & Astrin, 1981). There is no doubt that the expression of c-*myc* is enhanced in these tumours because the protein encoded by c-*myc* (a 58 000 dalton nuclear protein) is 20–100-fold more abundant in the tumour cells than in various normal sources (K. Alitalo, personal communication).

The scenario has now been extended to other tumours (see Table 4). In some instances, the activated genes are previously identified proto-oncogenes, in others they are genes that have been brought to light for the first time by insertional mutagenesis but which we can easily regard as candidates for proto-oncogenes. Activation is generally due to recent infection by a retrovirus, but there also exists at least one example where relocation of the provirus of an endogenous retrovirus has activated a proto-oncogene (c-*mos*) (Rechavi, Givol & Canaani, 1982). Activation of the proto-oncogenes by insertional mutagenesis has been attributed to either the promoters (Hayward *et al.*, 1981) or the enhancers (Payne *et al.*, 1982) of transcription carried by the LTRs of proviruses. Both explanations appear to be valid. It is also possible that insertion of proviral DNA could damage the coding unit of the activated cellular gene, but there is as yet no evidence for such damage.

Point mutations

The views of how damage to DNA might cause cancer have come full circle over the past decade. Large distortions of DNA structure – of the sort just reviewed – were first neglected, then earned their place in the sun and, for a brief moment, threatened to overshadow the role of point mutations. That threat vanished quickly with the discovery that single point mutations can account for the presence of active oncogenes in human tumours (Reddy *et al.*, 1982; Tabin *et al.*, 1982; Taparowsky *et al.*, 1982). The oncogenes were first detected by the ability of DNA from many human tumours (c. 20% of those tested to date) to transform cells in culture to neoplastic growth (Cooper, 1982; Weinberg, 1982). A substantial number of these oncogenes have now been identified; all but a few are members of the *ras* family of proto-oncogenes; and at least several owe their transforming activity to single point mutations that change one amino acid residue in the protein encoded by the gene (Reddy *et al.*, 1982; Tabin *et al.*, 1982; Taparowsky *et al.*, 1982; and R. Weinberg, M. Wigler & S. Aaronson, personal communications). These findings return us to retroviruses in at least two regards: it was from the study of retroviral oncogenes that the *ras* proto-oncogenes first came into view (Bishop, 1983*a*); and the transduced versions of *ras* that serve as viral oncogenes contain mutations analogous (albeit not necessarily identical) to those found in some of the oncogenes isolated from human tumours (Dhar *et al.*, 1982; Tsuchida, Ryder & Ohtsubo, 1982; Capon *et al.*, 1983; and R. Weinberg & M. Wigler, personal communications).

Table 5. *Progression: multiple genetic lesions in tumour cells, a premature synthesis*

Tumour	Genetic lesions[a]	
Avian B-lymphoma	c-*myc*	B-*lym*
Burkitt's lymphoma	c-*myc*	B-*lym*
Mouse plasmacytomas	c-*myc*	3T3+
Acute promyelocytic leukaemia (HL-60)	c-*myc*	c-*ras*
Human neuroblastoma	N-*myc*	N-*ras*
Chronic myelogenous leukaemia	22q⁻ (c-*abl*/c-*sis*)	3T3+
Mouse mammary carcinoma (MMTV)	*int*-1	Hu-*mam*
Abelson lymphoma (Ab-MuLV)	v-*abl*	3T3+

[a] The column on the left lists examples of gross chromosomal damage or viral oncogenes, the column on the right lists genes in tumour DNA that transform 3T3 mouse fibroblasts in culture.

MULTIPLE GENETIC LESIONS IN TUMORIGENESIS

Evidence of various sorts has long suggested that malignancies arise from a protracted sequence of events whose number remains in dispute – some observers say there are but two, others argue for more. The nature of the tumorigenic events is also obscure. The reductionist view holds that each step in tumorigenesis is a form of genetic damage, whereas opponents of this view argue that specific genetic lesions account for only part (or even none) of the tumorigenic scheme.

Recent findings with cellular oncogenes speak to these issues. The first clues came from B-cell lymphomas induced in chickens by avian leukosis virus. We now know of two oncogenes active in this tumour: c-*myc*, whose expression is enhanced by insertional mutagenesis – presumably the earliest event in the genesis of these tumours (see above); and a different gene, recognized by its ability to transform rodent cells in culture, unrelated to c-*myc*, and perhaps representing a later step in tumorigenesis (Cooper, 1982). If we are bold, analogous patterns can be perceived in other and varied neoplasms (Table 5). In each instance, a chromosomal translocation, gene amplification, gene activation by insertional mutagenesis, or the action of a viral oncogene is paired with the activity of a cellular oncogene that can transform cultured cells. These pairings were hardly in view before experimental evidence emerged to suggest that combinations of two oncogenes are necessary and sufficient to elicit the full-blown malignant phenotype. The mutant form of c-*ras* isolated from cells of a human bladder carcinoma (and analogous to mutant forms of *ras* found as viral oncogenes) elicits indefinite neoplastic growth in primary cultures of embryonic rodent cells only when used in concert with v-*myc*, c-*myc*, the gene encoding the large T antigen of polyoma virus, or the E1a region of the oncogenic domain in the adenovirus genome (R. Weinberg & M. Wigler, personal communications).

THE FUTURE?

It therefore appears that experimental models for multi-step carcinogenesis may be close at hand. With these models, and with parallel explorations of viral tumorigenesis, we can hope to address the following issues. (*a*) Are genetic lesions wholly accountable for carcinogenesis, or must we yet contend with more elusive events and influences? (*b*) What is the nature of the lesions that initiate tumorigenesis? Are they unique to initiation, or can they interchange with subsequent steps in the scheme, perhaps to give rise to different forms of tumours? Are they necessary steps, or can they on occasion be by-passed? (*c*) Is there a distinct class of end-stage or maintenance lesions, and if so, what is their character? (*d*) What happens during tumour progression? How many events must occur, and what is their nature? In particular, is progression driven by recurrent genetic damage, or are there in reality only two genetic lesions in all tumours – the lesion of initiation, and the lesion that sustains malignant growth in a manner akin to the action of viral oncogenes? (*e*) What is the nature of inherited susceptibility to cancer? Does it ever represent entry into the germ line of lesions afflicting proto-oncogenes? (*f*) What are the biochemical mechanisms by which oncogenes act? (*g*) Will we be able to parlay our new-found grasp on cancer genes into rational strategies for the prevention and therapy of human malignancy?

REFERENCES

ABRAMS, H. D., ROHRSCHNEIDER, L. R. & EISENMAN, R. N. (1982). Nuclear location of the putative transforming protein of avian myelocytomatosis virus. *Cell*, **29**, 427–39.
ALITALO, K., SCHWAB, M., LIN, C. C., VARMUS, H. E. & BISHOP, J. M. (1983). Homogeneously staining chromosomal regions contain amplified copies of an

abundantly expressed cellular oncogene (c-*myc*) in malignant neuroendocrine cells from a human colon carcinoma. *Proceedings of the National Academy of Sciences (USA)*, **80**, 1707–11.

BARKER, W. C. & DAYHOFF, M. O. (1982). Viral *src* gene products are related to the catalytic chain of mammalian cAMP-dependent protein kinase. *Proceedings of the National Academy of Sciences (USA)*, **79**, 2836–9.

BEUG, H., HAYMAN, M. J. & GRAF, T. (1982). Leukemia as a disease of differentiation: retroviruses causing acute leukaemias in chickens. *Cancer Surveys*, **1**, 205–30.

BISHOP, J. M. (1983*a*). Cellular oncogenes and retroviruses. *Annual Review of Biochemistry*, **52**, 301–54.

BISHOP, J. M. (1983*b*). Cancer genes come of age. *Cell*, **32**, 1018–20.

BISHOP, J. M. & VARMUS, H. E. (1982). Functions and origins of retroviral transforming genes. In *Molecular Biology of Tumor Viruses, Part III. RNA Tumor Viruses*, eds. R. A. Weiss, N. Teich, H. E. Varmus & J. M. Coffin, pp. 99–1108. Cold Spring Harbor: Cold Spring Harbor Press.

BOETTIGER, D. & DURBAN, E. M. (1980). Progenitor-cell populations can be infected by RNA tumor viruses, but transformation is dependent on the expression of specific differentiated functions. *Cold Spring Harbor Symposia on Quantitative Biology*, **44**, 1249–54.

BUNTE, T., GREISER-WILKE, I., DONNER, P. & MOELLING, K. (1982). Association of avian myelocytomatosis virus wild-type and mutant *gag-myc* proteins with chromatin. *European Molecular Biology Organization Journal*, **1**, 919–27.

CAPON, D. J., CHEN, E. Y., LEVINSON, A. D., SEEBURG, P. H. & GOEDDEL, D. V. (1983). Complete nucleotide sequences on the T24 human bladder carcinoma oncogene and its normal homologue. *Nature*, **302**, 33–7.

CHANG, E. H., GONDA, M. A., ELLIS, R. W., SCOLNICK, E. M. & LOWY, D. R. (1982). Human genome contains four genes homologous to transforming genes of Harvey and Kirsten murine sarcoma virus. *Proceedings of the National Academy of Sciences (USA)*, **79**, 4848–52.

COFFIN, J. M. (1979). Structure, replication, and recombination of retrovirus genomes: some unifying hypotheses. *Journal of General Virology*, **42**, 1–26.

COLLINS, S. & GROUDINE, M. (1982). Amplification of endogenous *myc*-related DNA sequences in a human myeloid leukaemia cell line. *Nature*, **298**, 679–81.

COOPER, G. M. (1982). Cellular transforming genes. *Science*, **217**, 801–6.

COOPER, J. A. & HUNTER, T. (1981). Four different classes of retroviruses induce phosophorylation of tyrosine present in similar cellular proteins. *Molecular and Cellular Biology*, **1**, 394–407.

DHAR, R., ELLIS, R. W., SHIH, T. Y., OROSZLAN, S., SHAPIRO, B., MAIZEL, J., LOWY, D. & SCOLNICK, E. (1982). Nucleotide sequence of the p21 transforming protein of Harvey murine sarcoma virus. *Science*, **217**, 934–7.

DONNER, P., GREISER-WILKE, I. & MOELLING, K. (1982). Nuclear localization and DNA binding of the transforming gene product of avian myelocytomatosis virus. *Nature*, **296**, 262–6.

DOOLITTLE, R. F., HUNKAPILLER, M. W., HOOD, L. E., DEVARE, S. G., ROBBINS, K. C., AARONSON, S. A. & ANTONIADES, H. N. (1983). Simian sarcoma virus *onc* gene, v-*sis*, is derived from the gene (or genes) encoding a platelet-derived growth factor. *Science*, **221**, 275–7.

FAVERA, R. D., WONG-STAAL, F. & GALLO, R. C. (1982). *onc*-gene amplification in the human promyelocytic leukemia cell line HL-60 and in primary leukemic cells of the same patient. *Nature*, **299**, 61–3.

FLINT, S. J. (1981). Transformation by adenoviruses. In *Molecular Biology of Tumor Viruses, Part II. DNA Tumor Viruses*, ed. J. Tooze, pp. 547–75. Cold Spring Harbor: Cold Spring Harbor Press.

GAY, N. J. & WALKER, J. E. (1983). Homology between human bladder carcinoma, oncogene product and mitochondrial ATP-synthase. *Nature*, **301**, 262–4.

GOLDFARB, M. P. & WEINBERG, R. A. (1981). Generation of novel biologically active Harvey sarcoma virus via apparent illegitimate recombination. *Journal of Virology*, **38**, 136–50.

GONDA, T. J., SHEINESS, D. K. & BISHOP, J. M. (1982). Transcripts from the cellular homologs of retroviral oncogenes: distribution among chicken tissues. *Molecular and Cellular Biology*, **2**, 617–24.

GOPINATHAN, K., WEYMOUTH, L., KUNKEL, T. & LOEB, L. (1979). Mutagenesis *in vitro* by DNA polymerase from an RNA tumor virus. *Nature*, **278**, 857–8.

GRUNWALD, D. J., DALE, B., DUDLEY, J., LAMPH, W., SUGDEN, B., OZANNE, B. & RISSER, R. (1982). Loss of viral gene expression and retention of tumorigenicity by Abelson Lymphoma cells. *Journal of Virology*, **43**, 92–103.

HAYWARD, W. S., NEEL, B. G. & ASTRIN, S. M. (1981). ALV-induced lymphoid leukosis: activation of a cellular *onc* gene by promoter insertion. *Nature*, **290**, 475–9.

JACOB, F. & MONOD, J. (1961). Genetic regulatory mechanisms in the synthesis of proteins. *Journal of Molecular Biology*, **3**, 318–56.

KLEIN, G. (1983). Specific chromosomal translocations and the genesis of B-cell-derived tumors in mice and men. *Cell*, **32**, 311–15.

KOLATA, G. (1983). Is tyrosine the key to growth control? *Science*, **219**, 377–8.

LANE, M.-A., NEARY, D. & COOPER, G. M. (1982). Activation of a cellular transforming gene in tumours induced by Abelson murine leukaemia virus. *Nature*, **300**, 659–61.

LEVIN, J. G. & SEIDMAN, J. G. (1979). Selective packaging of host tRNAs by murine leukemia virus particles does not require genomic RNA. *Journal of Virology*, **29**, 328–35.

LINIAL, M., MEDEIROS, E. & HAYWARD, W. S. (1978). An avian oncovirus mutant (SE2 1Q 1b) deficient in genomic RNA: biological and biochemical characterization. *Cell*, **15**, 1371–81.

LUCIW, P., BISHOP, J. M., VARMUS, H. E. & CAPECCHI, M. (1983). Location and function of retroviral and SV40 sequences that enhance biochemical transformation after microinjection of DNA. *Cell*, **33**, 705–16.

MULLER, R., SLAMON, D. J., TREMBLAY, J. M., CLINE, M. J. & VERMA, I. M. (1982). Differential expression of cellular oncogenes during pre- and postnatal development of the mouse. *Nature*, **299**, 640–4.

MUSCHINSKI, J. F., BAUER, S. R., POTTER, M. & REDDY, E. P. (1983*b*). Increased expression of *myc*-related oncogene mRNA characterizes most BALB/c plasmacytomas induced by pristane or Abelson murine leukaemia virus. *Proceedings of the National Academy of Sciences (USA)*, **80**, 1073–7.

MUSCHINSKI, J. F., POTTER, M., BAUER, S. R. & REDDY, E. P. (1983*a*). DNA rearrangement and altered RNA expression of the c-*myb* oncogene in mouse plasmacytoid lymphosarcomas. *Science*, **220**, 795–8.

NEEL, B. G., HAYWARD, W. S., ROBINSON, H. L., FANG, J. M. & ASTRIN, S. M. (1981). Avian leukosis virus-induced tumors have common proviral integration sites and synthesize discrete new RNAs: oncogenesis by promoter insertion. *Cell*, **23**, 323–34.

PAYNE, G. S., BISHOP, J. M. & VARMUS, H. E. (1982). Multiple arrangements of viral DNA and an activated host oncogene (c-*myc*) in bursal lymphomas. *Nature*, **295**, 209–17.

PAYNE, G. S., COURTNEIDGE, S. A., CRITTENDEN, L. B., FADLY, J. M. & VARMUS, H. E. (1981). Analysis of avian leukosis virus DNA and RNA in bursal tumors: viral gene expression is not required for maintenance of the tumor state. *Cell*, **23**, 311–22.

RECHAVI, G., GIVOL, D. & CANAANI, E. (1982). Activation of a cellular oncogene by DNA rearrangement: possible involvement of an IS-like element. *Nature*, **300**, 607–10.

REDDY, E. P., REYNOLDS, R. K., SANTOS, E. & BARBACID, M. (1982). A point mutation is responsible for the acquisition of transforming properties by the T24 human bladder carcinoma oncogene. *Nature*, **300**, 149–52.

ROSS, R. & VOGEL, A. (1978). The platelet-derived growth factor. *Cell*, **14**, 203–10.

ROWLEY, J. D. (1983). Human oncogene locations and chromosome aberrations. *Nature*, **301**, 290–1.

SCHIMKE, R. T. (1982). *Gene Amplification*. Cold Spring Harbor: Cold Spring Harbor Press.

SCHWAB, M., ALITALO, K., VARMUS, H. E., BISHOP, J. M. & GEORGE, D. (1983). A cellular oncogene (c-Ki-*ras*) is amplified, overexpressed, and located within karyotypic abnormalities in mouse adrenocortical tumour cells. *Nature*, **303**, 497–501.

SHILO, B. Z. & WEINBERG, R. A. (1981). DNA sequences homologous to vertebrate oncogenes are conserved in Drosophila melanogaster. *Proceedings of the National Academy of Sciences (USA)*, **78**, 6789–92.

SIMON, M. A., KORNBERG, T. B. & BISHOP, J. M. (1983). Drosophila possesses three loci related to the oncogene of Rous sarcoma virus and has tyrosine specific protein kinase activity. *Nature*, **302**, 837–9.

SWANSTROM, R., PARKER, R. C., VARMUS, H. E. & BISHOP, J. M. (1983). Transduction of a cellular oncogene: the genesis of Rous sarcoma virus. *Proceedings of the National Academy of Sciences (USA)*, **80**, 2519–23.

TABIN, C. J., BRADLEY, S. M., BARGMANN, C. I., WEINBERG, R. A., PAPAGEORGE, A. G., SCOLNICK, E. M., DHAR, R., LOWY, D. R. & CHANG, E. H. (1982). Mechanism of activation of a human oncogene. *Nature*, **300**, 143–9.

TAKEYA, T. & HANAFUSA, H. (1983). Structure and sequence of the cellular gene homologous to the RSV *src* gene and the mechanism for generating the transforming virus. *Cell*, **32**, 881–90.

TAPAROWSKY, E., SUARD, Y., FASANO, O., SHIMIZU, K., GOLDFARB, M. & WIGLER, M. (1982). Activation of the T24 bladder carcinoma transforming gene is linked to a single amino acid change. *Nature*, **300**, 762–5.

TEICH, N., WYKE, J., MAK, T., BERNSTEIN, A. & HARDY, W. (1982). Pathogenesis of retrovirus-induced disease. In *Molecular Biology of Tumor Viruses, Part III. RNA Tumor Viruses*, eds. R. A. Weiss, N. Teich, H. E. Varmus & J. M. Coffin, pp. 785–998. Cold Spring Harbor: Cold Spring Harbor Press.

TODARO, G. J. & HUEBNER, R. H. (1972). The viral oncogene hypothesis: new evidence. *Proceedings of the National Academy of Sciences (USA)*, **69**, 1009–15.

TOPP, W. C., LANE, D. & POLLACK, R. (1981). Transformation by SV40 and polyoma virus. In *Molecular Biology of Tumor Viruses, Part II. DNA Tumor Viruses*, ed. J. Tooze, pp. 205–96. Cold Spring Harbor: Cold Spring Harbor Press.

TSUCHIDA, N., RYDER, T. & OHTSUBO, E. (1982). Nucleotide sequence of the oncogene encoding the p21 transforming protein of Kirsten murine sarcoma virus. *Science*, **217**, 937–41.

TYLER-SMITH, C. & ALDERSON, T. (1981). Gene amplification in methotrexate-resistant mouse cells. I. DNA rearrangement accompanies dihydrofolate reductase gene amplification in a T-cell lymphoma. *Journal of Molecular Biology*, **153**, 203–18.

VARMUS, H. E. (1983*a*). Retroviruses. In *Transposable Elements*, ed. J. Shapiro, pp. 411–503. New York: Academic Press.

VARMUS, H. E. (1983*b*). Recent evidence for oncogenesis by insertion mutagenesis and gene activation. *Cancer Surveys*, **1**, 309–19.

VARMUS, H. E. & SWANSTROM, R. (1982). Replication of retroviruses. In *Molecular Biology of Tumor Viruses, Part III. RNA Tumor Viruses*, eds. R. A. Weiss, N. Teich, H. E. Varmus & J. M. Coffin, pp. 369–512. Cold Spring Harbor: Cold Spring Harbor Press.

VOGT, P. K. (1977). Genetics of RNA tumor viruses. In *Comprehensive Virology*, eds. H. Fraenkel-Conrat & R. R. Wagner, vol. 9, pp. 341–455. New York: Plenum Press.

WATERFIELD, M. D., SCRACE, G. T., WHITTLE, N., STROOBAUT, P., JOHNSSON, A., WASTESON, Å., WESTERMARK, B., HELDIN, C.-H., HUANG, J. S. & DUELL, T. F. (1983). Platelet-derived growth factor is structurally related to the putative transforming protein p28^{sis} of simian sarcoma virus. *Nature*, **304**, 35–9.

WATSON, R., OSKARSSON, M. & VANDE WOUDE, G. F. (1982). Human DNA sequence homologous to the transforming gene (*mos*) of Moloney murine sarcoma virus. *Proceedings of the National Academy of Sciences (USA)*, **79**, 4078–82.

WEINBERG, R. A. (1982). Oncogenes of spontaneous and chemically induced tumors. *Advances in Cancer Research*, **36**, 149–63.

WIERENGA, R. K. & HOL, W. G. J. (1983). Predicted nucleotide-binding properties of p21 protein and its cancer-associated variant. *Nature*, **302**, 842–4.

THE ROLE OF RECOMBINATION IN THE LIFE OF BACTERIAL VIRUSES

NEVILLE SYMONDS

School of Biological Sciences, University of Sussex, Brighton, UK

INTRODUCTION

Recombination is usually thought of as a mechanism whereby genetic diversity is accelerated by the intermixing of traits derived from different parents. This aspect of recombination certainly holds for bacterial viruses – or phages for short. However, in many phage systems the main relevance of recombination is something quite different. It plays a key role in fashioning some characteristic in the life-cycle of the phage, sometimes to the extent of being absolutely essential for virus multiplication. Just as there are a bewildering variety of life-styles amongst phages, there are also a wide variety of recombination mechanisms involved in these life-styles. In this paper I want first to describe in a qualitative way some examples of the dependency of phages on particular recombination systems, and then go on to discuss in some detail the particular case of recombination and phage Mu. This choice is partly personal, but also has a more general justification. Over the last five years or so a big change has begun to appear in attitudes to genetics. Previously, chromosomes were considered basically as static entitites whose genetic complement could be altered by recombination or mutation. However, it is now being realized that during development, and under stress, the genomes of organisms can undergo dynamic alterations leading to the switching on and off of relevant genes. These rearrangements are largely due to novel recombination systems such as transposition and site-specific recombination. An understanding of the basic mechanisms underlying these systems therefore becomes crucial. Phage Mu is perhaps the best model system in which to study these systems, as its whole life-cycle is based upon transposition, and it also contains one of the first site-specific recombination systems which have been shown to control gene expression.

Classification of recombination systems

In the broadest of terms, recombination can be said to take place when two parental DNA molecules (or two regions in the same molecule) interact and generate new molecules containing information derived from each parent. Historically, recombination has been divided into four classes. In general recombination, the recombination event is mediated by extensive homologous tracts of DNA. Site-specific recombination occurs only between specific (and short) DNA sequences; these sequences need not be identical. Transposition is a process where a DNA element (a transposon), initially present at one site in a genome, is subsequently found to be located at any one of a large number of different DNA sites within the same cell. Illegitimate recombination refers to recombination events which occur between DNA sequences which bear no known sequence relation to one another. Although this classification is an oversimplification, it is useful and we shall employ it in the discussion which follows of the involvement of recombination in the life-cycle of various phages.

Role of recombination in some phage systems

The following survey is not intended to be exhaustive, but it gives a broad picture of the ways recombination has been harnessed to solve specific phage problems.

Phage T4

T4 has a linear genome which is terminally redundant and circularly permuted. Soon after infection, DNA replication starts from a unique origin on the T4 genome and proceeds to the ends of the linear molecule where a single-stranded segment with a 3' terminus is left because of the inability of the T4 replisome to initiate the last Okazaki fragment. These single-stranded ends are extremely recombinogenic and proceed to invade the homologous region in another T4 genome (in multiple infection) or the homologous region at the other end of the same genome (in single infection). The recombination intermediates generated in this way have the structure of replication forks and act as secondary origins of T4 replication. This replication again proceeds to the ends of the molecules where the process is repeated, with the result that a branched network of replicating molecules with many growing forks is established (Luder

& Mosig, 1982). Finally, mature phage particles are packaged by a 'headful' mechanism from the long concatemers which have been formed. The secondary replication forks account for nearly all DNA replication in T4-infected cells. When any of the recombination enzymes are inactive, and all are phage coded, the T4 burst is reduced to less than one phage per cell. For T4, recombination (of the general type) is an essential component of normal replication.

Phage λ

λ codes for two types of recombination, each with a specific function. *Red* is a general recombination system which plays a role in replication analogous to that just discussed for recombination in T4. When a lytic cycle of phage growth is started, either by infection or induction, the λ genome initially exists as a monomeric circular molecule. DNA replication then proceeds bidirectionally from a primary origin, and a few rounds of so-called theta-replication occur, with the formation of more circular monomers. Mediated in some as yet unknown way by the *red* system, replication then changes to a rolling-circle mode, long concatemers of DNA are formed, and phage particles are fashioned by cutting these at particular sites. In the absence of active *red* enzymes, replication is limited to the theta-mode, no concatemers are produced and the λ burst drops to around 10 per cell, so again normal replication is crucially linked to a general recombination system (Enquist & Skalka, 1978).

Also in phage λ, a site-specific recombination system is operative which determines the integration and excision of the λ genome into (or out of) the host chromosome. Integration needs the presence of the *int* gene product together with certain host factors, and involves a reciprocal recombination event between two *att* sequences, one on λ and the other on the bacterial chromosome. Excision involves recombination between two different but related sequences at either end of a λ prophage and, in addition to *att*, needs another gene product from the *xis* gene (Gottesman & Weisberg, 1971). The *int* system was the first site-specific reaction to be reproduced *in vitro* and much of our present knowledge of the factors influencing this type of recombination system comes from these studies (Nash, 1975; Nash, 1981).

Phage P22

This is a temperate phage with a linear genome that is terminally redundant and circularly permuted. Upon infection the genome circularizes and this circular form is essential for both integration and replication. The circularization comes about by a generalized type of recombination occurring between the homologous regions at the ends of each genome. In rec^+ cells this recombination can be mediated by host enzymes, but P22 also circularizes efficiently in rec^- hosts due to the product of the P22 gene erf (Botstein & Matz, 1970). This is a case, therefore, where recombination is a necessary step in the phage life-cycle, but there are two alternative pathways to carry it out, one host-controlled and the other controlled by the phage.

Phage P1

P1 is a temperate phage with a large genome (90 kb) which is circularly permuted and has a terminal redundancy that covers about 20% of its length. Like P22, it is a generalized transducing phage. P1 is interesting for two reasons. Firstly, in the prophage state, it is not integrated but exists as an extra-chromosomal replicon at a level of 1–2 copies per cell. Secondly P1 codes for two important site-specific recombination systems. One of these systems controls the inversion of a genome region, G which, in turn, determines two alternative forms of the P1 host range. An almost identical system also exists in phage Mu and the details of this G inversion will be discussed when dealing with the properties of that phage. The second site-specific system is called lox. It consists of two elements, a lox site on the P1 genome and the product of a P1 gene called cre. The cre gene product stimulates efficient reciprocal recombination between two lox sites particularly when they are in the same molecule.

Two functions have been ascribed to the lox system. One is reminiscent of the erf systems in P22. Upon infection, the linear P1 genome has to circularize. Usually this is mediated by the host rec system by homologous recombination across the ends of the linear P1 molecule. In recA cells this is not possible but, in those phages having two lox sites in the redundant regions, the circularization can occur by site-specific recombination. Due to the large terminal redundancy in P1 this means about 20% of P1 particles are infective in recA cells, and explains why P1 lysates can be prepared which are able to transduce the recA phenotype (Sternberg et al., 1980). The other role ascribed to the lox system concerns copy-number control (Austin, Ziese & Sternberg, 1981). In order to maintain a stable prophage copy-number of 1–2, it is essential that there is some mechanism which ensures that at cell division each daughter cell receives a copy of the P1 prophage. In rec^+ cells, which are the normal hosts, recombination between the circular P1 prophage will always occur with the formation of dimers from two monomers. If such a recombination event occurs in a cell containing two P1 prophages just before cell division, then one of the daughter cells will receive a dimer, the other no P1 at all. As the fraction of cells which have lost the P1 prophage in a lysogenic culture is very small (less than one in ten thousand) any dimers formed in this way must be split into two monomers almost as soon as they are formed. It is here that *lox* is supposed to work. The P1 dimer contains two *lox* sites in the same molecule, which is exactly the substrate on which the site-specific *lox* recombination acts so efficiently, in this case breaking down the dimer to two monomers. As the reverse reaction (that of inducing recombination between two *lox* sites on different molecules) works much less efficiently, there is little tendency to form (or reform) dimers from monomers. This proposed liaison between site-specific recombination and copy-number control is an intriguing one and clearly has implications in the wider context of the inheritance of plasmids.

When examined carefully, virtually all phage systems can be shown to put recombination to some particular use. The examples that have been considered above are reasonably representative of the situation as it is known at present, but no doubt other aspects will turn up to add to the diversity in our knowledge of phage life-styles.

RECOMBINATION AND PHAGE MU

Mutation, transposition and the Mu lytic cycle

In the remainder of this paper, I want to discuss some of the problems posed by the behaviour of phage Mu that are linked to various types of recombination, in particular to transposition. An excellent and comprehensive review on Mu which looks at the subject from a broader point of view has recently appeared (Toussaint & Resibois, 1983).

Before discussing any specific problems it is necessary to have a nodding acquaintance with the Mu life-cycle and with some consequences that stem from it. Mu is a temperature phage so a lytic cycle of growth can be initiated either by infecting sensitive cells with phage particles or by inducing a Mu lysogen. Mu induction is not triggered by UV-irradiation or similar treatments and most experimental work employs Mu*cts* mutants which possess a temperature-sensitive mutation in the repressor gene. The normal host of Mu is *E. coli* K12 and cultures of the lysogen, *E. coli* K12 (Mu*cts*) are routinely grown at 32 °C and induced at 42 °C.

Superficially, the lytic cycle is similar whether it starts by infection or induction. Some early Mu enzymes are synthesized almost immediately. Synthesis of Mu DNA is first detected after about 10 min and continues throughout the rest of the latent period. Synthesis of late proteins (components of heads and tails) starts after approximately 20 min and the first infectious phage particles are formed at 30 min, being made then at a roughly constant rate until lysis occurs, liberating a burst of 100 phage particles. The complete cycle takes about 60 min.

The overall pattern shown in this cycle of early enzymes, specific DNA synthesis, late protein formation, linear production of phage and finally lysis is much the same as that shown by other phages. The unique character of Mu is related to two other characteristics. The first is that Mu can form lysogens by integrating at virtually any location in the bacterial chromosome. As these insertions inactivate the genes concerned, Mu is therefore a mutagen; 'Mu' in fact is an abbreviation of mutator (not an indication of classical scholarship as the author, who came into the field a bit late, thought for quite a while).

The other characteristic of Mu which unquestionably is related to the mutator phenotype is that Mu is the most potent transposon known. Indeed Mu is capable of transposing 100 times in 30 min and, as transposition is considered to be a replicative process, this degree of transposition is sufficient to account for all the copies of Mu generated during a lytic cycle. Uniquely among phages, it is therefore considered that *replication of Mu occurs by repeated transposition*.

The phage genome

DNA extracted from Mu phage particles consists of 37 kb of unique double-stranded DNA flanked by random sequences of bacterial DNA amounting to about 0.05 kb at one end (traditionally called the left end) and approximately 2 kb at the right end. The random

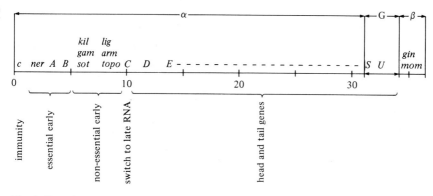

Fig. 1. Genetic map of phage Mu. Scale is in kilobases.

sequences of DNA at the end of the phage genomes (which can easily be visualized as 'split-ends' in the electron microscope by extracting this DNA and then separating and reannealing the strands) have a ready explanation in terms of the transposition scheme for Mu replication and the 'headful' mechanism of packaging used to form infectious phage particles. Each replication results in the insertion of a copy of Mu DNA at a new bacterial location. Maturation of these randomly integrated Mu genes into phage particles is then initiated by the action of a Mu gene which recognizes the left end of Mu then stretches its arm, so to speak, a defined distance to the left and institutes a double-strand break from which packaging of Mu DNA commences. Mu DNA is then stuffed into a phage head of defined size, which holds on average 39 kb of DNA. Each phage particle therefore contains different flanking sequences of bacterial DNA, with a small precise length attached to the left end of Mu and a much longer but variable length attached to the right end.

Genetic map of Mu

A simplified physical and genetic map of Mu is shown in Fig. 1. Historically, the genome has been divided into three regions, α, G and β. The α region, which comprises most of the genome, contains the genes controlling immunity, all the early functions, all the head functions and most of the tail functions. The G region contains the rest of the tail genes, while the β region contains genes affecting site-specific recombination and DNA modification.

The first gene on the left of the phage codes for Mu repressor and is read constitutively. Next comes *ner*, a negative regulator for transcription which is involved in the control both of repressor and also of the two essential early genes *A* and *B* which are the Mu genes intimately involved in transposition and are both essential for Mu replication. The *A* product is thought to recognize the ends of Mu and is considered to be the Mu transposase. The *B* product modifies *A* activity in some unknown way. Next follows a cluster of genes which are non-essential for phage production but all of which modify transposition, lysogeny or DNA synthesis. The product of *kil* is sufficient to kill the host even in the absence of replication, *gam* protects linear DNA from degradation by host exonucleases, *sot* supports transfection when present and expressed in recipient cells, *arm* causes amplification in Mu replication, *lig* and *topo* are apparently expressed constitutively and lead to partial complementation of bacterial and T4 mutants defective in ligase and topoisomerase.

The middle part of Mu is occupied by genes concerned with morphogenesis, the tail functions intruding into G. G is flanked by a 34 base-pair inverted repeat and is capable of inversion, this being catalysed by *gin* immediately to the right. The last gene is *mom*, a gene involved in modifying adenine residues contained in the sequence 5'C/G-A-C/G3'. This gene is only expressed after Mu induction and needs methylation of its promotor sequence to function.

Comparison between the genetic maps of Mu and λ

Mu and λ are both temperate phages, λ being about 25% larger (49 kb compared to 37 kb). Their properties for the most part are dissimilar. For example their mechanisms of genome integration, replication and packaging are quite distinct. However, as first pointed out by van de Putte *et al.* (1978), the genetic organization of the two phages show some startling similarities in the location, control and function of analogous genes. These similarities are highlighted in the genetic maps drawn in Fig. 2. It can be seen that the Mu map can be made to correspond with that part of the λ prophage map stretching from the *cI* repressor gene to the *b2* region. The first gene in both phages codes for repressor and is followed by a gene (*cro* or *ner*) which negatively controls early transcription. Then come, in each case, two essential genes that determine DNA replication, *O* and *P* for λ, *A* and *B* for Mu. The

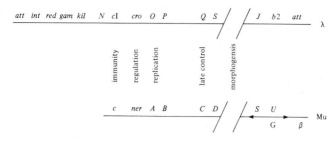

Fig. 2. Comparison between λ and Mu prophage maps. The maps are not drawn to scale, but genes with similar functions are placed in corresponding positions.

next 5 kb in Mu contain the non-essential early cluster which does not have an analogue at the corresponding position in λ (although interestingly λ does possess *kil* and *gam* genes but they are located to the left of *cI*). About 10 kb to the right of the repressor genes comes *Q* in λ and *C* in Mu, both involved in turning on late genes. Then, in both phages, there is a stretch of approximately 20 kb coding for morphogenetic components. At the far right finally comes the *b2* region in λ whose role is still a mystery, and the G and β regions in Mu.

It is hard to believe that these similarities are entirely fortuitous, although no plausible hypothesis has been proposed to explain them. On the practical side the search for analogies between λ and Mu genes having comparable map positions has given rise to some useful clues as to gene function.

GENERAL RECOMBINATION AND PHAGE MU

Mu does not code for any known system of general recombination but relies on host enzymes which it utilizes for (*a*) vegetative recombination between Mu genomes during the lytic cycle, (*b*) resolution of cointegrate structures formed during Mu transposition, and (*c*) excision of Mu prophage from lysogens.

Of these, the first two are straightforward examples of homologous recombination, but the involvement of general recombination in Mu excision poses a difficult problem. When Mu integrates into a gene it generates a small direct repeat of host DNA at the target site, as do other transposons. With Mu this repeat is of 5 base-pairs which then flank the prophage. Exact excision, meaning the restoration of the genes to its original configuration, therefore entails both removal of Mu and loss of one of the direct repeats. In order to observe such exact excisions at all with Mu, a roundabout path has to be followed. From the original insertion, a Mu mutant has to be isolated which is blocked both in DNA replication and in expression of the *kil* gene. Usually this type of 'X' mutant arises by insertion of an Insertion Sequence (IS-) element in the *B* gene which directly inactivates *B* and so stops Mu replication, and also acts in a polar manner to block transcription of *kil*. From cultures of these X lysogens, cells can then be recovered at a low frequency (10^{-6}) in which gene activity has been restored and which can be shown to have lost the Mu prophage. Analysis of these excisions has shown a requirement both for the *A* gene product and also for the general recombination functions of the host (Bukhari, 1975). How these two types of recombination work together is not known. Naively, one would think the *A* gene product is involved in excising the Mu prophage and the host genes in recombining the direct repeats and so deleting one of them. But five base-pairs is too small a segment for any known homologous recombination system to work on, so more investigations are required to elucidate the situation. Another unanswered question is whether with wild-type Mu exact excision is in fact a common event but is always associated with phage induction and cell death; and that in the formation of the X mutants (which are not inducible) this excision capacity has been lost.

SITE-SPECIFIC RECOMBINATION AND PHAGE MU: THE G SEGMENT AND FLIP-FLOP CONTROL OF HOST-RANGE

Investigations with the G segment of Mu over the last decade have unravelled an intriguing story with some unexpected twists. The existence of the invertible G segment was first deduced from electron-microscope photographs (Hsu & Davidson, 1972). When DNA was extracted from Mu phage generated by induction of lysogenic cultures, and the strands then subjected to separation and reannealing, 50% of the observed molecules contained a bubble approximately 3 kb long. It was inferred that Mu possessed a DNA sequence, called the G segment, which could exist in two orientations termed $G(+)$ and $G(-)$; and that equal numbers of phage particles in the induced lysates had G in either orientation. This idea was reinforced by the demonstration that in lysogenic cultures equal numbers of Mu prophage also had the G segment in either orientation. The mechanism underlying the existence of the two orientations was also pin-pointed by these electron-microscope studies when it was shown that there were short inverted repeats at the ends of G, so that recombination between these repeats could lead to inversion of the whole segment (Hsu & Davidson, 1974).

A puzzling result that turned up in these early investigations, which all used *E. coli* K12 as host bacteria, was that when the electron-microscope experiment was repeated with DNA from phage particles derived not from induced cells but from lytic cycles of phage growth, then no G bubbles were observed (Daniell, Boram & Abelson, 1973). Phages formed in this way therefore all had their G segments frozen in just one orientation, say G(+). The explanation for this result did not come until nearly five years later from two different types of investigation. In one, which used a modification of the classic single-burst technique (Symonds & Coelho, 1978), it was shown that approximately 50% of *E. coli* K12 (Mu*cts*) cells yielded no viable phage when induced. However if these non-producing cells were allowed to grow for several generations then induction of the resulting clones did lead to formation of viable phage. The inference from this experiment was that lysogens with Mu prophage in the G(+) orientation produce viable phage upon induction but those in G(−) do not; and that inversion of G proceeds at a measurable rate during bacterial growth. This suggested correlation between viability and G was only circumstantial, and definite proof of this point came from other studies which identified a gene called *gin* (short for G inversion), adjacent to G, which was essential for G inversion to occur and presumably coded for a site-specific recombination enzyme recognizing the inverted repeats. Mutants which locked the G region into either the G(+) or G(−) configuration could then be constructed and these mutants used to directly prove the result that only G(+) phage can grow in *E. coli* K12 cells (Kamp *et al.*, 1978).

The picture that emerged from these experiments was that as a lysogenic Mu culture grows from a single cell, inversion of the G segment occurs at a slow but steady rate (about 0.03 per cell per generation) due to a low level of constitutive expression of the *gin* gene. By the time 10^8 cells have been produced the G(+):(G−) ratio amongst the prophage has increased almost to unity so that, upon induction, roughly equal numbers of phage are formed having G in either orientation. When these phage are used to infect cells of *E. coli* K12, only the G(+) phage can grow and, as the inversion

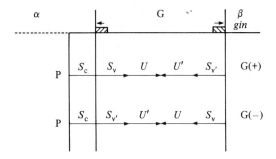

Fig. 3. Scheme for the organization of the G region in phage Mu. P represents the promotor for the *S* and *U* genes. The inverted repeat is represented by cross-hatching.

rate is so low, virtually all progeny phage from this lytic cycle retain the G(+) configuration. This interpretation of the electron microscope results was satisfying and was bolstered by other genetic data which showed that two tail-fibre genes, *S* and *U*, mapped at least partially in G (Howe, Schumm & Taylor, 1979) and so could only be expected to function and produce tail fibres with G in one particular orientation.

The next finding which gave importance to a phenomenon which had appeared, up to that time, merely to reflect a Mu idiosyncrasy was that G(−) phages, rather than being dead because they produced no tail-fibres, in fact produced a different set of tail fibres. These enabled them to absorb and grow in a range of hosts on which G(+) phages were impotent. These hosts included strains of *Citrobacter*, *Shigella* and *Erwinia* (van de Putte, Cramer & Giphart-Gassler, 1980). Clearly the inversion of G did not just turn off the *S* and *U* genes, it replaced them by another set (*S'* and *U'*). Details of the mechanism by which this occurs were then worked out and are shown in Fig. 3. In a manner somewhat reminiscent of that used to create diversity among immunoglobulins, the variations between the *S* and *S'* genes is brought about by joining a constant DNA sequence S_c adjacent to G with either of two different sequences, S_v or $S_{v'}$, located within G. In G(+), transcription starts at a promoter site to the left of G, proceeds through S_c and S_v and then through the adjacent *U* gene which lies wholly within G. Tail fibres corresponding to *S* and *U* are therefore produced. In G(−) the situation is reversed, the reading being $S_c + S_{v'}$, *U'* so that *S'* and *U'* tail fibres are made, but not *S* and *U* (Giphart-Gassler, Plasterk & van de Putte, 1982).

About the same time as this 'flip-flop' mode of controlling gene expression was being worked out in Mu, two other examples were reported whereby inactive genes are relocated by some type of recombination process to new sites where transcription can take place. These were phase-variation in *Salmonella* (Zieg & Simon, 1980) and mating-type in yeast (Hicks, Strathern & Klar, 1977). This recombinational control, which underlies the three examples mentioned above and is exemplifid by the variation in host ranges of Mu, is fundamentally different from control of protein synthesis by positive and negative feedback circuits. Such circuits are able to modulate the activity of genes and, in the extreme case, turn them off completely. Recombinational control, on the other hand, introduces a two-way switch into these circuits, allowing alternative pathways to be introduced into the system. This added flexibility is useful when only one of two possible phenotypes can be tolerated at a particular time. It is well known from experiments involving mixed infection of cells with wild-type phage and host-range phage mutants that progeny particles which receive a mixed complement of tail fibres acquired from different parents have the worst of all worlds, not being able to adsorb properly to any of the usual hosts. If the host-range of a phage, such as Mu, is to be extended by providing two distinct sets of tail fibres, then it is clearly advantageous that only one set or the other should be synthesized at any stage – a two-way switch is therefore appropriate.

The precise recombinational mechanism by which the switch between mating types is instituted in yeast, where some type of gene conversion seems to occur, is different from the inversion systems used to bring about phase variation in *Salmonella* or the alternation in host range shown by Mu (and also by phage P1 where a closely related system operates). All these inversion switches use site-specific recombination systems which act on short inverted repeats at either end of the relevant inverted region, and which are catalysed by genes immediately adjacent to it. Surprisingly, although the invertible regions in *Salmonella* and Mu are quite distinct, there is some sequence homology between their inverted repeats. The *gin* gene which mediates G inversion and the *hin* gene mediating inversion in *Salmonella* can complement each other (Szekely & Simon 1981). The known recombination switches of the inversion type therefore all seem to stem from some common origin. Another evolutionary oddity of the G system is that the emergence of two kinds of Mu phage with different host ranges would be expected to lead to an increase in the number of bacterial strains

which harbour Mu prophage. However only the original Mu isolate
detected in *E. coli* K12 has ever been reported. Whatever the
reason for this, the whole of the G region has been preserved in the
Mu phage we have, even though within K12 strains part of the
region (and the *gin* gene) could easily be deleted and the phage
locked in the (+) orientation. Perhaps there is still another twist to
come in the G story, by which G and *gin* are implicated in some
other aspect of the Mu life-cycle.

TRANSPOSITION AND MU

Integration as a transposition event

Upon infection, the DNA within a phage particle enters the cell and
the Mu gene becomes integrated into the host chromosome (or
other intracellular replicon). The flanking sequences of bacterial
DNA present in the phage DNA are lost in this process. The
integration is therefore akin to a transposition event where the
donor molecule (the DNA within the phage particle) is linear and
the target molecule (the bacterial chromosome) is circular. The
result of the integration is either the formation of a Mu lysogen or,
more often, the start of a lytic cycle which from that stage is
indistinguishable from a cycle started by induction.

The ability of Mu to integrate, and the relative frequency of the
lysogenic and lytic responses, is affected by the phage *A* and *B* genes
which control ordinary transposition. In the absence of either *A* or
B gene products no viable phage are formed so no successful lytic
responses are initiated. Lysogeny is also completely blocked with *A*
mutants, but *B* mutants can still lysogenize but with approximately
10-fold lower frequency (O'Day, Schultz & Howe, 1978). Some
preliminary experiments with host strains that contain cloned fragments of Mu indicate that the presence of excess amounts of *A* gene
product at the time of infection favours the lysogenic against the
lytic response.

A quite different factor has been suggested which could affect
integration: the replicative structure of the host chromosomes.
More specifically, the proposal is that Mu integrates at host replication forks. Two experiments have been reported bearing on this
idea, both utilizing synchronized cell cultures. In the first, which
relates to lysogeny, the frequency of bacterial mutants induced by

Mu infection was compared in exponentially and in synchronized growing cells. For genes near the bacterial origin, which were just being replicated in the synchronous culture at the time of infection, there were approximately six times as many induced mutants in the synchronized culture as in the control. This factor was less for other genes, becoming close to unity for genes near the terminus of host replication (Paolozzi, Jucker & Calef, 1978). A more searching experimental test, this time relating to integrations starting a lytic cycle, was performed using a system of five Hfr strains having different transfer origins. Initiation of DNA synthesis was synchronized in these strains, all of which had the same origin of replication, *oriC*. After 0, 8 and 15 min the cells were infected with Mu and subsequently mated with non-immune F⁻ recipient cells. Mating was interrupted at regular intervals and samples were assayed for infective centres. Analysis of the results showed that conjugal transfer of integrated Mu from infected Hfr cells was delayed when the chromosomal replication forks had not reached the Hfr transfer origin by the time mating mixtures were formed; this delay was increased in strains having transfer origins far from *oriC* (Fitts & Taylor, 1980).

Genetic experiments of this kind cannot *prove* that integrations occur at replication forks. The idea must however be taken seriously in connection with integration, and must also be borne in mind when considering the factors leading to the initiation of the more general type of transposition that occurs between circular replicons.

IS MU INTEGRATION CONSERVATIVE?

As the transposition reaction involved in integration involves a linear and a circular molecule rather than the two circular molecules involved in normal transposition, it is not obvious that the usual rules must apply. In particular the question can be asked as to whether integration is necessarily a replicative process, as only a single Mu genome needs to be integrated to form a lysogen or initiate an infectious cycle. This is unlike the situation occurring during the lytic cycle where transposition is necessarily replicative as it has to account for the synthesis of new Mu DNA. Two approaches have been made in an attempt to anwer this question. The more definitive (and difficult) is actually to follow the fate of the DNA in parental phage and determine whether or not the first integration

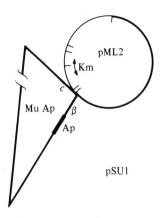

Fig. 4. Position of *Sma*I restriction sites in pSU1. All six sites are in pML2. The thick line represents the molecule active in transfection after complete restriction with *Sma*I. It consists of 37 kb of Mu DNA and about 9 kb of pML2.

event is conservative or semi-conservative. A less decisive but simpler experiment, which we shall discuss first, takes up the genetic point that if integration of an infecting Mu genome occurs in one semi-conservative event then, upon single infection, only genetic information on one strand should be transmitted to progeny phage. If infection is carried out using phage heterozygotes only pure bursts should therefore be observed.

A genetic experiment of this kind has recently been performed, but using transfection rather than phage adsorption as a starting point (Akroyd & Symonds, 1983). This enables the heterozygous molecules to be prepared by denaturation and reannealing, a device used previously in analogous studies with *Bacillus subtilis* phage and with phage λ. An initial difficulty which had to be overcome was that Mu DNA derived from phage particles and prepared in this way is not active in transfection (presumably because of the presence at either end of single-stranded 'split-ends' of bacterial DNA susceptible to degradation by host nucleases). In order to circumvent this difficulty it was necessary to find a source of Mu DNA in which the sequences at the ends of all genomes were identical. This was obtained by isolating linear DNA from the plasmid pSU1 (Fig. 4). This contains an insertion of the Mu derivative MuAp1 into the ColE1-type plasmid pML2. It can be seen from the figure that all the *Sma*I sites in pSU1 are in the pML2 moiety, complete restriction with this endonuclease yielding a linear molecule containing the Mu genome flanked now by two unique sequences of pML2 DNA amounting to about 0.05 kb at the c end of Mu and 9.0 kb at the β end.

Using this method of preparing linear Mu DNA it was then possible to perform the heterozygote experiment using DNA derived from pSU1 and from a derivative pSU1B^- in which an amber mutation had been introduced into the Mu B gene of pSU1 by genetic recombination. Linear DNA obtained from pSU1 and pSU1B^- was mixed in equimolar amounts, heat denatured and the strands allowed to reanneal at random. This reannealed DNA, which in principle consists of 50% B^+/B^- heterozygotes and 25% of both B^+/B^+ and B^-/B^- homozygotes, was then used to transfect competent cells which were plated on a Su$^+$ indicator before lysis. Individual plaques were then picked into broth and plated on a mixed indicator of Su$^+$ or Su$^-$ bacteria on which it was easy to distinguish wild-type and amB plaques. As a control for possible mixed indicator of Su$^+$ and Su$^-$ bacteria on which it was easy to DNA derived from pSU1 and pSU1B^- in which the strands had been separated and then reannealed *before* being mixed in equimolar amounts. The results of the experiment were quite unambiguous. Approximately 50% of competent cells infected with heteroduplex DNA regularly gave mixed bursts while the homoduplex controls averaged less than 10%.

Although this experiment was performed using transfection, the result can reasonably be expected to hold after phage adsorption. The conclusion from the experiment is therefore that the information of both strands of the infecting Mu DNA is transmitted with equal efficiency to progeny phage. This result can be reconciled with a semi-conservative mechanism of transposition by postulating that at least two copies of any infecting Mu DNA are integrated after infection. The simpler assumption however is to say that only a single integration event occurs and both parental strands integrate without any concomitant replication.

An isotope experiment to test for conservative or semi-conservative integration has been performed by Liebart, Ghelardini & Paolozzi (1982). The ingenious idea behind this work was to follow the integration of density-labelled Mu DNA not into the host chromosome, but into the multicopy plasmid pBR322 whose length (4 kb) is negligible compared to that of Mu. This enabled any Mu insertions into pBR322 to be isolated because they would be supercoiled and could be separated (along with pBR322 molecules) from other intracellular DNA by centrifugation in a caesium chloride ethidium bromide gradient. Then, as the density of a Mu insertion in pBR322 is essentially that of Mu itself, the question of whether the integrated Mu genome had undergone any replication could be decided by respinning the supercoiled DNA in a CsCl gradient. Experimentally host cells containing pBR322 were grown for 20 generations in a heavy (^{13}C, ^{15}N) hot (^{3}H) medium, and then suspended in buffer. These were then multiply infected with light ^{32}P-labelled phage. After 20 min, DNA was extracted from the infected cells and supercoiled molecules isolated. Part of the ^{32}P label from the parental Mu was found in this supercoiled fraction which was then respun in a CsCl gradient. All the ^{32}P counts were found to occur at a density corresponding to double-stranded light (that is unreplicated) DNA. Biological confirmation that these ^{32}P counts referred to Mu insertions in pBR322 was obtained by showing that DNA derived from the light bands in the density gradient had transforming ability both for plasmid and viral markers.

This isotope experiment provides clear evidence that infecting Mu DNA *can* integrate conservatively. There is the reservation, however, that the result only refers to conditions where phage adsorption was performed in buffer, conditions which would militate against reactions involving DNA synthesis. It is not certain that the result can be generalized to normal growth conditions, just as it is not certain that the results of the heterozygote experiment can be extended from transfection to infection. However although the experiments do not establish beyond doubt that Mu integration always occurs conservatively, taken together they do suggest very strongly the important conclusion that *a possible pathway of transposition open to phage Mu is a conservative one.*

Transposition between circular replicons

Phage Mu mediates the formation of an extensive array of gene rearrangements which includes replicon fusions (Toussaint & Faelen, 1973), deletions (Howe & Zipser, 1974; Faelen & Toussaint, 1978), and inversions (Faelen & Toussaint, 1980). Recently a number of models have been proposed which can explain the formation of all these rearrangements in terms of a single transposition scheme (for a summary of these see Bukhari, 1981). The first of these models is due to Shapiro (1979): when the donor and target molecules are on different circular genomes, this leads automatically

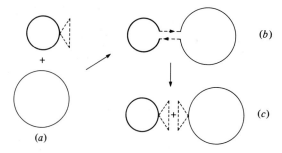

Fig. 5. Transposition scheme according to the Shapiro model. (*a*) Transposition between a donor replicon (thick circle) in which a transposon (dotted lines) is inserted and a recipient replicon (light circle) gives a cointegrate (*b*) which is resolved by recombination (*c*) to yield donor and recipient replicons both containing insertions of the transposon.

to the formation of a cointegrate structure where the two replicons are fused together by two copies of the transposon in direct repeat (Fig. 5). Reciprocal recombination between the two copies of the transposon then leads to a simple insertion in the target replicon which is considered to be the normal end product of transposition. In the Tn3 family of transposons such a reciprocal recombination system is encoded in the element, and there is convincing evidence that this two-step pathway for transposition is followed. As no analogous recombination system has been found in Mu it becomes important to determine whether simple insertion as well as cointegrates are obtained from Mu-mediated transposition events occuring between circular replicons.

One of the simplest systems with which to study this question (Fig. 6) uses cells having a donor replicon like pSU1, where the small plasmid pML2 (coding for Kanamycin-resistance, Km^R) contains an insertion of the Mu derivative MuAp1 (coding for Ampicillin-resistance, Ap^R). The target replicon is the sex-factor R388 (coding for Trimethoprin-resistance, Tp^R) which does not mobilize pML2 during conjugation. Donor cells containing these two plasmids are grown at low temperature, then Mu enzymes are induced by a temperature shift. After a defined interval these induced donor cells are mated with suitably marked recipient cells. Selection initially is for Streptomycin-resistance Str^R (to kill donor cells), Tp^R (to select transfer of R388) and Ap^R (to identify those recipient cells receiving a copy of Mu which, as pSU1 cannot transfer on its own, can only have been acquired by some kind of association of Mu with R388). These $Tp^R Ap^R$ colonies are then tested for their resistance to Km. The Km^R colonies are presumed to be cointegrates as they

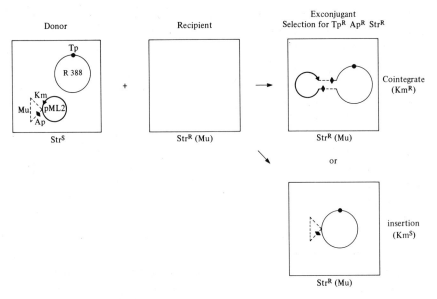

Fig. 6. Typical conjugation experiment used to detect cointegrates and insertions. Donor cells contain an insertion of MuAp1 (dotted lines) in pML2 (thick circle) as well as the sex-factor R388 (thin circle). Recipient cells are Str^R (for counter-selection) and lysogenic for Mu (to prevent zygotic induction). Location of genes conferring resistance to antibiotics: ♦ Ap^R, ▼ Km^R, ● Tp^R.

also have inherited pML2, while the Kanamycin-sensitive Km^S colonies are possibly due to insertions of Mu into R388. Further biochemical tests (molecular weight determinations, restriction analysis) and genetic tests (transformation experiments and reversion studies) are then carried out on DNA isolated from these two classes of recipient cells in order to check the identification with cointegrates or insertions (Maynard-Smith *et al.*, 1980). An entirely analogous genetic system uses the insertion of a MuKm^R phage in the plasmid pSC101 (coding for Tetracycline-resistance Tet^R) as a donor replicon and the sex factor F*lac*$^+$ as the target replicon (Chaconas, Harshey & Bukhari, 1980). In all experiments both donor and recipient cells are always *rec*A$^-$ to inhibit any resolution of cointegrates by host recombination systems.

These experimental systems can easily be adapted to study the effect of factors which might modify the transposition reaction. In particular, variants of the donor molecules have been constructed which lack functional *A* or *B* genes or which have abnormal Mu end sequences. The most common of these variants consist of 'mini-Mu' derivatives which retain the Mu ends but have deleted the *A* and *B* genes and most of the internal region of Mu. The level of the A and B gene products in the donor cell can then be varied by introducing Mu derivatives as prophage on the host chromosome or, in experiments which are just getting started, by introducing compatible plasmids into the donor cells on which the A and B genes have been cloned alongside controllable promotors.

The results of numerous experiments performed in various laboratories using these and other methodologies are completely in agreement that for any detectable Mu transposition to occur it is essential that both ends of Mu and the product of the Mu A gene must be present. Accord is not so complete about other aspects of the experiments, but a general consensus seems to be developing which may be summarized by saying that both cointegrates and insertions can be formed by Mu transposition between circular replicons, the relative frequency of the two depending on the ratio of A and B gene products in the cell. Cointegrates predominate at wild-type levels of A and B, but as the A/B ratio in cells increases the proportion of insertions increases. In the extreme case, when no B product is present, transposition still occurs but the overall rate is reduced to about 1% of normal (Coelho et al., 1980; Howe & Schumm, 1980; Chaconas et al., 1981; Toussaint & Faelen, 1981; Harshey, 1983).

Possible explanations for the finding of both cointegrates and insertions

As there are no known Mu enzymes which break down cointegrates there must be some other mechanism by which simple insertions are formed. Two types of explanation have been suggested to explain the experimental results, and at this stage it is impossible to say which is correct. Probably both come into play under different conditions. One possibility is that there are two distinct pathways by which transposition can occur with Mu: one leading to cointegrates, the other to insertions (e.g., cointegrates could reflect a semi-conservative pathway, insertions a conservative one). The other possibility is that there is just one basic pathway whose initial steps lead to a transposition intermediate than can be resolved in two ways, one yielding cointegrates the other insertions.

An interesting scheme embodying the second alternative has been proposed, prompted by an analysis of electron-microscope pictures of replicating Mu DNA. These pictures showed structures (which

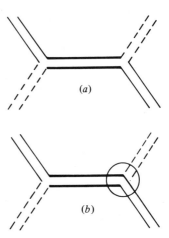

Fig. 7. The initial intermediate structures in transposition according to (a) the original model of Shapiro (two replication forks); and (b) the modification of Harshey & Bukhari (1981: one replication fork). Continuous lines: donor DNA containing the transposon (heavy lines). Broken lines: target molecule. The circle in (b) represents a protein complex and could be located at the replication fork. Normally the donor and target molecules would be on different circular genomes, or at different locations on the same circular genome.

were later shown to be true intermediates in Mu transposition) (Harshey, McKay & Bukhari, 1982), consisting of 'keys' having circles of variable size attached to tails of variable length. These structures are not predicted from the Shapiro model whose intermediate structure with two symmetric replication forks is illustrated in Fig. 7(a). In order to accommodate the key structures, Harshey & Bukhari (1981) suggested the alternative scheme with an intermediate essentially as depicted in Fig. 7(b). The main innovation is that it contains a single replication fork starting from one end of the transposon. The second end of the target molecule is left unligated until the last stage in the transposition process and, in the interim, is held by a protein complex which could be the replication complex itself. The extra flexibility of this model over the Shapiro one comes about because, when Mu replication has reached the other end of the genome, the intermediate can be resolved in two ways by introducing a further nick either (a) in the 'top' Mu strand, which allows ligations to occur that lead to the Shapiro junctions and so to cointegrates, or (b) in the 'bottom' Mu strand which lead to insertions.

CONCLUSIONS

Our understanding of Mu transposition is at an intriguing stage. No doubt, results from DNA sequence studies, enzyme isolations, and *in vitro* systems will be necessary before the precise roles of the Mu ends and the products of the *A* and *B* genes are known, and before one can tell whether the relationships which have been suggested between Mu transposition and replication forks and between the λ and Mu genetic maps are important clues or are just chance correlations. What does seem certain, even at this point, is that Mu can transpose in more than one way. This realization reflects a situation which holds in the context of recombination in general. From one point of view, all recombination systems have elements in common. Initially two sequences in the parental molecules come together, then an intermediate structure is formed, and finally resolution of this intermediate occurs with formation of two recombinants. However, although this unity is real enough, it can be misleading. The crucial point, which I hope has emerged to some extent in this paper, is that there is almost unlimited variation in the ways in which recombination can occur. It is diversity, not uniformity, which is important in trying to understand the role of recombination in the evolutionary scheme of prokaryotes and eukaryotes.

REFERENCES

AKROYD, J. & SYMONDS, N. (1983). Evidence for a conservative pathway of transposition of bacteriophage Mu. *Nature*, **303**, 84–6.

AUSTIN, S., ZIESE, M. & STERNBERG, N. (1981). A novel role for site-specific recombination in maintenance of bacterial replicons. *Cell*, **25**, 729–36.

BOTSTEIN, D. & MATZ, M. (1970). A recombination function essential to the growth of bacteriophage P22. *Journal of Molecular Biology*, **54**, 417–30.

BUKHARI, A. (1975). Reversal of mutator phage Mu integration. *Journal of Molecular Biology*, **96**, 87–99.

BUKHARI, A. (1981). Models of DNA transposition. *Trends in Biological Sciences*, **6**, 56–60.

CHACONAS, G., DE BRUIJN, F. J., CASADABAN, M. J., LUPSKI, J. R., KWOH, T. J., HARSHEY, R. M., DU BOW, M. S. & BUKHARI, A. I. (1981). *In vitro* and *in vivo* manipulations of bacteriophage Mu DNA: cloning of Mu ends and construction of mini-Mu's carrying selectable markers. *Gene*, **13**, 37–46.

CHACONAS, G., HARSHEY, R. M. & BUKHARI, A. I. (1980). Association of Mu-containing plasmids with the *E. coli* chromosome upon prophage induction. *Proceedings of the National Academy of Sciences (USA)*, **77**, 1778–82.

COELHO, A., LEACH, D., MAYNARD-SMITH, S. & SYMONDS, N. (1980). Transposition studies using a *ColE1* derivative carrying phage Mu. *Cold Spring Harbor Symposia on Quantitative Biology*, **45**, 323–8.

DANIELL, E., BORAM, W. & ABELSON, J. (1973). Genetic mapping of the inversion loop in bacteriophage Mu DNA. *Proceedings of the National Academy of Sciences (USA)*, **70**, 2153–5.

ENQUIST, L. & SKALKA, A. (1978). Replication of bacteriophage lambda DNA. *Trends in Biological Sciences*, **3**, 279–83.

FAELEN, M. & TOUSSAINT, A. (1978). Mu mediated deletions in the chromosome of *E. coli*. *Journal of Bacteriology*, **136**, 477–83.

FAELEN, M. & TOUSSAINT, A. (1980). Inversions induced by bacteriophage Mu-1 in the chromosome of *E. coli* K12. *Journal of Bacteriology*, **142**, 391–9.

FITTS, R. & TAYLOR, A. L. (1980). Integration of bacteriophage Mu at host chromosomal replication forks during lytic development. *Rroceedings of the National Academy of Sciences (USA)*, **77**, 2801–5.

GIPHART-GASSLER, M., PLASTERK, R. & VAN DE PUTTE, P. (1982). G inversion in bacteriophage Mu: a novel way of gene splicing. *Nature*, **297**, 339–42.

GOTTESMAN, M. & WEISBERG, R. (1971). Prophage insertion and excision. In *The Bacteriophage Lambda*, ed. A. D. Hershey, pp. 113–38. Cold Spring Harbor: Cold Spring Harbor Laboratories.

HARSHEY, R. (1983). A switch in the transposition products of Mu DNA mediated by proteins: cointegrates versus simple insertions. *Proceedings of the National Academy of Sciences (USA)*, **80**, 2012–16.

HARSHEY, R. & BUKHARI, A. (1981). A mechanism of DNA transposition. *Proceedings of the National Academy of Sciences (USA)*, **78**, 1090–4.

HARSHEY, R., McKAY, R. & BUKHARI, A. (1982). DNA intermediates in transposition of phage Mu. *Cell*, **29**, 561–71.

HICKS, J., STRATHERN, J. & KLAR, A. (1977). The cassette model of mating type interconversion. In *DNA insertion elements, plasmids and episomes*, ed. A. Bukhari, J. Shapiro & S. Adhya, pp. 457–62. Cold Spring Harbor: Cold Spring Harbor Laboratories.

HOWE, M. & SCHUMM, J. W. (1980). Transposition of bacteriophage Mu. Properties of λ phages containing both ends of Mu. *Cold Spring Harbor Symposia on Quantitative Biology*, **45**, 337–46.

HOWE, M. M., SCHUMM, J. W. & TAYLOR, A. L. (1979). The S and U genes of bacteriophage Mu are located in the invertible G segment of Mu DNA. *Virology*, **92**, 108–24.

HOWE, M. M. & ZIPSER, D. D. (1974). Host deletions caused by the integration of bacteriophage Mu-1. *American Society for Microbiology Abstracts*, No. V, 208, 235.

HSU, M. & DAVIDSON, N. (1972). Structure of inserted bacteriophage Mu-1 DNA and physical mapping of bacterial genes by Mu-1 DNA insertion. *Proceedings of the National Academy of Sciences (USA)*, **69**, 2923–7.

HSU, M. & DAVIDSON, N. (1974). Electron microscope heteroduplex study of the heterogeneity of Mu phage and prophage DNA. *Virology*, **58**, 229–39.

LIEBART, J., GHELARDINI, P. & PAOLOZZI, L. (1982). Conservative integration of bacteriophage Mu DNA into pBR322 plasmid. *Proceedings of the National Academy of Sciences (USA)*, **79**, 4362–6.

LUDER, A. & MOSIG, G. (1982). Two alternative mechanisms for initiation of DNA replication forks in bacteriophage T4: Priming by RNA polymerase and by recombination. *Proceedings of the National Academy of Sciences (USA)*, **79**, 1101–5.

KAMP, D., KAHMANN, D., ZIPSER, T., BROKER, L. & CHOW, L. (1978). Inversion of the G segment of phage Mu controls phage infectivity. *Nature*, **271**, 577–80.

MAYNARD-SMITH, S., LEACH, D., COELHO, A., CAREY, J. & SYMONDS, N. (1980). The isolation and characteristics of plasmids derived from the insertion of MupAp1 into pML2: their behaviour during transposition. *Plasmid*, **4**, 34–50.

NASH, H. (1975). Integrative recombination of bacteriophage lambda DNA *in vitro*. *Proceedings of the National Academy of Sciences (USA)*, **72**, 1072–6.

NASH, H. (1981). Integration and excision of bacteriophage λ: the mechanism of conservative site specific recombination. *Annual Review of Genetics*, **15**, 143–67.

O'DAY, K., SCHULTZ, D. & HOWE, M. M. (1978). A search for integration deficient mutants of bacteriophage Mu-1. In *Microbiology 1978*, ed. D. Schlessinger, pp. 48–51. Washington D.C.: American Society for Microbiology publications.

PAOLOZZI, L., JUCKER, R. & CALEF, E. (1978). Mechanism of phage Mu-1 integration: nalidixic acid treatment causes clustering of Mu-1 induced mutations near replication origin. *Proceedings of the National Academy of Sciences (USA)*, **75**, 4940–3.

SHAPIRO, J. (1979). Molecular model of the transposition and replication of bacteriophage Mu and other transposable elements. *Proceedings of the National Academy of Sciences (USA)*, **76**, 1933–7.

STERNBERG, N., HAMILTON, D., AUSTIN, S., YARMOLINSKY, M. & HOESS, R. (1980). Site-specific recombination and its role in the life cycle of bacteriophage P1. *Cold Spring Harbor Symposia on Quantitative Biology*, **45**, 297–309.

SYMONDS, N. & COELHO, A. (1978). Role of the G segment in the growth of phage Mu. *Nature*, **271**, 573–4.

SZEKELY, J. & SIMON, M. (1981). Homology between invertible deoxyribose nucleic acid sequence that controls flagella-phase variation in salmonella and deoxyribose nucleic acid sequences in other organisms. *Journal of Bacteriology*, **148**, 829–36.

TOUSSAINT, A. & FAELEN, M. (1973). Connecting two unrelated DNA sequences with a Mu dimer. *Nature New Biology*, **242**, 1–4.

TOUSSAINT, A. & FAELEN, M. (1981). Formation and resolution of cointegrates upon transposition of mini-Mu. In *Microbiology 1981*, ed. D. Schlessinger, pp. 69–72. Washington D.C.: American Society for Microbiology publications.

TOUSSAINT, A. & RESIBOIS, A. (1983). Phage Mu or 'How to use transposition as a life style'. In *Mobile Genetic Elements*, ed. J. Shapiro. New York, London: Academic Press.

VAN DE PUTTE, P., CRAMER, S. & GIPHART-GASSLER, M. (1980). Invertible DNA determines host specificity of bacteriophage Mu. *Nature*, **286**, 218–22.

VAN DE PUTTE, P., GIPHART-GASSLER, M., GOOSEN, T., VAN MEETERAN, A. & WIJFFELMAN, C. (1978). Is integration essential for Mu development? In *Integration and excision of DNA Molecules*, ed. P. Hofshneider & P. Starlinger, pp. 33–40. Berlin: Springer-Verlag.

ZIEG, J. & SIMON, M. (1980). Analysis of the nucleotide sequence of an invertible controlling element. *Proceedings of the National Academy of Sciences (USA)*, **77**, 4196–200.

THE MOLECULAR EVOLUTION OF VIRUSES

DARRYL REANNEY

Department of Microbiology, La Trobe University, Bundoora, Victoria 3083, Australia

INTRODUCTION

Because DNA and RNA encode a common genetic language, there is a tendency to think of the two types of polynucleotide as variants of the same thing. In fact, the chemical and functional distinctions between DNA and RNA are profound. Whereas DNA is almost always a monotonous double helix, RNA can assume a degree of topological complexity which rivals that of proteins (Kim *et al.*, 1974). Perhaps more significantly, RNA can catalyse reactions in the absence of proteins (Abelson, 1982; Kruger *et al.*, 1982; Zaug, Grabowski & Cech, 1983). This suggests that RNA is a dynamic molecule whose versatility we are only just beginning to appreciate (Lewin, 1982).

Chemical differences between DNA and RNA extend to their modes of replication. DNA molecules replicate by an energy-intensive, semi-conservative process; RNA molecules replicate by asymmetric transcription from one strand (even if that strand is part of a double helix). This distinction automatically denies RNA the possibility of using corrective processes to remove genetic lesions because known error-minimizing mechanisms normally use the information specified by one intact strand in a duplex molecule to repair damage to the other (see Loeb & Kunkel, 1982). *Because RNA lacks the proofreading and editing functions of DNA, the error level in RNA copying is between 100 000 and 10 000 000 times higher than that in DNA copying* (Kornberg, 1980; Holland *et al.*, 1982; Reanney, 1982) (see 'note' on p. 192). I shall argue that this high level of genetic 'noise' has moulded the evolution of RNA viruses in ways which rationalize many of their characteristic features. I shall also argue that overlapping genes, a pattern of genetic organization common to many DNA and RNA viruses, not only compacts the genetic information but also couples the expression of linked genes and so regulates the synthesis of viral proteins.

Error rates in RNA genome replication

The spontaneous rate of RNA genome mutation can be estimated in a number of ways. One relatively reliable way to estimate mutation rates is to use monoclonal antibodies to select for resistant phenotypes. The advantage of this technique is that monoclonal antibodies monitor nucleotide changes at defined locations in small sections of the genome. Prabhakar et al. (1982) prepared a panel of 18 monoclonal antibodies that neutralized the B4 strain of human Coxsackie virus. By dividing the titre of virus escaping neutralization by the titre before neutralization, mutation frequencies ranging from $10^{-4.0}$ to $10^{-5.8}$ were obtained (Prabhakar et al., 1982). Monoclonal antibodies have also been used to determine a major antigenic site for poliovirus neutralization (Minor et al., 1983). These studies confirm that the average frequency with which resistant mutants appear under monoclonal selection is about 10^{-4} (although a somewhat higher figure was obtained for the parental Leon strain than for its Sabin type 3 derivative). In the case of the attenuated Sabin 1 poliovirus, the complete nucleotide sequence is available (Nomoto et al., 1982). Comparison with the sequence of the parental (Mahoney) strain shows that 57 nucleotide substitutions had occurred in a genome of 7441 bases. This represents a change in 0.77% of the entire genome. While the frequency with which mutations were introduced into the wild-type genome cannot be calculated because the number of generations needed to generate the attenuated strain is not known, the observed difference supports the concept that poliovirus RNA replicases are very error-prone.

Another method for picking up variations in the nucleotide codes of some RNA viruses is to examine the migration patterns of their segmental RNAs on gels. This is especially applicable to the reo- and rotavirus groups whose genomes consist of 10–12 individual RNA segments. Separation techniques can detect genome segments which differ in about 15 base pairs (i.e., 10 000 daltons) (see Sabara et al., 1982). Application of this technology to the rotaviruses has shown that individual virus isolates contain multiple subpopulations (Sabara et al., 1982; see also Hrdy, Rosen & Fields, 1979). A similar study on the avian reoviruses revealed marked polymorphism of migration pattern of any individual double-stranded RNA segment among isolates of the same serotype as well as among different serotypes (Gouvea & Schnitzer, 1982). Similar variation was observed some time ago with the RNA of wound tumour virus

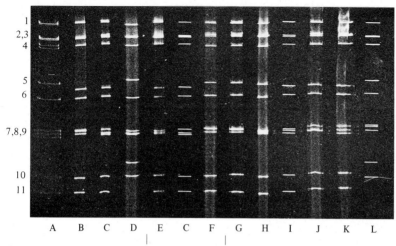

Fig. 1. Variation in rotavirus RNAs: a comparison of electrophoretic patterns of human rotavirus electropherotypes A–L recognized in Central Australia in 1976–9. Migration was from top to bottom and genome segments are numbered in descending order of size. The variant RNA molecules result from mutations or reassortment or both. Photograph kindly provided by Dr R. Schnagl.

(Reddy & Black, 1974). It seems likely that a continuous spectrum of mutationally altered RNA molecules exists in nature and that many of the observed polymorphisms represent only the small fraction of RNAs (or RNA combinations) that survive the editing of natural selection (Fig. 1).

An interesting analysis of variations in the genomes of human and animal rotaviruses has been performed using 'Northern blots'. These hybridization data indicated that different human rotavirus electropherotypes showed 'a remarkable degree of sequence diversity' (Street *et al.*, 1982). Antigenic drift (i.e., mutations in the genes coding for target proteins) was inferred from the fact that the homology demonstrated under conditions of low hybridization stringency was absent under more stringent conditions (Street *et al.*, 1982). A comparison of human and animal rotaviral RNAs showed that where an 8% mismatch was tolerated, few segments exhibited homology (Schroeder *et al.*, 1982). This low order of genetic relatedness was attributed to an interchange of segmental RNAs as well as to cumulative changes in nucleotide sequence (Fig. 1). The demonstration of substantial heterogeneity in the structural glycoprotein (VP7) gene of simian rotavirus SA11 (Estes *et al.*, 1982) reinforces the idea that these genomes are highly mutable.

The influenza viruses exhibit notorious variability but whether this variability is due to mutation (drift), reassortment, the re-emergence of old strains or combinations of these factors remains unclear (Palese & Young, 1982). Recent sequence data provide direct evidence of the mutability of influenza virus genomes. Bishop *et al.* (1982) report that the 1968 influenza virus isolate (A/NT/60/68) differs in RNA segment 3 by 159 point mutations from a 1934 strain (A/PR/8/34). Similar results were obtained by comparing segment 5 of the two viruses (Bishop *et al.*, 1982). While these data give no indication of authentic mutation rates because there is no information as to the number of divisions the isolates have passed through since their divergence from an inferred common ancestor, the 7.3% difference between the sequences of segment 3 points to a high degree of genetic noise in the replicative system. More meaningful information as to mutation rates comes from a comparative study of sequence variation at the 3' end of the neuraminidase gene from 39 type A influenza viruses (Blok & Air, 1982). This study showed very high rates of base substitution at particular positions (in general, the region around 150 bases from the 5' end seemed more mutation-prone than other regions sequenced to date). There is no reason to single this region out as a 'hot spot'. Since this region is in the 'stalk' segment of the neuraminidase molecule it is probably not under antigenic selection and so can retain and express a degree of variability which would, in the absence of selection, be a feature of the whole genome.

Evidence that the influenza viruses are not hypermutable has come from a comparative study by Portner, Webster & Bean (1980) in which monoclonal antibodies were used to assess the frequency of evolution of resistant phenotypes in influenza A viruses as well as in Sendai virus and vesicular stomatitis virus. The calculated 'mutation' rate for all three viruses was $10^{-4.5}$–$10^{-4.7}$, which is very similar to the value already given for the picornaviruses. In view of this consistency, a figure of $10^{-4.5}$ will be accepted as the *measured* index of mutability per nucleotide in the genomes of *all* RNA viruses. The *actual* mutation frequency is certain to be somewhat higher. Nucleotide substitutions in position 3 of codons do not normally result in amino acid substitutions and hence are silent in terms of their effect on phenotype. Changes in the first position of codons often exchange functionally related amino acids and this minimizes the chances of detecting phenotypic variation. Also, high error rates make it difficult to determine whether a given phenotype is a result of one mutational event or several (Holland *et al.*, 1982). Working on the assumption that the genome target for monoclonal antibodies is never more than 10 nucleotides and assuming an equiprobability of base changes in all positions, Holland *et al.* (1982) calculated that a 10 kb RNA, with an error rate of $10^{-4.5}$, would average an overall mutation frequency of $10^{-1.5}$. Put another way, well over 10% of all genomes should carry errors. This figure seems very reasonable on the basis of the available data.

Perhaps the most definitive study of mutation rates in an RNA virus is that of Domingo *et al.* (1978) on the ribophage Qβ. By subjecting the empirical data to a mathematical analysis they arrived at an error rate per nucleotide per doubling of 3×10^{-4} in this RNA virus. There seems no reason not to extrapolate this value to other RNA viruses.

This high level of intrinsic noise means that it is impossible to specify the genome of an RNA virus except in a probablistic sense. Eigen & Schuster (1977) point out that RNA genomes exist in nature not as uniquely defined nucleotide codes but as populations of different sequences clustered round a weighted average. They call this spread-out distribution a *quasi species* (Eigen & Schuster, 1977). The quasi species concept has been experimentally verified for the ribophage Qβ by Domingo *et al.* (1978). By digesting Qβ RNAs with RNase T1 and by cloning viable mutants, they were able to show (*a*) that about 15% of clones from a multiply passaged population had fingerprint patterns different from those of total population RNA and (*b*) that viable clones differed by 1–2 base changes from the average sequence of the 'wild-type'. Competition experiments between the wild-type and variant RNAs showed that the wild-type sequence predominated because it grew faster than its mutant derivatives. Thus, the Qβ population exists as a *dynamic equilibrium* in which the identity of the wild-type is maintained only because of strong selection against a continuously appearing spectrum of mutants.

This thesis predicts that a change in selective pressure should shift the equilibrium in favour of a pre-existing mutant. This has been shown experimentally by subjecting small, self-replicating RNAs *in vitro* to different environmental stresses such as low concentrations of agents, like ethidium bromide and RNase, which damage the RNAs or interfere with their multiplication. The resistant molecules which emerge are always changed in a few defined nucleotide positions (Spiegelman, 1970; Eigen *et al.*, 1981).

Table 1. *Probability of occurrence of errors in RNA and DNA viral genomes*[a]

Number of total errors	Probability of occurrence	
	RNA	DNA
0	0.2592	0.9999955
1	0.3500	4.5×10^{-6}
2	0.2363	1.0×10^{-11}
3	0.1063	1.5×10^{-17}
4	0.0359	
5	0.00969	
6	0.00218	Infinitesimal
7	0.000420	
8	0.0000708	

[a] Derivation: consider a quasi-species of viruses. Two factors affect their distribution: (*a*) $1 - q$, the error rate per nucleotide per generation, which tends with each new generation to spread the distribution of sequences to one of complete randomness; and (*b*) a restorative force based on selection against mutants which are disadvantaged with respect to perfect copies, due to slower rates of replication or other factors. The counter-balancing of the two effects creates an equilibrium situation. It is not an absolute equilibrium however and very rare favourable mutations might completely change the behaviour of the system.

For the basis of comparison, assume the restorative force is neglible. For RNA: $1 - q = 3 \times 10^{-4}$ and DNA: $1 - q = 10^{-9}$ let size of virus (n) = 4500 nucleotides. Then each new replication may be considered a set of Bernoulli trials, so that

$$b(k;n,q) = \binom{n}{k}(1 - q)^k q^{n-k}$$

is the probability of replication producing k errors.

It is hard to overstate the importance of this spread of sequence options for the evolution of RNA viruses. Consider two populations of phage particles of equivalent genome sizes, population A consisting of DNA phage, population B of RNA phage. The error rate of genome replication in the DNA phages is 10^{-9}–10^{-10} (Kornberg, 1980), that in the RNA phages, 3×10^{-4} (see above). An evolutionary consequence of this near millionfold difference in relative noise levels is that *an RNA genome, in the absence of counteracting selection, can accumulate in 33 generations a degree of genetic variation that an equivalent DNA genome under identical conditions would take 10 000 000 generations to achieve* (Table 1). This is a dramatic illustration of the way in which the high noise level in the RNA replicative system speeds up the rate of RNA virus evolution and telescopes the time-scale needed to evolve new species.

Evolutionary consequences of high noise levels in RNA genomes

Small genome sizes

The amount of information a self-copying system can reproducibly maintain is inversely proportional to the error rate of the copier process. This relationship can be formally given as

$$\nu\text{max} = 1n\sigma/1 - q$$

where νmax is the maximum information capacity of the system for a given error rate, $1 - q$ is the average error rate per digit (nucleotide) and σ is the selective superiority of the wild-type to the average of its mutant distribution (Eigen & Schuster, 1977). An accurate value for $1 - q$ can be fed into this equation from the preceding discussion of error rates. If one selects a σ value of 200, then the upper limit for any RNA genome works out at about 17 600 nucleotides. This is somewhat smaller than the size reported for the largest continuous RNA genome known: that of the coronavirus, avian infectious bronchitis virus (Lomniczi & Kennedy, 1977). (Their value of 23 000 bases or 8×10^6 daltons may err on the high side since most coronaviruses have genome sizes of $6-7 \times 10^6$ daltons (Matthews, 1979).). There may be a certain circularity in this derivation since the value selected for σ can be adjusted to give a νmax value that corresponds with the biological data; although, since σ is a logarithmic function, the variations may be rather slight. However it is a fact that the average size of RNA viral genomes is very much smaller than that of DNA viral genomes (Reanney, 1982). This suggests that there are restraining factors operating on RNA genes that do not affect DNA genes to the same degree, and it seems reasonable to identify the 10^5-10^7-fold difference in replicative error rates between the two classes of polynucleotide as one of the most 'significant' of these restraining factors (see Eigen & Schuster, 1977).

Other limiting factors should also be considered. While a variety of processing and corrective enzymes have the ability to cleave DNA, purely degradative DNases are rather rare because DNA, the permanent repository of genetic information in cells, is not normally put at risk by endogenous factors. By contrast, active cells maintain a high mRNA turnover and RNases are plentiful in the cytoplasm. A single-stranded RNA genome multiplying in the cytoplasm is a potential target for normal degradative enzymes and these can destroy infectivity with a single endonucleolytic cut. Since the kinetics of RNA–nuclease interactions follow the laws of probability, the smaller an RNA, the longer its predicted 'half-life' in an RNase-rich environment. In addition to limiting sizes, the abundance of RNases may also have shaped the *character* of RNA genomes. For example, the tendency of single-stranded RNAs to fold into compact, largely double-helical formats (see Min Jou *et al.*, 1972) may be viewed, in part, as an adaptive response to hazards in the immediate cellular environment. It is also possible that the fully double-helical character of reovirus RNAs is related to the need to protect their relatively large genomes against the possibility of nucleolytic degradation. It is perhaps worth stressing that double-helical reovirus RNAs are not functionally equivalent to duplex DNAs since the latter can be unwound whereas the former cannot.

Another point is that the vicinal 2′ hydroxyl groups in RNA labilize RNA molecules at physiological pH to the extent that RNA is substantially more prone to hydrolysis at this pH than DNA (J. Langridge, personal communication). A short RNA has a longer projected survival time at alkaline pH than a long RNA does because random hydrolysis of the 3′—5′ phosphodiester bonds is a length-related function (i.e., if there is a certain probability of hydrolysis in a given time interval, it follows that the greater the number of bonds - the longer the molecule - the greater the chances of hydrolytic self destruction). RNA viruses may have compensated for these liability factors by magnifying the number of progeny RNAs generated from any input copy. This is also a size-related strategy since the *shorter* the initiating strand, the *larger* the population of replicas that can be generated from it using a *fixed* pool of precursor elements.

In my view then, the relatively small genome sizes of most RNA viruses are direct functions of (*a*) the high error rates of RNA replication (*b*) the chemical lability of 3′—5′ bonds due to the destabilizing effect of the 2′ hydroxyl of the ribose sugar at alkaline pH and (*c*) the susceptibility of single-stranded RNA to degradative enzymes which are abundant in most replicative sites. Other factors such as susceptibility to ultraviolet (UV) light and hydrogen peroxide may make a minor contribution.

Genome segmentation

Genome size is not the only feature of RNA viruses affected by noise levels. One way to increase the reliability with which a unit of information is sent through a noisy channel is to divide the information into smaller subunits whose senses are checked independently (Shannon, 1949). In the light of this premise, one can argue that the genome *segmentation* which characterizes many RNA viruses (Matthews, 1979) is a strategy which *protects* viral RNAs from the harmful consequences of the high level of noise in their replicative machinery (Reanney, 1982; Pressing & Reanney, 1983). UV inactivation kinetics (target theory) provide a model for this kind of protective effect. Consider an idealized situation in which UV dosage is calibrated so that it introduces one lethal 'hit' per molecule in a large population of RNAs, all of length x. If each RNA is divided into two equal-sized modules A and B of length $x/2$ then the same UV dosage will most likely not inactivate the whole population. This is because a 'hit' in module A need not affect the integrity of the information in module B and vice versa. The logic is similar when mutations introduced by an error-prone polymerase are not a result of outside factors (Reanney, 1982).

However, a fundamental distinction must be made between what happens when the modular RNAs are packaged in one capsid (monocompartment viruses) and what happens when each RNA is separately accommodated in a discrete particle (multicompartment viruses). Consider the above example again and assume that the error rate of the replicase which reproduces the two equal length modules A and B is such that 50% of progeny carry lethal errors. The following combination of modules are possible, where letters with asterisks indicate lethally damaged modules, while letters without asterisks indicate viable modules.

Case 1	Case 2
Monocompartment	Multicompartment

In the first case each pair of modular RNAs can be accommodated in one capsid so a single particle can initiate infection. However this strategy limits the number of particles carrying

Table 2. *Divided genomes in RNA viruses*

Virus[a]	Number of modular RNAs	Total MW ($\times 10^6$)	Host
Strategy 1: Monocompartment viruses			
Cysto[b]	3	10.4	Bacteria
Reo[b]	10–12	12–20	Animals, plants
Influenza	8	5	Animals
Bunya	3	5.5	Animals
Arena	5 (3 of host origin)	?	Animals
Tomato spotted wilt	4	7.5	Plants
Strategy 2: Multicompartment viruses			
Penicillium chrysogenum group	3	?	Fungi
Penicillium stoloniferum group	2	?	Fungi
Nepo	2	4.6	Plants
Pea enation mosaic	2	3	Plants
Como	2	2.4	Plants
Tobra	2	3.8	Plants
Cucumo	3	3.2	Plants
Bromo	3	2.8	Plants
Ilar	3	2.7	Plants
Alfalfa mosaic	3	2.6	Plants
Hordei	2–4	?	Plants

[a] Names recommended by Matthews (1979) have been used, the suffix 'virus' or 'viridae' being omitted for simplicity.
[b] Double-stranded RNA; MWs for these viruses should be divided by two to give values which can be compared with the single-strand RNAs.

acceptable copies of the genetic information to 25%. In this instance, unless there is some preferential association of non-damaged modules (see below), a segmented genome has no advantage over a non-segmented one. In the second case 50% of the population of progeny particles carry viable copies of units of the genetic information. This gives the multicompartment strategy a distinct survival advantage, so long as the transmission of the various encapsidated RNAs remains random and independent. This requirement is met in nature.

In the real world, however, the advantage gained in terms of increased reliability of information transmission due to separate packaging is offset by the need to infect a sensitive cell with two (or more) particles. There is thus a 'pendulum' effect in which increased fidelity of information transmission must be weighed against decreased fidelity of host-to-host transmission. Clearly, the multicompartment strategy will only be favoured in situations where the chances of successful host-to-host transmission with independently encapsidated RNAs are high enough to ensure a reliable flow of viral information. It may be no accident that multicompartment viruses are as common in plants as they are rare in animals (Table 2). Because they are sessile and photosynthetic, plants can attain very high population densities denied to animals which must move to obtain food. This crowding together of susceptible hosts must improve the likelihood of viral dissemination. Also the multiple mechanisms developed by flowering plants to attract insects open up a variety of channels for reliable host-to-host passage by means of vectors (Nahmias & Reanney, 1977).

In the case of monocompartment viruses, segmentation will not confer an advantage if the modular RNAs associate at random. If a consistent bias in favour of well-copied sequences over miscopied sequences can be demonstrated, then segmentation in this case also emerges as a favoured strategy.

Does such a bias exist? Consider the reoviruses. During the maturation of this group of viruses *each* particle accumulates the *correct* combination of 10–12 *different* modular RNAs. The mechanism of this selective assembly is believed to be a set of RNA–RNA or RNA–protein interactions that operate while the RNAs are single-stranded (Silverstein, Christman & Acs, 1976). Lane (1979) has proposed a co-operative process in which the successful completion of one step is required for the next step to proceed. Whatever the details, it is obvious that this assembly process is based on a number of highly *specific* interactions among the components of the divided genome or between these components and one or more proteins. The basis of this specificity must reside in a defined nucleotide sequence at some point, or points, in each RNA (this remains true even if topology is also a factor). Given this situation, it follows that a badly copied sequence will have a lower chance of entering a nascent assembly complex than an accurately copied one. For each individual RNA, the chance of 'rejection' may be very small (since it requires one or more errors in a target area which could be as short as 10 nucleotides) but the 10–12 steps needed to assemble the whole virus offer a hierarchy of rejection possibilities. One may infer then that the multi-step nature of reovirus development constitutes a crude form of molecular 'proofreading' (Reanney, 1982) which gives selection some control over the quality of the sequences used to make viable viruses.

In the case of multicompartment viruses, the problem of infecting one cell with several different RNAs limits the number of modular RNAs per virus to 2–3 (Table 2). For monocompartment viruses, if the above argument holds good, the higher the number of modular RNAs the greater the degree of discrimination in favour of well-copied sequences that can be achieved (Pressing & Reanney, 1983). Thus it is not surprising that monocompartment viruses contain relatively high numbers of modular RNAs (Table 2). It is possible to show that the improved fidelity of information transfer achievable with highly divided genomes allows RNA molecules to exceed significantly the νmax value for nondivided genomes (Pressing & Reanney, 1983). Accordingly the reoviruses, which have the biggest genome *sizes* of any RNA virus ($12–20 \times 10^6$), contain the largest *number* of genomic segments (10–12).

Diploidy in the retroviruses

The thrust of the preceding argument is that genome segmentation is favoured where error rates are high because small modules of information have a better chance of passing undamaged through a noisy channel than larger ones. A different strategy to combat noise is to *repeat* the information i.e., to send the message more than once (see Reanney, MacPhee & Pressing, 1983). This strategy of genetic redundancy has been followed by most forms of life which have large genomes. Diploidy, for example, protects the information in eukaryote genomes by representing each gene twice.

Genetic redundancy is found in only one group of RNA viruses – the retroviruses. The genome of a retrovirus exists as *two* equivalent RNAs encapsidated in a common envelope. The 'alleles' of the retroviral genome are linked by base-paring at their 5' ends to give an inverted dimer structure. Transcription is initiated from tRNA primers located close to these 5' ends. The specific pairing between the redundant RNAs provides the basis for a degree of discrimination against miscopied sequences, as in the case of the reoviruses. Whether this explanation *per se* is sufficient to account for the diploid character of retroviruses is dubious. What is beyond dispute is that the DNA phase of the retrovirus life cycle means that retroviruses, alone among viruses, alternate reversibly between a high noise level (the free RNA genome) and a low noise level (the integrated DNA copy). This means that retroviral genes can use the low-error copying systems of chromosomal genes to preserve their preselected identities while retaining the ability to spread their survival options across a wide distribution of variant sequences whenever they revert to an extracellular state. Retroviruses have the best of both worlds.

Recent data make it clear that alterations between high and low fidelity systems are not confined to viral genes. The presence of poly A tails at the 3' ends of the *Alu* family of repetitive sequences and various pseudogenes suggests that many cellular sequences participate in a flow of information from RNA to DNA and perhaps back again (Sharp, 1983). These oscillations may have had important consequences for the evolution of eukaryotic genomes. For example, the removal of introns via RNA processing may have helped to create less fragmented genes from more highly divided ones (Reanney, 1983).

This point has some important consequences for evolutionary calculations which use sequence data. A molecular comparison between a rat α-tubulin gene and its corresponding pseudogene, for example, showed that the pseudogene differed from its inferred ancestor gene by 38 point substitutions in 1646 positions (Lemischka & Sharp, 1982). This corresponds to a 2.3% divergence. Since the RNA precursor of the pseudogene may have passed through a (large) number of transcriptions before becoming fixed in its present chromosomal locus, it seems unwise to use uncritically the number of mutations in a pseudogene to estimate the time of origin of the gene, as some authors have done (see Lemischka & Sharp, 1982; Sharp, 1983). As Table 1 shows, a molecular 'clock' based on RNA runs at a much faster rate than one based on DNA.

Overlapping genes

Given that a variety of factors limit the sizes of RNA genomes (preceding discussion) it follows that any strategy which compresses the genetic information into a smaller 'space' will have a survival advantage. This requirement is met by overlapping genes.

At first sight, the presence of overlapping genes in RNA viruses seems to generate more problems than it solves. A high error rate means that when a given gene is read in two different frames, an error acceptable in frame 1 (e.g., one which occurs in position 3 of a codon) may become unacceptable in frame 2 (if it now lies in position 2 of a codon). However, since overlapping genes are common in RNA viruses, the average number of errors is evidently not high enough to outweigh the other advantages of small genomes. As discussed, when energy reserves or precursors are limiting, a short RNA generates more copies of an initiating strand than a large one; hence the *greater* the *coding* capacity of the RNA, the *smaller* its *physical* size need be.

I wish to show here that in addition to compacting the genetic information, overlapping genes also serve a different function, which may be just as important for their survival fitness. Overlapping genes fall into four distinct categories:

(1) Structural genes may be read in different frames; this occurs in RNA viruses such as MS2 and f2 (Beremand & Blumenthal, 1979) and in DNA viruses such as ϕX174 (Fiddes & Godson, 1979) (see Reanney, 1982).

(2) Structural genes may be read in the same phase but generate different products by the use of varying stop and start signals. This occurs in several RNA viruses such as $Q\beta$ and turnip yellow mosaic virus as well as in DNA viruses such as ϕX174 (see Reanney, 1982).

(3) Transcriptional signals (promotors, attenuators or terminators) may overlap coding regions. This strategy mixes the rate-promoting and read-out aspects of genes. This type of genetic organization is found in the DNA phages fd and ϕX174 and in the *ampC/frd* region of the *E. coli* chromosome (Grundström & Jaurin, 1982).

(4) Sequences on complementary strands of bihelical DNA may be transcribed in both directions. This occurs in the *cro* and *c*II genes of coliphage λ (Honigman *et al.*, 1976).

Consider first viral genes which are read in different frames. Overlapping genes occurs in segments 7 and 8 of the composite influenza genome (Lamb & Lai, 1980, 1981). Segment 8 generates two distinct proteins NS_1 and NS_2 by treating an internal sequence as an exon in the first instance but as an intron in the second (Lamb & Lai, 1981). The resulting mRNA contains a short 5' sequence, read in the same frame as before, ligated to a relatively long sequence read in a different frame. As a result of the splicing event, the production of the NS_1 protein from the unprocessed RNA and the NS_2 protein from the processed RNA may be staggered in time, NS_2 appearing only after NS_1 has accumulated. Alternatively, sequence overlap may influence the relative amounts of the two gene products. In this example then the economy of genetic organization achieved by overlapping genes may have the added bonus of allowing their expression to be *regulated* in a sensitive way.

More convincing evidence for the idea that out-of-phase sequence overlaps have been exploited for gene regulation has come from recent work on the ribophage, MS2. MS2 contains four genes coding for (in sequence from the 5′ end) a maturation protein, the viral coat protein, a lysis (L) protein and the viral replicase. The L protein overlaps the coat and replicase cistrons in the + 1 reading frame. Kastelein *et al.* (1982) have shown that the L cistron is translated by that fraction of ribosomes that arrive out of frame from the upstream coat protein gene. The two UAA codons which are positioned in the + 1 and − 1 frames just prior to the L cistron elicit peptide chain termination and this evidently allows reinitiation at the beginning of the L message in an area normally blocked to ribosomes. In this case the tight coupling of the coat and lysis genes by use of shared sequences guarantees that an infected cell cannot be lysed unless enough coat proteins have been made to produce a good crop of progeny viruses.

Another aspect of the overlapping gene phenomenon is exemplified by the organization of the *Nu*3 and *C* genes in phage λ. These genes are translated in the same reading frame so that the *Nu*3 protein is identical to the carboxy-terminal third of the *C* protein (Shaw & Murialdo, 1980). Because the *Nu*3 and *C* products are essential for the construction of the phage prohead, it has been suggested that homologous structural domains which result from shared amino acid sequences may be the basis of the functional interaction between these proteins during viral assembly (Shaw & Murialdo, 1980). It seems likely that in-phase sequence overlaps represent a general strategy for generating proteins which *conserve* some amino acid sequences and *vary* others (Reanney, 1982). This strategy is not restricted to viruses having been found at the *che*A locus of *E. coli* (Smith & Parkinson, 1980).

How are these overlaps achieved? In the case of viruses such as adenovirus and influenza, which have a nuclear phase in their replication cycle, multiple mRNAs are produced from a common precursor RNA by one or more RNA-splicing events (Chow & Broker, 1978; Lamb & Lai, 1980, 1981). In all other cases, alternative outcomes are introduced into the translation of individual mRNAs by the calibrated use of variant stop and start signals. Perhaps the most striking example of this is the *G* gene of phage φX174 in which a single nucleotide sequence is evidently read into four different polypeptides (Pollock, Tessman & Tessman, 1978). These proteins result from the use of internal initiation signals and from the presence of suppressor tRNAs which neutralize appropriately positioned UAG or UGA codons (Geller & Rich, 1980). UAG and UGA codons will always be present at variable positions in a population of genetic RNAs because self-replicating RNAs are so inaccurately copied. One is tempted to conclude that multiple-option genes have been closer to the rule than the exception during (RNA) virus evolution but that 'overlaps' have been stabilized by selection only in those cases where the coupling confers survival advantages.

A detailed analysis of the use to which UAG and UGA codons have been put by selection can be made using the recently elucidated sequence of tobacco mosaic virus (TMV) RNA (Goelet *et al.*, 1982). Examination of this sequence reveals three major open reading frames. The first AUG codon initiates an open reading frame which continues until residue 3417. This codes for a protein of M_r 125 941. The UAG codon which terminates this gene may be suppressed by amber tRNAs to yield a read-through protein of M_r183 253 (Pelham, 1978). The end of the reading frame for the M_r 183 253 protein overlaps (by five codons) a gene read in a different phase that encodes a protein of M_r 29 987. These three proteins are known to be produced during TMV infection (Beachy & Zaitlin, 1977; Pelham, 1978). In addition, at least five other 3' coterminal RNAs are synthesized during infection. Comparison of the peptide maps of several proteins made *in vitro* by packaged subgenomic mRNAs shows that the proteins have overlapping carboxy-termini (Goelet *et al.*, 1982). These data suggest that a variety of proteins may be generated from an average-sized RNA genome by exploiting the apparent reluctance of the eukaryotic release factor to recognize the UAG triplet (Goelet *et al.*, 1982). Viral genomes enriched in amber triplets may thus offer possibilities for the regulation of gene expression not readily available to the mRNAs of their cellular hosts.

Further evidence that sequence overlaps are used for regulatory purposes comes from an examination of the genes for the β and ε subunits of the spinach chloroplast ATPase (Zurawski, Bottomley & Whitfield, 1982). These data indicate that the two genes are cotranscribed from a dicistronic mRNA with a four base pair overlap between the *STOP* codon of the β subunit gene and the *START* codon of the ε subunit gene (Zurawski *et al.*, 1982). It has been suggested that this arrangement serves as a translational coupling mechanism (Oppenheim & Yanofsky, 1980) to ensure equimolar synthesis of the two subunits. Whether this type of coupling is a feature of virus evolution remains to be determined.

The most unexpected type of overlap occurs when complementary sequences from each strand of a (potentially) double-helical DNA are transcribed. The advantages of this type of organization are obscure. Some speculations, however, can be made. The DNA-containing geminiviruses have divided genomes consisting of two similar-sized, single-stranded, covalently closed, circular molecules. The nucleotide sequence of the geminivirus, cassava latent virus, has recently been determined (Stanley & Gay, 1983). Examination of this sequence shows that both genomic DNAs contain overlapping regions in their + and − strands (Stanley & Gay, 1983). If bidirectional transcription of a duplex DNA (1) starts from a common origin as suggested, then rightwards transcription gives a potential mRNA almost immediately whereas leftwards transcription does not reach this complementary region until the enzyme has travelled about two-thirds of the way round the molecule. Use of overlapping sequences in this system may thus ensure a *kinetic delay* between the appearance of transcriptional products from each *complementary* strand.

Another possible use of complementary sequences arises from the fact that RNA sequences so transcribed can base-pair. Since 'incompatibility' in bacterial plasmids appears to be based, at least in part, on interference between complentary RNAs (Lacatena and Cesarini, 1981) it seems possible that such a mechanism could be exploited to protect a cell harbouring a given virus from superinfection with another virus of the same kind.

CONCLUSIONS

Recent discoveries have sharpened our perception of the differences between DNA and RNA. In particular, it is obvious that RNA genes face much greater survival problems than DNA genes. The most significant problem is the low copying fidelity of RNA as opposed to DNA genomes. The main proposition of this paper has been that this 10^5–10^7-fold difference in error rates has moulded the evolution of RNA viruses in a number of characteristic ways. The high noise level in combination with other factors has placed an ill-defined upper limit on the sizes of self-replicating RNAs and favoured the evolution of composite viruses in which the genetic information is divided between discrete RNA subspecies. The advantages of this genome segmentation follow largely from the fact that a small unit of information has a higher chance than does a large unit of passing undamaged through a noisy channel (Reanney, 1982). 'Noise' in this context refers not only to copying errors but also to the destructive influences of RNases, UV, hydrogen peroxide, mutagenic chemicals and any other factors which damage RNA molecules in a manner which obeys the laws of 'target theory'.

This last point raises an important evolutionary issue. Damage due to mutagenic chemicals or UV constitutes a source of *external* noise. Chemical labilities and copying errors constitute a source of *intrinsic* noise since they are inbuilt features of the RNA molecules themselves or of their copying machinery. One cannot say therefore that genome size, for example, results from natural selection due to *environmental* pressures. This would imply that the external habitat of the virus favours small genomes whereas in fact the critical limitations come from within the self-replicating system itself. In an important sense then this breaks with the traditional neo-Darwinian concept of selection and proposes a novel explanation for the design of the genetic machinery in this group of viruses. The implications of this 'intrinsic' form of selective pressure for general biology have been explored elsewhere (Reanney *et al.*, 1983).

Note

1. The term *mutation.* is now one of the most confusing in the literature. Its proper scientific meaning is a *heritable* alteration in the genetic material. However this definition means that a spectrum of events which affect the genome – from the deamination of cytosine to uracil to the formation of thymine-to-thymine dimers– should be described as *premutational lesions* rather than as mutations *sensu stricto*. Sloppiness over the use of the term mutation leads to such confusing statements as 'the role of DNA repair is to eliminate harmful mutations' (since mutations are heritable as normal base pairs they cannot, by definition, be eliminated).

Throughout this paper I make frequent use of the term *noise*. This word has a precise meaning in information theory (Shannon, 1949) where it is used to refer to any factor or factors which diminish the integrity of information or the fidelity of information transfer. Noise can thus be used to describe mutations, premutational lesions or errors in the decoding machinery which result in the incorporation of 'wrong' amino acids. The level of *noise* as opposed to the level of *mutation* is high in *all* cells. The error rate in DNA synthesis in *E. coli* cells containing a DNA polymerase 1 defective in its repair properties is about 10^{-3} per nucleotide per generation (Eigen & Schuster, 1977). This value defines the *intrinsic* error rate in enzymatic DNA synthesis before proofreading mechanisms intervene. We obtain a fidelity figure of 10^{-10} for DNA copying *only because we see noise in modern cells through the screen erected by subsequent corrective events.* My point is simply that RNA genes cannot use these restorative mechanisms in their replicative processes.

I am grateful to Dr Jeff Pressing for the mathematical derivation of Table 1.

REFERENCES

ABELSON, J. (1982). Self-splicing RNA. *Nature*, **300**, 400–1.

BEACHY, R. N. & ZAITLIN, M. (1977). Characterization and *in vitro* translation of the RNAs from less-than-full-length, virus-related, nucleoprotein rods present in tobacco mosaic virus preparations. *Virology*, **81**, 160–9.

BEREMAND, M. N. & BLUMENTHAL, T. (1979). Overlapping genes in RNA phage: a new protein implicated in lysis. *Cell*, **18**, 257–66.

BISHOP, D. H. L., JONES, K. L., HUDDLESTON, J. A. & BROWNLEE, G. G. (1982). Influenza A virus evolution: complete sequences of influenza A/NT/60/68 RNA segment 3 and its predicted acidic P polypeptide compared with those of influenza A/PR/8/34. *Virology*, **120**, 481–9.

BLOK, J. & AIR, G. M. (1982). Sequence variation of the 3' end of the neuraminidase gene from 39 influenza type A viruses. *Virology*, **121**, 211–29.

CHOW, L. T. & BROKER, T. R. (1978). The spliced structures of adenovirus 2 fiber message and the other late mRNAs. *Cell*, **15**, 497–510.

DOMINGO, E., SABO, D., TANIGUCHI, T. & WEISSMAN, C. (1978). Nucleotide sequence heterogeneity of an RNA phage population. *Cell*, **13**, 735–44.

EIGEN, M., GARDINER, W., SCHUSTER, P. & WINKLER-OSWATITSCH, R. (1981). The origin of genetic information. *Scientific American*, **244**, 78–96.

EIGEN, M. & SCHUSTER, P. (1977). The hypercycle, Part A. *Naturwissenschaften*, **64**, 541–65.

ESTES, M. K., GRAHAM, D. Y., RAMIG, R. F. & ERICSON, B. L. (1982). Heterogeneity in the structural glycoprotein (VP7) of simian rotavirus SA11. *Virology*, **122**, 8–14.

FIDDES, J. C. & GODSON, G. N. (1979). Evolution of the three overlapping gene systems in G4 and φX174. *Journal of Molecular Biology*, **133**, 19–43.

GELLER, A. I. & RICH, A. (1980). A UGA termination suppression tRNA^{Trp} active in rabbit reticulocytes. *Nature*, **283**, 41–6.

GOELET, P., LOMONOSSOFF, G. P., BUTLER, P. J. G., AKAM, M. E., GAIT, M. J. & KARN, J. (1982). Nucleotide sequence of tobacco mosaic virus RNA. *Proceedings of the National Academy of Sciences (USA)*, **79**, 5818–22.

GOUVEA, V. S. & SCHNITZER, T. J. (1982). Polymorphism of the migration of double-stranded RNA genome segments of avian reoviruses. *Journal of Virology*, **43**, 465–71.

GRUNDSTRÖM, T. & JAURIN, B. (1982). Overlap between *ampC* and *frd* operons on the *Escherichia coli* chromosome. *Proceedings of the National Academy of Sciences (USA)*, **79**, 1111–15.

HOLLAND, J., SPINDLER, K., HORODYSKI, F., GRABAU, E., NICHOL, S. & VANDE POL, S. (1982). Rapid evolution of RNA genomes. *Science*, **215**, 1577–85.

HONIGMAN A., HU, S.-L., CHASE, R. & SZYBALSKI, W. (1976). 4S *oop* RNA is a leader sequence for the immunity-establishment transcription in coliphage λ. *Nature*, **262**, 112–16.

HRDY, D. B., ROSEN, L. & FIELDS, B. N. (1979). Polymorphism of the migration of double-stranded RNA genome segments of reovirus isolates from humans, cattle and mice. *Journal of Virology*, **31**, 104–11.

KASTELEIN, R. A., REMAUT, E., FIERS, W. & VAN DUIN, J. (1982). Lysis gene expression of RNA phage MS2 depends on a frameshift during translation of the overlapping coat protein gene. *Nature*, **295**, 35–41.

KIM, S., SUSSMAN, J. L., SUDDATH, F. L., QUIGLEY, G. J., McPHERSON, A., WANG, A. H. J., SEEMAN, N. C. & RICH, A. (1974). The general structure of transfer RNA molecules. *Proceedings of the National Academy of Sciences (USA)*, **71**, 4970–4.

KORNBERG, A. (1980). *DNA Replication*. San Francisco: W. H. Freeman & Co., 724 pp.

KRUGER, K., GRABOWSKI, P. J., ZANG, A. J., SANDS, J., GOTTACHLING, D. E. & CECH, T. R. (1982). Self-splicing RNA: Autoexcision and autocyclization of the ribosomal RNA intervening sequence of tetrahymena. *Cell*, **31**, 147–57.

LACATENA, R. M. & CESARENI, G. (1981). Base pairing of RNA I with its complementary sequence in the primer precursor inhibits ColE1 replication. *Nature*, **294**, 623–31.

LAMB, R. A. & LAI, C.-J. (1980). Sequence of interrupted and uninterrupted mRNAs and cloned DNA coding for the two overlapping nonstructural proteins of influenza virus. *Cell*, **21**, 475–85.

LAMB, R. A. & LAI, C.- J. (1981). Interrupted mRNA(s) and overlapping genes in influenza virus. In *The Replication of Negative Strand Viruses*, ed. D. H. L. Bishop & R. W. Compans, pp. 251–9. New York: Elsevier.

LANE, L. C. (1979). The RNAs of multipartite and satellite viruses of plants. In *Nucleic Acids in Plants*, vol. 2, ed. T. C. Hall & J. W. Davies, pp. 65–110. Boca Raton: CRC Press.

LEMISCHKA, I. & SHARP, P. A. (1982). The sequences of an expressed rat α-tubulin gene and a pseudogene with an inserted repetitive element. *Nature*, **300**, 330–5.

LEWIN, R. (1982). RNA can be a catalyst. *Science*, **218**, 872–4.

LOEB, A. A. & KUNKEL, T. A. (1982). Fidelity of DNA synthesis. *Annual Review of Biochemistry*, **51**, 429–57.

LOMNICZI, B. & KENNEDY, I. (1977). Genome of infectious bronchitis virus. *Journal of Virology*, **24**, 99–107.

MATTHEWS, R. E. F. (1979). Classification and nomenclature of viruses. *Intervirology*, **12**, 129–296.

MIN JOU, W., HAEGEMAN, G., YSEBAERT, M. & FIERS, W. (1972). Nucleotide sequences of the gene coding for the bacteriophage MS2 coat protein. *Nature*, **237**, 82–8.

MINOR, P. D., SCHILD, G. C., BOOTMAN, J., EVANS, D. M. A., FERGUSON, M., REEVE, P., SPITZ, M., STANSWAY, G., CANN, A. J., HAUPTMAN, R., CLARKE, L. D., MOUNTFORD, R. C. & ALMOND, J. W. (1983). Location and primary structure of a major antigenic site for poliovirus neutralization. *Nature*, **301**, 674–9.

NAHMAIS, A. J. & REANNEY, D. C. (1977). The evolution of viruses. *Annual Review of Ecology and Systematics*, **8**, 29–49.

NOMOTO, A., OMATA, T., TOYODA, H., KUGE, S., HORIE, H., KATAOKA, Y., GENBA, Y., NAKANO, Y. & IMURA, N. (1982). Complete nucleotide sequence of the attenuated poliovirus Sabin 1 strain genome. *Proceedings of the National Academy of Sciences (USA)*, **79**, 5793–7.

OPPENHEIM, D. S. & YANOFSKY, C. (1980). Translational coupling during expression of the tryptophan operon of *Escherichia coli*. *Genetics*, **95**, 785–95.

PALESE, P. & YOUNG, J. F. (1982). Variation of influenza A, B and C viruses. *Science*, **215**, 1468–73.

PELHAM, H. R. B. (1978). Leaky UAG termination codon in tobacco mosaic virus. *Nature*, **212**, 469–71.

POLLOCK, T. J., TESSMAN, I. & TESSMAN, E. S. (1978). Potential for variability through multiple gene products of bacteriophage ϕX174. *Nature*, **274**, 34–7.

PORTNER, A., WEBSTER, R. G. & BEAN, W. J. (1980). Similar frequencies of antigenic variants in sendai, vesicular stomatitis and influenza A viruses. *Virology*, **104**, 235–8.

PRABHAKAR, B. S., HASPEL, M. V., McCLINTOCK, P. R. & NOTKINS, A. L. (1982). High frequency of antigenic variants among naturally occurring human Coxsackie B4 virus isolates identified by monoclonal antibodies. *Nature*, **300**, 374–6.

PRESSING, J. & REANNEY, D. C. (1983). Divided genomes and intrinsic noise. *Journal of Molecular Evolution*, in press.

REANNEY, D. C. (1982). The evolution of RNA viruses. *Annual Review of Microbiology*, **36**, 47–73.

REANNEY, D. C. (1983). Genetic noise in evolution. *Nature*, in press.

REANNEY, D. C., MacPHEE, D. G. & PRESSING, J. (1983). Intrinsic noise and the design of the genetic machinery. *Australian Journal of Biological Sciences*, **36**, 77–91.

REDDY, D. V. R. & BLACK, L. M. (1974). Deletion mutations of the genome segments of wound tumor virus. *Virology*, **61**, 458–73.

SABARA, M., DEREGT, D., BABIUK, L. A. & MISHRA, V. (1982). Genetic heterogeneity within individual bovine rotavirus isolates. *Journal of Virology*, **44**, 813–22.

SCHROEDER, B. A., STREET, J. E., KALMAKOFF, J. & BELLAMY, A. R. (1982). Sequence relationships between the genome segments of human and animal rotavirus strains. *Journal of Virology*, **43**, 379–85.

SHANNON, C. E. (1949). The mathematical theory of communication. In *The Mathematical Theory of Communication*, ed. C. E. Shannon & W. Weaver. Urbana, Illinois: University of Illinois Press.

SHARP, P. A. (1983). Conversion of RNA to DNA in mammals: *Alu*-like elements and pseudogenes. *Nature*, **301**, 471–2.

SHAW, J. E. & MURIALDO, H. (1980). Morphogenetic genes *C* and *Nu*3 overlap in bacteriophage λ. *Nature*, **283**, 30–5.

SILVERSTEIN, S. C., CHRISTMAN, J. K. & ACS, G. (1976). The reovirus replicative cycle. *Annual Review of Biochemistry*, **45**, 375–408.

SMITH, R. A. & PARKINSON, J. S. (1980). Overlapping genes at the *cheA* locus of *Escherichia coli*. *Proceedings of the National Academy of Sciences (USA)*, **77**, 5370–4.

SPIEGELMAN, S. (1970). Extracellular evolution in replicating molecules. In *The Neurosciences*, ed. F. O. Schmitt, pp. 927–45. New York: Rockefeller University Press.

STANLEY, J. & GAY, M. R. (1983). Nucleotide sequence of cassava latent virus DNA. *Nature*, **301**, 260–2.

STREET, J. E., CROXSON, M. C., CHADDERTON, W. F. & BELLAMY, A. R. (1982). Sequence diversity of human rotavirus strains investigated by northern blot hybridization analysis. *Journal of Virology*, **43**, 369–78.

ZAUG, A. J., GRABOWSKI, P. J. & CECH, T. R. (1983). Autocatalytic cyclization of an excised intervening sequence RNA is a cleavage–ligation reaction. *Nature*, **301**, 578–83.

ZURAWSKI, G., BOTTOMLEY, W. & WHITFIELD, P. R. (1982). Structures of the genes for the β and ε subunits of spinach chloroplast ATPase indicate a dicistronic mRNA and an overlapping translation stop/start signal. *Proceedings of the National Academy of Sciences (USA)*, **79**, 6260–4.

MECHANISMS OF VIRUS–HOST INTERACTIONS

BERNARD N. FIELDS

Department of Microbiology and Molecular Genetics, Harvard Medical School and Department of Medicine (Infectious Disease), Brigham and Women's Hospital, Boston, MA 02115, USA

INTRODUCTION

The dramatic advances in understanding the detailed stages of virus replication in infected cells are largely a result of the ease of culturing cells *in vitro*, and the ability to perform chemical analysis of such cell cultures with defined media. The infection of an eukaryotic host is a much more complex process and thus has been more difficult to elucidate in precise biochemical terms. However, by separating out the individual stages involving the interaction of viruses with their hosts, it has become feasible to begin to determine the precise mechanisms involved.

Our own studies have relied heavily on one model system, the infection of mice by the mammalian reoviruses (Fields, 1982). The mammalian reoviruses contain a segmented double-stranded RNA genome and have been subjected to extensive genetic analyses (Fields, 1981). Hence, this system has been extensively used to define the genetic strategies and the functions of individual viral proteins in viral pathogenesis.

In this paper, I would like to outline the several stages of viral pathogenesis, citing recent data which have provided insights into possible mechanisms determining virus virulence. I will stress studies with lytic viruses, particularly the mammalian reoviruses, and will not include studies with tumour viruses.

ENTRY INTO THE HOST

Viruses enter the mammalian host primarily via the gastrointestinal tract (e.g., polio), the respiratory tract (e.g., measles), direct inoculation by arthropods into the blood stream (equine encephalitis), or by other breaks in the skin (rabies) (Fenner *et al.*, 1974). Following introduction into these primary sites, viral replication takes place in primary target cells and, following amplification, there begins the process of spreading to more distal targets. Very little is known concerning molecular mechanisms during the initiation of virus infection. It is probable, however, that the interaction of viruses with specific receptors on the surface of different epithelial cells or endovascular cells (see discussion below) plays a central role at this stage in determining the localization of virus infection (Holland & McLaren, 1961). For example, certain viruses (i.e., rotaviruses) infect primarily the intestinal absorptive cells, resulting in direct epithelial injury (Schreiber, Blacklow & Trier, 1973). By contrast, the reoviruses bypass the intestinal barrier by binding to the surface of intestinal M cells (specialized lymphoepithelial cells that are located over Peyer's patches), moving through the M cells in intracellular vesicles, and finally being released to the antilumenal surface where they are presented to various cellular elements (primarily macrophages and lymphocytes) (Wolf, et al., 1981). Thus, the initial pattern of virus–host interactions is determined by the nature of the cells that are permissive to the virus and thus, to a large extent, by the nature of viral recognition on the surface of infected cells. It is quite likely that most instances of specific virus–cell interactions involve the binding of virus to specific cell surface receptors (see below). It remains to be determined whether cells that function to transport viruses (such as M cells) have specific receptors or, alternatively, non-specifically transport a wide variety of infectious agents and other non-infectious materials.

In addition to early interactions with selected cells, the nature of the body fluids in the anatomic location of viral entry (e.g., respiratory secretion, gastrointestinal fluids, blood serum), determines in part whether a productive infection will be established. Recent studies with the reoviruses, for example, have indicated that one of the surface polypeptides (the μ1C polypeptide) is the target of proteolytic enzymes (Rubin & Fields, 1980). Loss in infectivity in some isolates results from digestion of the μ1C polypeptide by gastrointestinal proteases. Resistance to proteases in vivo thus allows primary infection to occur and ultimately determines whether successful initiation of systemic infection will occur. It is possible, but still somewhat controversial, that the influenza virus neuraminidase similarly plays a primary role at this stage of infection, allowing virus to reach the primary site of multiplication successfully (Bucher & Palese, 1975). Although analogous specific proteins have not been recognized for other viruses, it is likely that most, if not all, viruses have surface proteins that play similar roles in entry and spread. Thus, the nature of these body fluids at surfaces, as well as throughout the host, represents an important host component involved in the mechanism of interaction with an infecting viral pathogen.

HOW DO VIRUSES SPREAD FROM THE SITE OF PRIMARY MULTIPLICATION?

An infecting virus that has successfully withstood the hostile environment at the portal of entry must traverse body fluids as well as cellular and membranous barriers. The three major pathways of spread in the host are (*a*) bloodstream, (*b*) nerves, (*c*) lymphatics (Johnson & Griffin, 1978). Although the pathways of spread by these routes have been well described, especially using the detection of viral antigen by fluorescent antibody techniques, molecular mechanisms involved in these stages of transport have not been proposed.

The nature of the dramatically different strategies chosen by those viruses utilizing viraemia (such as picornaviruses, reoviruses) or retrograde neural spread (rabies, ? herpes simplex) suggests possible differences in mechanism. For example, viruses transported from sites of primary multiplication into the blood could move freely in plasma (polioviruses) or attached (via surface receptors, either specific or non-specific) to cellular elements. Thus, a virus that uses the haematogenous route must resist serum factors (complement, salts, etc.) and be transported either in the fluids or cellular components of the blood. The specific mechanisms of entry into the host and spread from the primary site to more distant targets have not been determined.

CELL TROPISM

The ultimate pattern of disease within the host is determined to a considerable degree by the localization of viruses within particular organs or cells within those organs. Certain viruses are highly localized in certain cell types (rabies and reovirus type 3 in neurones, reovirus type 1 in ependyma), while others appear to be more widely distributed (herpes simplex encephalitis, measles in

sub-acute sclerosing panencephalitis, SSPE). Studies using the
model of the mammalian reoviruses have indicated that the component of the virus responsible for selective localization in the CNS is
the viral haemagglutinin (HA) interacting in a highly specific
manner with cell-surface receptors (Weiner *et al.*, 1977; Weiner,
Powers & Fields, 1980). Thus, the reovirus type 3 HA interacts with
receptors on neurones. The details of this reaction have been
studied using monoclonal antibodies. Three structurally and functionally discrete epitopes are present on the T3 HA (Burstin,
Spriggs & Fields, 1982). One epitope was identified as the site that
interacts with neutralizing antibody (NT epitope). The red blood
cell binding site was physically distinct. Analyses using the monoclonal antibodies to block a property or alternatively, to select HA
mutants, have indicated that the NT epitope is the site responsible
for recognizing cytotoxic T lymphocytes (CTLs) and neurones
(Finberg, Spriggs & Fields, 1982; Spriggs, Bronson & Fields, 1983).
In fact, HA mutants that are selected as resistant to NT antibody are
attenuated and show restricted tropism (Spriggs & Fields, 1982;
Spriggs *et al.*, 1983). Thus, it is feasible, by selectively altering
regions of the viral attachment protein (HA), to alter the extent of
tissue injury.

Recent studies with bunyaviruses have also indicated a role for the
surface glycoprotein in tropism (Beatty *et al.*, 1982). Additional
studies with monoclonal antibodies directed against a surface glycoprotein of the rabies virus have shown that selective mutations result
in attenuation in a fashion similar to that demonstrated by studies
with the reoviruses (Coulon *et al.*, 1982). Studies with these rabies
variants have revealed that the attenuated strains are altered by a
single amino acid change from the virulent parental virus (Dietzchold *et al.*, 1983).

Another example of how the viral HA (as well as other viral
surface glycoproteins) can affect the capacity of viruses to produce
disease has been shown by studies with myxoviruses and paramyxoviruses. The HA and other surface glycoproteins are responsible for
adsorption to cells. With both viruses, the two surface proteins exist
as precursor molecules that are cleaved by host proteases to active
forms (Klenk, 1980). Differences in susceptibility of the glycoproteins to proteases have been shown to be important in host range
and pathogenicity. Avirulent strains have limited host range and
most cell systems are nonpermissive, i.e., they produce non-
infectious virus, containing uncleaved glycoproteins. With virulent

strains, the proteases of most host cells cause cleavage of the glycoproteins and allow the virus to undergo multiple cycles of replication. Thus, the combination of the viral glycoproteins with specific host proteases helps determine the outcome of the infection. Host factors are also important in determining in which cells the virus localizes. These host factors include genetic factors that affect the capacity of cells of certain species to allow permissive virus growth (Rosenstreich et al., 1980). In addition, age is a well-known determinant of virus growth (Griffin et al., 1974). In neither of these instances is the biochemical mechanism known for these striking effects.

It should be pointed out that it is quite probable that receptor sites for viruses on the cell surface do not function solely as receptors for a particular virus. The probability of a cell generously providing receptors for viruses which inflict damages and death is low. The existence of cell-surface binding sites that serve both as phage or colicin receptors and as components of the uptake systems for low molecular weight compounds (Vitamin B12) and maltose or iron has been demonstrated in several bacterial systems. Thus phages $\phi80$, T5 and colicin M all appear to bind to a specific surface receptor for the siderophore (ferrichrome) and mutants lacking the receptors ($tonA^-$) are both phage-resistant and unable to utilize ferrichrome (Meager & Hughes, 1977).

INTERACTION OF VIRUSES WITH THE IMMUNE SYSTEM

Following entry into a host, viruses encounter a number of barriers and host defences. While anatomical and other barriers play critical roles in defence against viruses, it is quite apparent that the host immune system is a complex and highly developed network that has evolved in part to defend the host against foreign invaders. There are at least four distinct but somewhat interrelated components of host defence: interferons, macrophages, humoral immunity and cellular immunity. In this section, I will briefly outline some of the functions of each of the components, stressing recent data bearing on specificity of recognition and function.

Interferons are a family of molecules synthesized in cells infected with viruses and other molecules which induce an antiviral state in uninfected cells (Stewart, 1979). This antiviral state displays broad antiviral activity against a variety of RNA and DNA viruses. There are at least three different antigenic types of interferons (IFN-α, leukocyte; IFN-β, fibroblast; IFN-γ, T lymphocyte). It is now clear that there are at least two IFN-β genes and perhaps a family of 12–18 IFN-α genes (Allen & Fantes, 1980). The complexity of these host components is still poorly understood. The mechanisms of action of interferons is also not well understood but it is clear that they induce a number of enzymes in cells (including a protein kinase, an endoribonuclease, an oligoadenylate synthetase) which results in degradation of single-stranded RNA and inhibition of protein synthesis – in part by phosphorylating elF-2, a factor required for initiating protein synthesis (Baglioni, 1979). Interferons have been synthesized in large amounts and purified. It is quite clear that the interferons not only affect virus growth but also have a number of effects on cells and may play a role in regulating the immune system.

Macrophages play a critical role in the early response to viruses. The capacity of certain viruses to grow in, or be killed by, macrophages is an important determinant of the outcome of virus infections (Mims, 1964). Ia-bearing macrophages play a critical role in presenting antigens to the immune system. This is also true for viral antigens. For example, the generation of reovirus-specific CTLs is dependent on the presence of such antigen-presenting cells (Letvin, Kauffman, & Finberg, 1981). The precise manner by which antigen is processed and presented by the macrophage and the precise biochemical mechanisms involved in killing viruses, is not understood.

Antibodies are thought to protect against virus infection by combining with the surfaces of viruses and preventing their attachment to susceptible cells (neutralization), by combining with the viral surface, making them susceptible to complement-mediated or cell-mediated lysis, or by combining with viral antigens to form antigen–antibody complexes (Cooper, 1979; Mandel, 1979). Recent studies using genetic approaches and monoclonal antibodies have begun to provide more detailed information about the nature of the target site responsible for virus neutralization. For example, genetic studies with reovirus type 3 have indicated that the viral HA is the major type-specific neutralization antigen (Weiner & Fields, 1977). Using monoclonal antibodies, it is clear that there are 3 epitopes on the HA, one of which is the neutralizing site (Burstin et al., 1982). For influenza virus, the situation is somewhat more complex in that four sites may play a role in neutralization (Webster & Laver, 1980; Gerhard et al., 1981). Hence, clearly different strategies are used by different viruses but, in each case studied, specific regions on one (or two) viral surface proteins are involved.

Cellular immunity is thought to help defend against virus infections by generating various T cells that recognize viral components (Zinkernagel, 1979). Cytotoxic T lymphocytes are capable of lysing virus-infected cells containing viral antigens on their surface. Other T cell populations such as helper T cells and T cells that mediate delayed-type hypersensitivity can be stimulated to differentiate and produce lymphokines following interactions with specific viruses. The lymphokines produce a variety of effects, including activating macrophages for their virus-killing function. While little is known concerning the precise regulation of the cellular immune system, a great deal has been discovered about specific recognition of viruses by immune T cells. Immune T cells are specific for virus and for glycoproteins encoded in the major histocompatibility complex (MHC: reviewed in Zinkernagel, 1979). Thus in mice, it was originally discovered that immune T cells generated in the response to lymphocytic choriomeningitis (LCM) virus infection interacted with target cells that were both infected with the same virus and were at least partially histocompatible (Zinkernagel & Doherty, 1974). This MHC requirement has been seen for a number of other viral systems. The nature of the viral component involved in recognition has, in general, indicated that one surface protein – the virus HA – is involved (reovirus, influenza virus). Other proteins, however – such as the M protein of influenza virus – may also show some reactivity. Interestingly, with reovirus type 3, the major viral determinant that is recognized by CTLs is the NT epitope (Finberg et al., 1982).

Suppressor T (Ts) cells are an additional type of T cell that appears following virus infection and may have profound impact on virus–host interactions. For the reoviruses, such Ts cells appear following intravenous or peroral inoculation of UV-inactivated viruses (Greene & Weiner, 1980; Rubin et al., 1981). Such Ts cells can abrogate delayed-type hypersensitivity responses.

Although there have been many studies on lymphocytes in vitro, studies in vivo have not been as extensively performed. It is quite clear that antibody is more important in protecting against certain viruses (such as picornaviruses and certain togaviruses) while effector T lymphocytes play a major role in infections with pox and herpes viruses (Allison, 1974). The precise role of various lymphocyte sub-populations in host defence remains to be elucidated.

HOW DO VIRUSES INJURE CELLS?

The outcome of a virus infection can vary from highly cytocidal and destructive, to virtual absence of detectable effect (Tamm, 1975). Viral proteins have been shown to shut off host cell protein synthesis rapidly (polio, herpes). In other instances, this effect on host cell metabolism may occur after a delayed period and not be as efficient (paramyxoviruses). With certain viruses (rabies) no effects may be detected while, in LCM, vital cell functions were not affected but 'luxury functions' were reduced (Oldstone, Perrin & Welsh, 1976). Viral proteins have been found which cause rounding up of cells in culture and have been called 'toxins' (e.g., adenovirus fiber proteins). In all instances, it is clear that there remains little fundamental insight into the mechanism of injury. Recently, data have been generated to indicate more precisely the cellular target responsible for some of these effects on protein synthesis through modification of the 'cap' binding protein (Trachsel et al., 1980). In spite of such isolated examples, the viral components responsible for cell injury, how they mediate injury (on the surface of the cells or in a particular intracellular compartment) are poorly understood.

The nature of cellular injury has recently been studied with the mammalian reoviruses. In particular, the effect on the cytoskeleton has been studied in some detail. The major finding is that the vimentin filaments are strikingly disrupted, suggesting that this plays a critical role in producing cytopathic effects (Sharpe, Chen & Fields, 1982). In addition, studies on different reovirus serotypes, differing in their capacity to inhibit host RNA and protein synthesis, have indicated that a single viral gene (S4) is the responsible component (Sharpe & Fields, 1982).

In addition to direct viral injury, it has been demonstrated in several animal models that virus-initiated, immune-dependent damage can occur. Such models include infections by arenaviruses of which LCM virus is an example. Immunosuppression of T cell responses in adults or infected neonates prevents the development of CNS disease. In addition, disease is transferable by immune T lymphocytes but not by immune serum. The presumed mechanism is destruction of infected CNS tissue by immune lymphocytes (Nathanson et al., 1975). In another model, murine encephalomyelitis virus (Theiler virus) is associated with an initial virus-induced flaccid paralysis secondary to grey matter involvement. In animals which recover, there is persistence of virus and then primary demyelination in the spinal cord. The latter effects can be prevented by immunosuppression (Lipton & Del Canto, 1976). Well described clinical examples of virus-associated, immune-mediated damage include post-infectious or post-vaccinial encephalomyelitis and 'virally triggered' damage to peripheral nerves (Guillain-Barré). The mechanisms of injury in these instances are unknown. It is possible that shared antigenic structures between virus and portions of the central nervous system, or alternatively, imbalances of immunoregulation caused by the virus, could be related to these immunological events.

Studies with the reoviruses have indicated one possible way in which such sharing may occur. The neutralization epitope of the reovirus type 3 HA is thought to bind to cell-surface receptors. This epitope is defined by the idiotype (antigen binding site) of certain monoclonal antibodies (Burstin, et al., 1982). Anti-idiotypic antibodies have been isolated which bind in a manner similar to the reovirus HA (Nepom et al., 1982). These anti-idiotypic antibodies bind to, and thus could lead to injury in, the same uninfected cells as those to which reovirus type 3 attaches – notably, the antibodies bind to neuronal and lymphoid cells. It is likely that similar sharing of antigenic structures will be found for a number of infectious agents.

WHY DO SOME VIRUSES PERSIST?

Numerous theories have been advanced to explain the capacity of a number of different virus infections to persist in the host, often in the face of a highly effective immune response. In the case of DNA viruses and RNA viruses that contain reverse transcriptase, DNA copies of the viral genome are produced which can be integrated into the host cells as 'cellular' genes (retroviruses) or which can replicate as extrachromosomal episomes (herpes viruses). Thus, such viruses can exist in incomplete or latent forms and can periodically reactivate complete virus when the host immune response fails. Herpes simplex and varicella virus reside in sensory ganglia in the central nervous system, presumably as such incomplete virions. The retrovirus visna has been recovered from the choroid plexus of infected animals and undergoes sequential changes during the persistent infection, resulting in antigenic mutants that are no longer neutralized by the animals' sera (Narayan, Griffin & Chase, 1977).

Other mechanisms invoked to explain the persistence of normally lytic RNA viruses include mutation, defective interfering (DI) particles, and resistance to interferon (Stevens, Todaro & Fox, 1978). Temperature-sensitive (ts) and other virus mutants have been recovered from numerous persistently infected cell lines *in vitro* (Preble & Younger, 1975) and have been shown to alter disease *in vivo* (Fields, 1972). Data from experiments with the reoviruses suggest that a specific mutation in the viral gene product responsible for inhibition of protein synthesis (*S4* gene) is the critical mutation responsible for initiating a persistent infection (Ahmed & Fields, 1982). Experiments with ts mutants of reovirus *in vivo* indicate that mutants can attenuate an acute efficient lytic infection and allow illness to develop in a more progressive fashion (Fields, 1972). Interestingly, there is an altered host response to the M protein in SSPE (Hall, Lamb & Choppin, 1979; Wechsler, Weiner & Fields, 1979). In addition, there appears to be decreased production of M protein in the brain of SSPE patients (Hall & Choppin, 1979).

DI particles have also been suggested to play a role in persistent infection (Holland *et al.*, 1980). Experiments *in vivo* in mice suggest that acute illness may be modified by DI particles but do not clearly establish a role in persistence.

CONCLUSIONS

In this paper, I have outlined certain features of viral pathogenesis, particularly stressing recent insights into possible mechanisms. I have been selective in order to illustrate the advantages of using a well defined model system. The reoviruses are clearly not the only model system, but they provide a useful framework for approaching the problem of defining virulence and virus–host interactions at a genetic and biochemical level (Fields & Greene, 1982).

To summarize, the encounter between a virus and its host involves a highly complex series of interactions. The end-result of these interactions may result in no impact on the host or, alternatively may result in a severe, or even lethal, infectious disease. The overall process of infection may be divided into a series of stages involving entry into the host, multiplication at a primary site, spread throughout the host, and tropism involving different target tissues. We have been interested in defining the precise viral and host genetic and molecular determinants responsible for each of these stages of pathogenesis. Towards this end, we have utilized the mammalian reoviruses as a model system. The advantage of this system is that there are three serotypes with diverse biological properties. Due to the segmented nature of the viral genome, the reoviruses readily undergo genome segment reassortment following mixed infection. Using such 'intertypic' reassortants, we have been able to identify the viral components responsible for determining the specific features of several stages of virus–host interactions. In each instance, one of the components of the outer capsid has been found to be essential. The outer capsid consists of three polypeptides: the $\sigma1$ polypeptide (the viral HA), the $\mu1C$ protein, and the $\sigma3$ protein.

The viral HA is responsible for cell and tissue tropism and for specificity in cellular and humoral immunity. The $\mu1C$ protein determines the response to proteases, viral clearance, and the capacity to grow in intestinal and nervous tissue. The $\sigma3$ protein is responsible for inhibiting host cell protein and RNA synthesis and plays a critical role in the establishment of persistent infection.

In addition to the three outer capsid polypeptides, the $\lambda2$ core protein, although based in the viral core, is also exposed to the outside of the virus. We have recently found that it is the essential viral component responsible for generating deleting mutants.

These studies with the mammalian reoviruses have led to certain general principles concerning the genetic basis of viral pathogenesis. First, while virus virulence is clearly multigenic, each viral component plays a unique role in the infectious process. Understanding the specific individual genes responsible for different stages of pathogenesis has allowed us to carry out more detailed biochemical and immunological analyses on the exact mechanism. Second, mutations in each of the critical genes can lead to attenuation at different steps during the infectious process. For example, mutations in the receptor interacting protein (HA protein) can selectively alter tropism; mutations in the protease sensitive gene ($\mu1C$ polypeptide) can decrease yields below lethal levels; and mutations in the gene responsible for inhibiting host cell RNA and protein synthesis ($\sigma3$ polypeptide) can lead to viruses with reduced capacity to injure cells.

Thus, it is clear that viruses interact with hosts in a predictable fashion, that such interactions can be studied in a manner analogous to simpler more manipulable genetic systems and finally, that modern technologies as they are applied to the problems of virus–host interactions, will further enhance our understanding of viral pathogenesis.

REFERENCES

AHMED, R. & FIELDS, B. N. (1982). Role of the S4 gene in the establishment of persistent reovirus in L cells. *Cell*, **28**, 605–12.

ALLEN, G. & FANTES, K. H. (1980). A family of structural genes for human lymphoblastoid (leukocyte-type) interferon. *Nature*, **287**, 408–11.

ALLISON, A. C. (1974). Interactions of antibodies, complement components, and various cell types in immunity against viruses and various pyogenic bacteria. *Transplantation Review*, **19**, 3–55.

BAGLIONI, C. (1979). Interferon-induced enzymatic activities and their role in the antiviral state. *Cell*, **17**, 255–64.

BEATTY, B. J., MILLER, B. R., SHOPE, R. E., ROZHON, E. J. & BISHOP, D. H. L. (1982). Molecular basis of bunyavirus per os infection of mosquitoes: role of the middle-sized RNA segment. *Proceedings of the National Academy of Sciences (USA)*, **79**, 1295–7.

BUCHER, D. & PALESE, P. (1975). The biologically active proteins of influenza virus: neuraminidase. In *The Influenza Viruses and Influenza*, ed. E. D. Kilbourne, pp. 83–123. New York: Academic Press.

BURSTIN, S. J., SPRIGGS, D. R. & FIELDS, B. N. (1982). Evidence for functional domains on the reovirus type 3 hemagglutinin. *Virology*, **117**, 146–55.

COOPER, N. R. (1979). Humoral immunity to viruses. *Comprehensive Virology*, **15**, 123–70.

COULON, P., ROLLIN, P., AUBERT, M., & FLAMAND, A. (1982). Molecular basis of rabies virus virulence. I. Selection of avirulent mutants of the CVS strain with anti-G monoclonal antibodies. *Journal of General Virology*, **61**, 97–100.

DIETZCHOLD, B., WUNNER, W. H., WIKTOR, I. J., LOPES, A. D., LAFON, M., SMITH, C. L. & KOPROWSKI, H. (1983). Characterization of an antigenic determinant of the glycoprotein that correlates with pathogenicity of rabies virus. *Proceedings of the National Academy of Sciences (USA)*, **80**, 70–4.

FENNER, F., MCAUSLAN, B. R., MIMS, C. A., SAMBROOK, J. & WHITE, D. O. (1974). *The Biology of Animal Viruses*. New York: Academic Press.

FIELDS, B. N. (1972). Genetic manipulation of reovirus type 3: a model for alteration of disease? *New England Journal of Medicine*, **287**, 1026–33.

FIELDS, B. N. (1981). Genetics of reovirus. *Current Topics in Microbiology and Immunology*, **91**, 1–24.

FIELDS, B. N. (1982). Molecular basis of reovirus virulence. *Archives of Virology*, **71**, 95–107.

FIELDS, B. N. & GREENE, M. I. (1982). Genetic and molecular mechanisms of viral pathogenesis: implications for prevention and treatment. *Nature*, **300**, 19–23.

FINBERG, R., SPRIGGS, D. R. & FIELDS, B. N. (1982). Host immune response to reovirus: CTLs recognize the neutralization domain of the viral hemagglutinin. *Journal of Immunology*, **129**, 2235–8.

GERHARD, W., YEWDELL, J., FRANKEL, M. E. & WEBSTER, R. (1981). Antigenic structure of influenza virus hemagglutinin defined by hybridoma antibodies. *Nature*, **290**, 713–16.

GREENE, M. I. & WEINER, H. C. (1980). Delayed hypersensitivity in mice infected with reovirus. II. Induction of tolerance and suppressor T cells to virus specific gene products. *Journal of Immunology*, **125**, 283–90.

GRIFFIN, D. E., MULLINIX, J., NARAYAN, O. & JOHNSON, R. T. (1974). Age dependence of viral expression: comparative pathogenesis of two rodent-adapted strains of measles virus in mice. *Infection and Immunity*, **6**, 690–5.

HALL, W. W. & CHOPPIN, P. W. (1979). Evidence for the lack of synthesis of the M polypeptide of measles virus in brain cells of SSPE. *Virology*, **99**, 443–50.

HALL, W. W., LAMB, R. A. & CHOPPIN, P. W. (1979). Measles and subacute sclerosing panencephalitis virus proteins: lack of antibodies to M protein in patients with subacute sclerosing panencephalitis. *Proceedings of the National Academy of Sciences (USA)*, **76**, 2047–51.

HOLLAND, J. J., KENNEDY, I. T., SENLER, B. L., JONES, C. L., ROUX, L. & GRAKAU, E. A. (1980). Detective interfering RNA viruses and the host-cell response. *Comprehensive Virology*, **16**, 137–92.

HOLLAND, J. J. & McLAREN, L. C. (1961). The location and nature of enterovirus receptors in susceptible cells. *Journal of Experimental Medicine*, **114**, 161–71.

JOHNSON, R. T. & GRIFFIN, D. E. (1978). Pathogenesis of viral infections. In *Handbook of Clinical Neurology*, vol. 34, ed. P. J. Vinken & G. W. Bruyn, pp. 15–37. Amsterdam: North Holland.

KLENK, H. D. (1980). Viral glycoproteins: initiators of infection and determinants of pathogenicity. In *The Molecular Basis of Microbial Pathogenicity*, ed. H. Smith, J. J. Skehel & M. J. Turner, pp. 55–66. Weinheim: Verlag Chemie.

LETVIN, N. L., KAUFFMAN, R. S. & FINBERG, R. (1981). T lymphocyte immunity to reovirus: Cellular requirements for generation and role in clearance of primary infections. *Journal of Virology*, **127**, 2334–9.

LIPTON, H. L. & DEL CANTO, M. C. (1976). Theiler's virus-induced demyelination: prevention by immunosuppression. *Science*, **192**, 62–4.

MANDEL, B. (1979). Interactions of viruses with neutralizing antibody. *Comprehensive Virology*, **15**, 37–106.

MEAGER, A. & HUGHES, R. C. (1977). Virus receptors. In *Receptors and Recognition*, vol. 4, ed. P. Cuatrecasas & M. F. Greaves, pp. 141–95. London: Chapman & Hall.

MIMS, C. A. (1964). Aspects of the pathogenesis of virus diseases. *Bacteriological Review*, **28**, 30–71.

NARAYAN, O., GRIFFIN, D. E. & CHASE, J. (1977). Antigenic shift of visna virus in persistently infected sheep. *Science*, **197**, 376–8.

NATHANSON, N., MONJAN, A. A., PANITCH, H. S., JOHNSON, E. D., PETURSSON, G. & COLE, G. A. (1975). Virus-induced cell-mediated immunopathological disease. In *Viral Immunology and Immunopathology*, ed. A. Notkins, pp. 357–91. New York: Academic Press.

NEPOM, J. T., WEINER, H., DICHTER, M., SPRIGGS, D., GRAMM, C., POWERS, M. L., FIELDS, B. N. & GREENE, M. I. (1982). Identification of a hemagglutinin specific idiotype associated with reovirus recognition shared by lymphoid and neural cells. *Journal of Experimental Medicine*, **155**, 155–67.

OLDSTONE, M. B. A., PERRIN, L. H. & WELSH, R. M. (1976). Potential pathogenic mechanisms of injury in amyotrophic lateral sclerosis. In *Amyotrophic Lateral Sclerosis*, ed. J. M. Andrews, R. T. Johnson, M. A. B. Brazier, pp. 251–62. New York: Academic Press.

PREBLE, O. T. & YOUNGER, J. S. (1975). Temperature-sensitive viruses and the etiology of chronic and inapparent infections. *Journal of Infectious Disease*, **131**, 467–73.

ROSENSTREICH, D. L., O'BRIEN, A. D., GROVES, M. G. & TAYLOR, B. A. (1980). Genetic control of natural resistance to infection in mice. In *The Molecular Basis of Microbial Pathogenicity*, ed. H. Smith, J. J. Skehel, & M. J. Turner, pp. 101–14. Weinheim: Verlag Chemie.

RUBIN, D. H. & FIELDS, B. N. (1980). The molecular basis of reovirus virulence: the role of the M2 gene. *Journal of Experimental Medicine*, **152**, 853–68.

RUBIN, D., WEINER, H. L., FIELDS, B. N. & GREENE, M. I. (1981). Immunologic tolerance following oral administration of reovirus: requirement for two viral gene products for tolerance induction. *Journal of Immunology*, **127**, 1697–701.

SCHREIBER, D. S., BLACKLOW, N. R. & TRIER, J. S. (1973). The mucosal lesion of the proximal small intestine in acute infectious non-bacterial gastroenteritis. *New England Journal of Medicine*, **288**, 1318–23.

SHARPE, A. H., CHEN, L. B. & FIELDS, B. N. (1982). The interaction of mammalian reoviruses with the cytoskeleton of monkey kidney CV-1 cells. *Virology*, **120**, 399–411.

SHARPE, A. H. & FIELDS, B. N. (1982). Reovirus inhibition of cellular RNA and protein synthesis: role of the S4 gene. *Virology*, **122**, 381–91.

SPRIGGS, D. R., BRONSON, R. & FIELDS, B. N. (1983). Hemagglutinin variants of reovirus type 3 show altered tropism. *Science*, **220**, 505–7.

SPRIGGS, D. R. & FIELDS, B. N. (1982). Generation of attenuated reovirus type 3 strains by selection of hemagglutinin antigenic variants. *Nature*, **297**, 68–70.

STEVENS, J. G., TODARO, G. J. & FOX, C. F. (1978). *Persistent Viruses*. New York: Academic Press.

STEWART, W. E. II (1979). *The Interferon System*. Weinheim: Springer-Verlag.

TAMM, I. (1975). Cell injury with viruses. *American Journal of Pathology*, **81**, 163–77.

TRACHSEL, H., SONENBERG, N., SHATKIN, A. J., ROSE, J. K., LEONG, K., BERGMAN, J. E., GORDON, J. & BALTIMORE, D. (1980). Purification of a factor that restores translation of vesicular stomatitis virus mRNA in extracts from polio-virus-infected Hela cells. *Proceedings of the National Academy of Sciences (USA)*, **77**, 770–4.

WEBSTER, R. G. & LAVER, W. G. (1980). Determination of the number of nonoverlapping antigenic areas on Hong Kong (1T3N2) influenza virus hemagglutinin with monoclonal antibodies and the selection of variants with potential epidemiological significance. *Virology*, **104**, 139–48.

WECHSLER, S. L., WEINER, H. L. & FIELDS, B. N. (1979). Immune response in subacute sclerosing panencephalitis: reduced antibody response to the matrix protein of measles virus. *Journal of Immunology*, **123**, 884–9.

WEINER, H. L., DRAYNA, D., AVERILL, D. R. & FIELDS, B. N. (1977). Molecular basis of reovirus virulence: roles of the S1 gene. *Proceedings of the National Academy of Sciences (USA)*, **74**, 5744–8.

WEINER, H. L. & FIELDS, B. N. (1977). Neutralization of reovirus: the gene responsible for the neutralization antigen. *Journal of Experimental Medicine*, **146**, 1303–10.

WEINER, H. L., POWERS, M. L. & FIELDS, B. N. (1980). Reovirus virulence and central nervous system cell tropism: absolute linkage to the viral hemagglutinin. *Journal of Infectious Disease*, **141**, 609–16.

WOLF, J. L., RUBIN, D. H., FINBERG, R., KAUFMAN, R. S., SHARPE, A. H., TRIER, J. S. & FIELDS, B. N. (1981). Intestinal M cells: a pathway for entry of reovirus into the host. *Science*, **212**, 471–2.

ZINKERNAGEL, R. M. (1976). Cellular immune response to viruses and the biologic role of polymorphic major transplantation antigens. *Comprehensive Virology*, **15**, 171–204.

ZINKERNAGEL, R. M. & DOHERTY, P. C. (1974). Restriction of *in vitro* T cell mediated cytotoxicity in lymphocytic choriomeningitis within a syngeneic or semiallogeneic system. *Nature*, **248**, 701–2.

VIRUSES AND HUMAN CANCER

ROBIN A. WEISS

Institute of Cancer Research, Chester Beatty Laboratories, Fulham Road, London SW3 6JB, UK

INTRODUCTION

Viruses are not generally cited as important environmental causes of human cancer, yet as carcinogens they play a much larger role than, say, industrial chemicals. To take two examples: hepatocellular carcinoma is a major source of cancer mortality on a world-wide scale and nasopharyngeal carcinoma is the commonest fatal cancer amongst the southern Chinese. These cancers, as well as several other types of malignancy, almost certainly depend on a transmissible virus as a crucial factor in their development. The control of viral infection, by immunization for instance, would make a major impact on global cancer incidence.

Several kinds of virus can cause cancer in humans and animals. These viruses induce malignancy by different mechanisms, and interact with different environmental and host cofactors. Most human tumour viruses have been identified only after epidemiological studies indicated the importance of a transmissible agent, and before related viruses of animals came to light (an exception being the T-cell leukaemia retrovirus). In some cases it has been difficult to prove the oncogenicity of these viruses because malignancy is a rare consequence of infection occurring many years after initial infection, and because susceptible animals have not been available for experimental studies. Paradoxically, the two classes of human virus that are most highly oncogenic in animals, namely adenoviruses and the polyoma virus BK, have not been definitely associated with any human malignancy. It is not easy, however, to establish the involvement of a virus in human cancer. While the persistence and expression of viral genes in tumour cells provides a strong indication of a direct role in the maintenance of the malignant state, some viruses might conceivably act as 'hit-and-run' carcinogens, and others might persist as innocent passengers. Generally speaking, human viruses parasitize the majority of their hosts with minimum morbidity. Neither virulent infection nor neoplasia are common sequelae;

they typically result from predisposing factors, of which some kind of cellular immune dysfunction is the most notable condition.

The large investment of research effort over the last 25 years into the study of animal tumour viruses has greatly enhanced our understanding of the molecular and cellular processes involved in malignant transformation. For example, the elucidation of retrovirus oncogenes originally derived from the host genome has revealed a set of genes that may play a role in many human cancers which do not result from virus infection. This fascinating story is told by J. M. Bishop (this volume). By contrast, the contribution of animal virus studies to understanding human viral carcinogenesis has been relatively meagre. In this paper, I shall briefly review the natural history of human cancers believed to have a viral aetiology, and indicate some of the complexities of viral carcinogenesis.

ONCOGENIC HUMAN VIRUSES

Table 1 lists eight kinds of virus which are thought to act as aetiological agents in human cancer or which have been shown to possess oncogenic properties under experimental conditions. These potentially oncogenic viruses represent all the major virus families containing DNA, except the pox group, but only one RNA virus, a retrovirus. This reflects the families of viruses associated with cancer in animals, though pox viruses should then be added to the catalogue.

Three of the viruses listed Epstein-Barr virus (EBV), cytomegalovirus (CMV) and herpes simplex virus (HSV) belong to the herpesvirus family. Compelling evidence of cancer aetiology is confined to Epstein-Barr virus for nasopharyngeal carcinoma, with a very close association between EBV and lymphomas and between CMV and Kaposi's sarcoma. It now seems doubtful whether herpes simplex virus type II is aetiologically associated with cervical cancer. A fourth human herpes virus, varicella-zoster virus (VZV), has not been associated with malignant disease, although it is an ubiquitous infection causing chicken pox. Like HSV, after primary infection VZV remains latent throughout life in sensory nerve ganglia, with sporadic activation to cause shingles.

From time to time the question has been raised whether oncogenic animal viruses might cause tumours in humans, but no convincing evidence has accrued to support this notion. Butchers'

Table 1. *Human viruses with oncogenic properties*

Virus		Associated human neoplasm	Predisposing cofactor
Hepadna	Hepatitis B	Hepatocellular carcinoma	Alcohol, aflatoxin?
Herpes	Epstein-Barr	Burkitt's lymphoma Immunoblastic lymphoma Nasopharyngeal carcinoma	Malaria Immunodeficiency HLA genotype
	Simplex, types I, II	Cervical neoplasia?	
	Cytomegalo	Kaposi's sarcoma Prostatic neoplasia?	Immunodeficiency, HLA genotype
Retro	T-cell leukaemia	Adult T-cell leukaemia–lymphoma	
Papova	Papilloma	Warts Cervical neoplasia Skin cancer	Genetic disorders, sunlight
	Polyoma, JC, BK	Neural tumours?	
Adeno	Types 2, 5, 12	None	

warts are caused by a human rather than a bovine papilloma virus. Certain murine and feline leukaemogenic retroviruses can infect human cells *in vitro*, but serological evidence of natural infection is lacking so far. Human T-cell leukaemia–lymphoma virus (HTLV) may have evolved from a simian prototype but it is clearly transmitted by human-to-human infection, rather than as a zoonosis. The consequences of iatrogenic infection by the simian polyoma virus (SV40) as a contaminant of polio virus is still being evaluated and will be discussed under human polyoma viruses.

Each of the oncogenic human viruses will now be briefly reviewed.

Hepatitis B virus

The involvement of hepatitis B virus (HBV) in the development of hepatocellular carcinoma was first suggested in the 1950s (see Blumberg & London, 1981) and has been considerably strengthened in subsequent studies (reviewed by Szmuness, 1978; Zuckerman, 1982). More recently, related hepatitis viruses of ducks, woodchucks and ground squirrels have been shown to induce liver cancer in infected animals (Summers & Mason, 1982), and these will provide useful models for viral liver carcinogenesis.

Epidemiologically, primary liver cancer is most prevalent in those parts of the world and those communities where HBV is most common (Beasley, 1982), so the virus, rather than a cofactor, determines 'endemic' areas of cancer incidence. Nevertheless, only a small minority of patients with chronic HBV infection develop liver cancer. Early studies relied on the presence of HBV surface ('Australia') antigen as evidence of chronic infection. However, except for a small number of persistent HBV carriers, most patients eventually clear surface antigen from the plasma, and the only serological evidence of infection (which may remain latent in the liver) is the presence of antibody to viral core antigen. A detailed study of over 400 cases of hepatocellular carcinoma in Japan (Okuda et al., 1982) showed that, while fewer than 10% of patients secreted surface antigen, more than 90% expressed core antibodies, considerably strengthening the association between HBV infection and this malignancy. Longitudinal serological studies in Japan (Obata et al., 1980), Taiwan (Beasley et al., 1981) and Alaska (Heyward et al., 1982) indicate that HBV infection may occur as little as 10 years before tumour presentation, although the time lapse is often much longer, and infection is frequent in infancy.

The HBV genome can be demonstrated in the tumour cells (Summers et al., 1978) including integrated genomes (Shafritz et al., 1981). Presumably HBV DNA is also present in many non-malignant and pre-malignant cells in chronically infected liver. Hepatocellular carcinoma without apparent HBV infection occurs with a non-endemic, low incidence. If (like EBV-negative Burkitt's lymphoma) these tumours prove to lack viral DNA sequences, they may represent an alternative pathway of carcinogenesis.

More detailed studies are needed on the role of possible cofactors to HBV in hepatocellular carcinoma. For example, aflatoxins are known to act as potent liver carcinogens in rats; aflatoxins are produced by moulds growing on poorly stored grain and groundnuts, which form a substantial part of the human diet in some areas where HBV infection is prevalent (Lutwick, 1979). On the other hand, aflatoxin contamination of food is infrequent in Greece and also in Alaska, although both HBV infection and liver cancer occur at relatively high incidence in these countries (Trichopoulos, 1981; Heyward et al., 1982). Aflatoxin may act independently of HBV and be more relevant to HBV-negative liver cancer. Alcohol is also recognized as a weakly linked cofactor (Trichopoulos, 1981), possibly acting as a tumour promoter rather than as a primary carcinogen. Ohnishi, *et al.*, (1982) have reported that the mean age of presentation with primary liver cancer of HBV-positive patients was 9 years younger for heavy drinkers than for non-drinkers. Another possible cofactor that merits investigation is the Delta agent, a satellite virus associated with HBV in some patients. Delta virus increases the severity of HBV acute infection (Smedile *et al.*, 1982), but its relationship (if any) to hepatocellular carcinoma has not been studied.

Although primary liver cancer is relatively rare in the West, it is one of the most frequent carcinomas world-wide, particularly around the Pacific and Indian Ocean basins. We may therefore consider HBV to be a major environmental carcinogen. The development of mass immunization will be of great importance in preventing the morbidity of acute HBV infection and in reducing the incidence of liver cancer.

Epstein-Barr virus

Epstein-Barr virus (EBV) is the cause of infectious mononucleosis, as well as being strongly associated with three kinds of malignancy, Burkitt's lymphoma (Burkitt, 1962), immunoblastic lymphoma in immunodeficient patients, and nasopharyngeal carcinoma (reviewed by Epstein & Achong, 1979; de Thé, 1980). EBV is an example of an ubiquitous human herpes virus which usually infects infants with sub-clinical effects. Some B cells, and also probably some epithelial cells in the throat, remain chronically but latently infected throughout life, with sporadic activation of virus production. It is those few individuals not infected as infants who are at risk of developing infectious mononucleosis as adolescents or young adults, when infected with high virus doses in the saliva of their partners. Patients who have suffered infectious mononucleosis are at three-fold greater risk in the ensuing 5 years of developing Hodgkin's lymphoma, but it is not thought that EBV itself is causally related to Hodgkin's disease.

EBV was first discovered in Burkitt's lymphoma (BL) cells grown in culture (Epstein, Achong & Barr, 1964), and almost all African cases of BL harbour the virus and express its nuclear antigen in the tumour cells. The endemic areas where the incidence of Burkitt's lymphoma is remarkably high are tropical Africa and New Guinea.

In contrast to the geographic distribution of hepatocellular carcinoma, the areas endemic for BL reflect the presence of a major cofactor, holoendemic malaria, rather than the distribution of the putative causative virus. Longitudinal studies in Uganda indicate that susceptible children become infected very early in life, and it is thought that the effect of chronic malarial infestation is immunosuppressive for cytotoxic T-cells directed to EBV-induced cell surface antigens. Humoral (IgG) immunity to viral capsid antigens remains elevated in children at risk for Burkitt's lymphoma (de Thé, 1980). Children (largely boys) are at greatest risk at 6–9 years of age and the tumour most frequently develops in the jaw.

Most individuals have persistent EBV-infected B cells which are potentially 'immortal' when grown in culture. Experimental infection of umbilical cord B cells by EBV induces indefinite proliferation, showing that EBV can transform cells *in vitro* (Miller, 1980). Immortalization of B cells by EBV, however, is not sufficient to render these cells 'fully malignant'; they seldom grow as progressive tumours in nude mice whereas BL cells do so readily. Further changes in BL cells which appear to be closely associated with the development of a truly malignant phenotype are the specific chromosome translocations (Nilsson, 1982) that bring the *myc* oncogene on chromosome 8 into juxtaposition with actively expressed immunoglobulin genes on chromosomes 14, 2 or 22 (Klein, 1983; Bishop, this volume).

Burkitt's lymphoma cells do not invariably carry evidence of EBV infection. In the endemic area in tropical Africa, 98% of BL contain EBV in the lymphoma cells. BL also occurs sporadically as a much rarer malignancy outside the endemic areas, and the proportion of EBV-positive malignancies varies greatly from one geographic region to another. In the USA, only 15–20% of cases have EBV in the BL cells although more than 70% of the BL patients have been infected with EBV. The 2% incidence of EBV-negative BL in the African endemic area actually reflects the same absolute incidence in the population as BL in the USA. In North Africa, at least 90% of the 'sporadic' BL cases are EBV-positive, but the overall incidence is not accurately known; BL in this region might reasonably be considered semi-endemic (G. Lenoir, personal communication).

It therefore appears that EBV-independent BL occurs throughout the world in a non-clustered way. EBV-negative BL is histopathologically indistinguishable from the EBV-positive disease, and has the same specific chromosome translocations. However, EBV-negative BL presents in all age groups and is not typically located in the jaw. The occurrence of EBV-negative BL raises the question whether the virus plays any direct role in oncogenesis, other than expanding the population of target cells for malignant conversion dependent on chromosome translocation (Klein, 1983). However, the ability of EBV to immortalize cells *in vitro* or *in vivo* strongly suggests that EBV has a direct effect on an early step in neoplastic transformation, although viral transformation apparently is not the only pathway to BL oncogenesis.

Immunoblastic lymphoma (reticulum cell sarcoma) develops in patients as a result of deficient cellular immune responses to EBV-transformed cells. As discussed below, this type of tumour is prevalent in immunosuppressed recipients of organ transplants (Kinlen, 1982; Penn, 1982) and also in individuals with certain inherited immunodeficiencies (Purtilo, 1981). The emergence of a monoclonal tumour frequently occurs after polyclonal proliferation of EBV-infected lymphoblasts (Hanto *et al.*, 1982). The polyclonality of the initial immunoblastic proliferation (resembling infectious mononucleosis) has led some investigators to question whether this disease should be regarded as a true malignant lymphoma; suffice it to say that the neoplasia in polyclonal or monoclonal manifestation is frequently fatal in immunodeficient patients.

Nasopharyngeal carcinomas (NPC) of the undifferentiated type are also strongly associated with EBV (de Thé, 1980). This tumour is particularly prevalent in southern provinces of China and among ethnic southern Chinese communities throughout southeast Asia. Studies of Chinese patients in Singapore (Simons *et al.*, 1978) show an association of the disease with certain HLA haplotypes. The tumour also occurs at a relatively high incidence in the southern Mediterranean and in East Africa, particularly in Uganda and southern Sudan. The age-specific incidence of NPC in Chinese has a single broad peak between 45 and 54 years, whereas in North Africa it is bimodal, some 20% cases presenting at 10–19 years and the remainder mainly at 50–60 years.

The association between NPC and EBV emerged serendipitously when Old *et al.* (1966) observed high serum antibody titres to the EBV antigens in some NPC patients used as controls to BL sera. Although Henderson *et al.* (1976) have questioned the relationship between NPC and EBV on the basis of serology, case-control studies within defined geographical areas indicate the presence of significantly elevated titres of serum antibodies to capsid antigen, early antigen and nuclear antigen in NPC patients (de Thé, Lavoué & Muenz, 1978). Furthermore, the presence of anti-EBV IgA in saliva is restricted to NPC patients (Henle & Henle, 1976), although many healthy EBV carriers release virus and occasionally IgG into oral secretions. Salivary IgA has been developed as a useful marker in mass screening for early NPC (Zeng et al., 1980).

EBV DNA and nuclear antigen were first detected in the carcinoma cells by Klein et al. (1974). It has since been found that all cases of undifferentiated NPC, irrespective of ethnic origin, express EBV nuclear antigen, and the virus clearly must be an essential factor in this type of nasopharyngeal carcinogenesis. Direct transformation in vitro of epithelial cells by EBV, however, has not been achieved.

Herpes simplex virus

Herpes simplex virus (HSV) has been intensively studied as a potentially oncogenic virus. HSV type I is a very common orolabial infection which gives rise to cold sores. The virus becomes latent in sensory (trigeminal and facial) ganglia and may break out into new cold sores at times of stress or sunburn. HSV type II is a genital infection which is becoming increasingly frequent. HSV-I also occurs genitally as a sexually transmitted infection and either type can cause painful lesions in both sexes. Although evidence of HSV-II infection is frequently found in cervical cancer patients, a careful analysis of British women attending venereal clinics shows that HSV is not more closely associated with neoplasia than are the other sexually transmitted infections (Franceschi et al., 1983).

HSV-I and -II genome fragments are able, at very low efficiency, to transform hamster and human fibroblasts in culture into cells resembling neoplastic cells (Duff et al., 1974), and a specific transforming gene may have been identified (Camacho & Spear, 1978). RNA transcripts specific to HSV have been found in cervical intraepithelial neoplasia (CIN) or invasive carcinoma cells (Eglin et al., 1981), yet it has proved very difficult to detect integrated or free HSV DNA in the same tissues, even using as a probe the transforming fragment of DNA identified by Camacho & Spear. This has led Galloway & McDougall (1983) to support a 'hit-and-run' theory of HSV oncogenesis which does not require persistence of viral DNA in the carcinoma cells. Zur Hausen (1982) has further postulated that HSV may act synergistically with papilloma virus in causing cervical cancer. However, it is simpler to conclude, in agreement with epidemiological findings (Franceschi *et al.*, 1983), that HSV does not play a significant role in cervical cancer.

Human cytomegalovirus

Cytomegalovirus (CMV) is a herpes virus that can cause damaging congenital and perinatal infection (Stagno *et al.*, 1982), and a severe form of mononucleosis in immunosuppressed patients. In most healthy individuals, however, it occurs like so many chronic virus infections as a life-long commensal entity. Human CMV infection occurs world-wide, varying between 20% and 98% in different communities.

Experiments with partially inactivated CMV show that it can transform human fibroblasts into a neoplastic phenotype (Geder, Kneider & Rapp, 1977), though the transforming genes remain to be identified. Related animal CMV strains are not known to be naturally oncogenic. In humans, CMV is strongly associated with Kaposi's sarcoma (Giraldo *et al.*, 1975; Boldogh *et al.*, 1981) and may also be involved in neoplasia of the prostate (Boldogh *et al.*, 1983).

Kaposi's sarcoma is a rare malignancy in the west, occurring mainly in patients of Ashkenazi Jewish or of Mediterranean (southern Italy and Greece) origin (Giraldo *et al.*, 1975), but an aggressive form of the tumour is relatively common in tropical Africa (Slavin, Cameron & Singh, 1969). As Boldogh *et al.* (1981) have detected CMV antigens, RNA, DNA and virus particles in many Kaposi's sarcoma tissues *in situ*, it is generally thought that CMV is aetiologically associated with this tumour. Further studies are needed to investigate the prevalence of CMV infection in normal tissues, including endothelial cells from which Kaposi's sarcoma cells are probably derived. The clustering of Kaposi's sarcoma incidence is not so much related to virus prevalence as to unknown factors probably including deficient immune responses and the genetic constitution of the host population.

A highly malignant form of Kaposi's sarcoma has recently become evident as a frequent consequence of acquired immune deficiency syndrome (AIDS). This disease has arisen as a new epidemic among male homosexuals with multiple sexual partners (Drew *et al.*, 1982; Waterson, 1983) and is now spreading to female prostitutes, recipients of blood transfusions or blood products

(Ragni *et al.*, 1983), Haitian immigrants (Viera *et al.*, 1983), and other groups. AIDS patients may be especially susceptible to certain CMV-infected cells emerging as Kaposi's sarcomas (see discussion of immunodeficiency below). Kaposi's sarcoma in AIDS presents as diffuse, multicentric lesions. Costa & Rabson (1983) have pointed out that it might represent a benign, polyclonal proliferative disease of the endothelium, comparable to the progressive, EBV-induced, infectious mononucleosis seen in immunosuppressed patients, from which malignant clones might later emerge. On the other hand, Kaposi's sarcoma may be clonal from its first appearance, and might even be acquired as a homograft as discussed later. Patients with Kaposi's sarcoma, whether sporadic or as a result of AIDS, have a higher frequency of DR5 histocompatibility antigens than ethnically matched healthy individuals or AIDS patients without the sarcoma (Friedman-Kien *et al.*, 1982). One might also postulate that African Kaposi's sarcoma could be a manifestation of AIDS, endemic in a region where the opportunistic infections and other symptoms might not be recognized or recorded.

The agent causing AIDS remains to be identified; it is probably not CMV itself, although active CMV infection may exacerbate the disease. Recent publications (Essex *et al.*, 1983; Gallo *et al.*, 1983) suggest a link between AIDS and a retrovirus related to HTLV (described below). At the time of writing, however, it is not clear whether the retrovirus is the aetiological agent or an opportunistic infection.

Retroviruses

Retroviruses have long been recognized as the causative agents of leukaemias and lymphomas in numerous vertebrate species ranging from fish to apes, and also of mammary carcinoma in mice. Yet despite intensive research and many premature claims, human oncogenic retroviruses remained elusive entities until very recently (Weiss, 1982*a*). Three classes of human retrovirus have been described. First, there exist DNA elements in the human genome which are structurally related to endogenous proviruses of other mammalian species. These are expressed in the placenta, ovarian follicles and possibly in proliferative diseases of the haematopoietic system, but they have not been causally related to leukaemia or other malignancies. Second, exogenous foamy viruses are wide-spread in humans and monkeys as latent, non-pathogenic infections of the central nervous system. Third, convincing evidence has accrued for the involvement of an infectious C-type retrovirus in a form of mature T-cell leukaemia.

Adult T-cell leukaemia–lymphoma (ATLL) was first identified in southwestern Japan (Uchiyama *et al.*, 1977; Hanaoka, Takatsuki & Shimoyama, 1982) as a malignancy of mature T-cells distinguishable from classical mycoses fungoides/Sezary's syndrome in the aggressive course of the disease, visceral involvement, and frequent hypercalcaemia. More recently, further endemic areas have been discovered, notably the Caribbean basin and Central America following the appearance of ATLL in British patients born in the West Indies (Blattner *et al.*, 1982; Catovsky *et al.*, 1982). A C-type retrovirus known as human adult T-cell leukaemia virus (HTLV-I or ATLV) appears to be causally associated with this disease. The virus was first isolated by Poiesz *et al.* (1980, 1981) in lymphoma cell lines derived from American patients and subsequently isolates were characterized in Japan (Yoshida, Miyoshi & Hinuma, 1982). HTLV-I is distinct from all previously isolated animal viruses, although it is distantly related to bovine leukaemia virus. The US and Japanese HTLV isolates appear to be members of the same virus family (Popovic *et al.*, 1982). Our own studies of HTLV envelope antigens and of neutralizing antisera in ATLL patients (Nagy *et al.*, 1983; Clapham, Nagy & Weiss, 1984) show that the British, US and Japanese viruses belong to one major serotype. A second serotype, HTLV-II, readily distinguishable from HTLV-I, has been found in a single American patient with a T-cell variant of hairy cell leukaemia (Kalyanaraman *et al.*, 1982). There is also serological and virological evidence of viruses related to HTLV in old world monkeys and apes.

Various retroviruses are transmitted through contagion, insect vectors, congenital infection, milk transmission, or through endogenous Mendelian proviral elements in the germ line of the host (Weiss, 1982*b*). The mode of transmission of HTLV is not yet known. It is not an endogenous genome in man (Reitz *et al.*, 1981) and therefore must be acquired by infection, probably by the transfer of infected cells. As with the other human tumour viruses, malignancy is a rare and late consequence of infection. In the endemic areas of Japan, the incidence of infection can be as high as 25% (Hinuma *et al.*, 1982). The number of seropositive individuals increases linearly with age throughout life and is frequently clustered in families (Tajima *et al.*, 1982). Where one spouse is a carrier, spread to the other appears to be more frequent in wives than husbands, suggesting male-to-female sexual transmission. Clinical presentation with the malignancy is more prevalent in the summer, suggesting some seasonal cofactor such as mosquito-borne micro-filariasis (Tajima *et al.*, 1981). But, like hepatocellular carcinoma (and in contrast to Burkitt's lymphoma), the geographical clustering of ATLL reflects the incidence of primary virus infection rather than a secondary cofactor.

HTLV-related viruses may be associated with a broader range of T-cell tumours than those strictly classified as adult T-cell leukaemia–lymphoma including cases arising outside the known endemic areas. A small proportion of cases of classical mycoses fungoides/Sezary's syndrome and other mature T-cell malignancies have integrated HTLV DNA sequences in their tumour cells, including uncultivated peripheral blood cells (Wong-Staal *et al.*, 1983). Detailed surveys of mature T-cell malignancies are being undertaken to elucidate the significance of HTLV in atypical disease.

As mentioned in relation to Kaposi's sarcoma, a virus similar to HTLV-I has been found in several AIDS patients (Essex *et al.*, 1983; Gallo *et al.*, 1983) and should be considered as a possible aetiological agent. HTLV could plausibly cause neoplasia of a T-cell subset in one patient and induce lymphocytopaenia or T-cell aplasia in another, by analogy to avian and feline leukosis viruses which are known to cause either lymphoma or lymphoid atrophy.

Like EBV, HTLV has the capacity to transform lymphocytes, usually T-cells, into 'immortal' cell lines following infection *in vitro* (Miyoshi *et al.*, 1981; Popovic *et al.*, 1983). Cell-free preparations have very low infectivity and best results have been obtained by cocultivation of virus-producing cells with T-cells derived from umbilical cord, bone marrow, or peripheral blood. There is concern that HTLV may be unwittingly transmitted through transfusion of blood or blood products (Miyoshi *et al.*, 1982; Saxinger & Gallo, 1982).

Human papillomaviruses

There are several distinct strains of human papilloma virus (HPV) and more strains probably remain to be discovered and charac-terized (Zur Hausen & Gissmann, 1980). Considerable diversity of antigens and DNA sequences exists between strains so that a general probe for all HPVs does not exist. As with HBV, culture systems for HPV replication have not yet been developed, and this has hampered experimental investigation. With the introduction of gene-cloning methods, HPV DNA sequences can now be propagated. Suitable reagents for probing natural infection are becoming available.

Different HPV strains are associated with different types of wart. HPV 1, 2 and 4 are found in common warts, and HPV 3 in flat warts. HPV 5 and occasionally HPV 3 are found in epidermodysplasia verruciformis (Orth *et al.*, 1978; Green *et al.*, 1982), a hereditary disease leading to the development of multiple skin warts and depressed immune functions (Jablonska, Dabrowski & Jakubowicz, 1972; Prawer *et al.*, 1977). Genital warts (condylomata acuminata are florid growths associated with HPV 6 and 11 (Gissmann *et al.*, 1983) and some oro-laryngeal warts may also be associated with these HPV strains.

Papillomas (warts) are proliferative lesions in epithelia which usually remain benign. Occasionally papillomas progress to form malignant carcinomas (Zur Hausen & Gissmann, 1980). This has not been observed for common and flat warts in healthy individuals, but is frequently seen in patients with epidermodysplasia verruciformis (Jablonska *et al.*, 1972; Ostrow *et al.*, 1982). Malignant conversion has also been observed with orolaryngeal and genital warts (Gissmann *et al.*, 1983). Jarrett and colleagues (1978) have described viral papillomatosis of the alimentary tract of cattle in which a tendency to malignant conversion appears to be greatly increased in animals feeding on bracken fern which contains chemical carcinogens.

Papilloma viruses have emerged as the most likely primary cause of carcinoma of the uterine cervix. Epidemiological studies strongly indicated that a sexually transmissible, probably infectious agent, is a factor in cervical cancer (Kessler, 1976), although a particular virus was not consistently identified. The incidence of cervical cancer is correlated with promiscuous behaviour, including that of the male sexual partner (Skegg *et al.*, 1982). For many years, most emphasis has been placed on herpes simplex type II, but more recently, human papilloma viruses have attracted increasing attention. The recognition of 'flat' cervical warts (Meisels *et al.*, 1981; Reid *et al.*, 1982) has indicated that venereal papilloma virus infection may be far more frequent than formerly thought. Recent epidemiological evidence shows that CIN and invasive carcinoma are more strongly linked with papilloma virus than with any other known venereal infection (Franceschi *et al.*, 1983). HPV DNA (McCance *et al.*, 1983) and viral antigens (Walker *et al.*, 1983) have been detected in CIN and viral DNA has also been detected in malignant carcinoma (Green *et al.*, 1982; Gissmann *et al.*, 1983). The viruses were identified as HPV 6 and HPV 11 (Gissmann *et al.*, 1983; McCance *et al.*, 1983). New strains designated HPV 16 and 18, related to HPV 6 and 11, may be more strongly associated with cervical carcinomas (J. Durst, L. Gissmann, H. Ickenburg & H. Zur Hausen, personal communication).

Human polyomaviruses

Polyomaviruses are small DNA viruses structurally related to papillomaviruses. The best known members of this subfamily are murine polyoma virus, which can give rise to multiple kinds of tumours in mice, and simian vacuolating virus 40 (SV40), which is not known to cause cancer in its natural host, the rhesus monkey, but is highly oncogenic when inoculated into newborn rodents. The molecular biology of polyoma virus and SV40 has been very extensively studied as have the cells transformed by these agents.

Humans are frequent hosts to polyoma viruses related to SV40. The best known strains are BK and JC, named after the initials of the patients from whom they were first isolated. BK virus was detected in the urine of a renal transplant patient (Gardner *et al.*, 1971) and JC virus from the brain of a patient with progressive multifocal leukoencephalopathy (PML) following Hodgkin's disease (Padgett *et al.*, 1971). JC is aetiologically related to PML while BK is not causally associated with disease (Howley, 1980; Padgett, 1980). Both BK and JC are not uncommon active infections in the urinary tract of renal transplant patients. Most children and adults develop antibodies to human polyoma viruses indicating that infection is widespread.

Human polyoma viruses induce malignant sarcomas and neural tumours when inoculated into newborn hamsters and rats. Infection of rodent cells *in vitro* also frequently leads to stable neoplastic transformation. It remains controversial whether BK and JC are oncogenic in humans (Wold *et al.*, 1978; Howley, 1980; Padgett, 1980). Large T antigen and DNA sequences have occasionally been detected in urogenital tumours, meningiomas, gliobastomas and melanomas, but further investigation has not shown these viruses to be consistently or frequently associated with these tumours. However, further studies would be useful to see whether a minority of such tumours have a viral aetiology, particularly melanomas which arise after immunosuppression.

SV40 has been unwittingly inoculated into thousands of human infants as a contaminant of early batches of the Sabin and Koprowski live poliovirus vaccines. It is ironic that regulations prohibiting the propagation of vaccine strains in human cancer cells, such as HeLa, led to the widespread adoption of primary monkey kidney cell culture in which SV40 was latent. An American study of a cohort of children receiving contaminated poliovirus vaccine has not shown any increase in cancer incidence up to the age of 19 years (Mortimer et al., 1981). Unfortunately the study has been discontinued. Another study of recipients of SV40 in vaccine in eastern Europe suggests a higher than expected incidence of melanoma (Geissler et al., 1980), but these data are based on very small numbers of tumours so far.

Human adenoviruses

Finally, mention should be made of a group of human viruses which have oncogenic properties under experimental conditions, but are not linked epidemiologically with human malignancies. Human adenoviruses are common upper respiratory tract viruses, of which types 2, 12 and 18 readily cause malignant tumours in hamsters and rats and can transform human and rodent cells in culture (Trentin, Yabe & Taylor, 1962; MacAllister et al., 1966). With the discovery that human adenoviruses are oncogenic in rodents, plans to develop mass immunization with live or killed virus were promptly dropped. The molecular biology of adenoviruses has been extensively studied and specific transforming genes have been isolated and cloned (Flint, 1980). Several searches for DNA sequences specific to adenovirus genes in human malignancies have not been rewarded with success. If human adenoviruses were categorized as chemicals rather than microorganisms, their oncogenicity in hamsters and rats would place them at the top of the list of potent environmental carcinogens, although they are apparently non-malignant in humans.

MECHANISMS OF VIRAL ONCOGENESIS

Cell transformation

How viruses transform cells into a neoplastic state is not yet understood clearly, although remarkable advances have recently been made in probing the molecular genetics of animal tumour viruses. With the DNA tumour viruses, full replication of the virus is cytopathic so that the infected cell is destined to die rather than transform to malignancy. However, replication is frequently incomplete, either because the infected cell is of a type that does not support the complete viral life cycle or because the virus has lost genetic information essential for later stages of replication. With the RNA retroviruses, full replication is not usually cytocidal, so the infected cell may support virus production as well as becoming neoplastic. Nevertheless, many retrovirus infections are non-productive, and the most highly oncogenic retrovirus are defective for viral replication owing to the substitution of oncogenes in place of genes essential for virus production. Viruses could induce neoplasia by a variety of mechanisms.

Non-specific tissue damage inducing proliferation. The chronic cytopathic effects of virus infection may stimulate regeneration or hyperplasia in neighbouring uninfected cells which eventually become neoplastic. One can envisage such events, say in liver tissue infected with hepatitis B virus, but as only a few cytopathic viruses are oncogenic, it is more likely that neoplastic transformation is the result of a specific viral effect. In most cases the viral genome (or part of it) can be found in each of the tumour cells, implying that specific viral genetic information is important in viral oncogenesis.

'Hit-and-run' mutagens. In cases where the viral genome cannot be found in the cancer cells, it is thought that pre-cancerous cells ancestral to the tumour have been infected by the virus and that the virus has caused mutations – just as physical and chemical carcinogens do – without persisting in the malignant clone. Such a model has recently been proposed for herpes simplex virus in cervical carcinogenesis (Galloway & McDougall, 1983). Lymphomas in cats and cattle, while strongly linked epidemiologically with retrovirus infection, are occasionally found to be virus-negative, possibly because the viral genome did not persist in the cellular clone that became malignant.

Tumour viruses as insertional mutagens. The DNA of retroviruses, hepadnaviruses and some papovaviruses integrates into host chromosomal DNA. Integration itself is a recombinational event that causes changes in the organization of host genetic information. The site of viral integration into host DNA has been studied extensively with these viruses. In most cases it appears to be largely random, that is, the viral genome integrates at a different site, usually in a different chromosome for each infected cell. Some of these integration events will be mutagenic, and in rare cases could lead to neoplasia, analogous to mutagenesis by chemical carcinogens. Integrating polyoma viruses, for example, are mutagenic and also induce chromosome breaks and rearrangements (Geissler *et al.*, 1980), whereas the related papillomaviruses do not appear to integrate into the host genome. With avian retroviruses causing malignant bursal lymphomas, there is evidence that the integration of the viral genome adjacent to the host c-*myc* proto-oncogene enhances or activates the expression of the cellular genes, analogous to the translocation of the same gene in Burkitt's lymphoma (Hayward, Neal & Astrin, 1981; Bishop, this volume).

Viral oncogenes. Some acutely oncogenic retroviruses carry oncogenes of host origin in place of the virus' normal replication genes. These genes are not required for viral replication, indeed they usually render the virus defective, but they induce rapid cell transformation and malignancy (Bishop, this volume). The products of different retrovirus oncogenes have different properties, exerting their neoplastic effect at different sites in the cell. For instance, the *src* product is a tyrosine phosphokinase, the *ras* product a GTP-binding protein and the *myc* product a nuclear protein. With the exception of avian reticuloendotheliosis virus, oncogene-carrying retroviruses are not transmitted from host to host in nature; epidemiologically speaking, their occurrence is trivial, yet it is difficult to exaggerate their importance in illuminating the molecular biology of neoplasia.

In contrast to retroviruses, papovaviruses and adenoviruses code for viral proteins required for early phases of viral replication which also play a role in cellular transformation; that is to say their 'oncogenes' are essential for virus propagation. The best understood oncogene product is the large T-antigen of SV40, a nuclear antigen that forms a complex with a normal cellular protein. The complex binds to DNA and probably initiates DNA synthesis. Specific transforming genes in HBV and herpes viruses have not been so clearly defined or their functions analysed but it is expected that they also code for essential viral proteins. Viral oncogenes were initially defined by genetic analysis of viruses which lost transforming properties and by the selection of conditional mutants for cell transformation. Cells transformed by viral oncogenes rely on the continued expression of these genes for the malignant phenotype to be maintained.

Thus, different tumour viruses probably induce malignant transformation by different molecular mechanisms. In human cancer, transformation by viruses is not understood at the molecular level for any malignancy. For example, it has not yet been determined whether the human retrovirus HTLV carries a viral oncogene, consistently activates a cellular proto-oncogene, or transforms cells by some other method.

Multistage oncogenesis

Cancer is the end result of a series of stepwise changes in a lineage of cells leading to a fully malignant phenotype (reviewed by Peto, 1978; Whittemore, 1978). In this respect, virus-induced tumours are no different from other tumours, except for neoplasms induced by some oncogene-bearing retroviruses, where the newly introduced oncogene(s) itself is the result of changes in structure and regulation which have taken place before infection.

An analysis of bursal lymphomagenesis in chickens exemplifies the multistep nature of viral oncogenesis (Neiman et al., 1980). In chickens inoculated with avian leukosis virus 1 week after hatching, the probability of death from lymphoma increases progressively at approximately the sixth power of time in birds known to be susceptible to the disease. Moreover, a genetically 'resistant' strain of birds with a different major histocompatibility haplotype exhibited the same increase in mortality with time except that the constant relating probability of death to time was about ten-fold lower. This observation indicates that, in resistant birds, the rate at which events leading to death from lymphoma differs rather than the number of events required. One early event must be the infection of target cells in the bursa; another probably early event is the integration of a provirus near the c-*myc* host proto-oncogene (Hayward et al., 1981) and a third, later event is an alteration in an unlinked proto-oncogene, B-*lym*, (Goubin et al., 1983). In Burkitt's lymphoma following EBV infection, activation of the human homo-logues of c-*myc* and B-*lym* also occur, though possibly in reverse order to avian lymphomagenesis as the chromosome translocation leading to c-*myc* activation appears to take place as a late event.

In humans with genetic predispositions to viral or non-viral carcinogenesis, it remains to be determined whether the increased incidence is a result of one or more 'events' being inherited or of an accelerated rate of events occurring. Genetic defects of DNA repair such as xeroderma pigmentosum probably increase the rate of mutational events, while other genetic disorders such as familial polyposis coli or retinoblastoma may represent the inheritance of one of the essential steps in the pathway to malignancy.

Risk factors in viral oncogenesis

In the foregoing discussion, human malignancies have been described for which virus infection appears to be a necessary com-ponent in multifactorial oncogenesis. Exposure to the virus, then, is clearly the primary risk factor. Because infection with some of the viruses, e.g., EBV and CMV, affects a substantial proportion of the human population, whereas the associated cancers are relatively rare, secondary risk factors may play a determining role in the incidence of the disease. As already discussed in relation to EBV, this is exemplified by a secondary environmental factor, namely holoendemic malarial infestation, for Burkitt's lymphoma in tropi-cal Africa and New Guinea, and the secondary host factor, namely HLA haplotype, for nasopharyngeal carcinoma amongst the south-ern Chinese (Simons *et al.*, 1978). Ito (1983) has postulated that chemicals acting as promoter cocarcinogens are synthesized by bacteria in the jaw during the replacement of milk teeth by permanent dentition. These promoters which activate EBV may then lead to the final transformation of B cells, accounting for the localization of BL in the jaw with peak incidence during tooth replacement.

It is often difficult to apportion quantitative risks to individual factors, especially when they act synergistically, and epidemiologi-cal studies remain the best approach to this problem. It is not clear that HSV plays any role in HPV-induced cervical carcinoma, as proposed by Zur Hausen (1982), or to what extent dietary carci-nogens and promoting agents, such as aflatoxins or more probably alcohol, are involved in HBV-associated hepatocellular carcinoma.

Skin cancer might be attributable to papilloma viruses (Spradbrow, Beardmore & Francis, 1983), with exposure to ultra-violet light as an environmental cocarcinogen, enhanced susceptibility to papillomavirus transformation (epidermodysplasia verruciformis) as a genetic factor (xeroderma pigmentosum is probably not associated with HPV), and immunosuppression as a predisposing pathophysiological factor.

Immunodeficiency and human cancer

Burnet's (1969) hypothesis that malignant cell clones continuously arise in the body but are quickly eliminated by immune surveillance mechanisms has been largely discredited because immunodeficient mice and humans do not show a markedly higher incidence of most types of cancer. However, those cancers which clearly have a viral aetiology do occur at a significantly higher rate in immunodeficient hosts, and impairment of the immune system may be considered the most important risk factor in human viral carcinogenesis. Many viruses which usually remain as life-long latent infections in healthy individuals frequently become activated following immunosuppression, e.g., VZV, EBV and polyoma BK.

For potentially oncogenic viruses, reactivation may present further opportunities for the infection of target cells for transformation, and (probably of greater importance) for the emergence of neoplastic cells expressing viral antigens which would normally be eliminated by cytotoxic T-cells. For instance, Moss, Rickinson & Pope (1978) have shown that T-lymphocytes from healthy EBV-positive individuals will inhibit the growth of EBV-transformed cells *in vitro*, and cell-mediated immunity to EBV is impaired in renal transplant patients (Crawford *et al.*, 1981). The immunoblastic lymphomas in patients with iatrogenic immunosuppression and in genetic X-linked immunodeficiency (Purtilo, 1981) contain EBV DNA and antigens. A clonal tumour frequently emerges following polyclonal B-cell hyperplasia in these patients (Hanto *et al.*, 1982) and in other cases of persistent infectious mononucleosis (Abo *et al.*, 1982). As a high proportion of immunoblastic lymphomas in transplant patients appear shortly after immunosuppressive treatment (Kinlen, 1982), most of the oncogenic steps must have taken place before immunosuppression commences.

It has been argued that the great majority of human malignancies do not have a viral aetiology because their incidence is not increased in immunodeficient patients (Cairns, 1978). Conversely, those cancers for which the frequency is elevated in immunosuppressed patients should be scrutinized for oncogenic viruses (Kinlen, 1982; Purtilo, 1982). They include non-Hodgkin lymphoma (mainly immunoblastic lymphoma), Kaposi's sarcoma, hepatocellular carcinoma, cervical carcinoma, squamous- and basal-cell carcinomas of the skin and melanoma (Kinlen, 1982; Penn, 1982). All of these malignancies have already been discussed in relation to oncogenic viruses. Relatively few transplant patients develop liver cancer (Penn, 1982) although there is nonetheless a 38-fold relative risk (Kinlen, 1982) and an association with HBV infection (Schröter et al., 1982). HPV is associated with cervical neoplasia in women with impaired immunity (Shoki-Tabibzadeh et al., 1981). The epidermal ncers and melanoma merit further investigation, particularly in ㅤtion to papovaviruses. However, it has proved difficult to obtain ㅤㅤaccurate assessment of the risk of skin cancer in transplant ㅤㅤents as the incidence varies greatly in the normal population ㅤㅤording to exposure to ultra-violet radiation in sunlight (Kinlen, ㅤㅤ2). It is noteworthy that 10 out of 14 cutaneous melanomas ㅤㅤdied by Greene, Young & Clark (1981) in renal transplant patients arose from precursor naevi. If the human polyoma viruses, JC and BK, are aetiologically involved in neural tumours, one would expect to see the incidence of gliomas and meningiomas rise in transplant patients. While activation of BK and JC viruses is frequently encountered in immunosuppression, the tumours supposedly associated with them do not appear to be significantly increased.

In AIDS, Kaposi's sarcoma and lymphoma are the most frequent tumours observed (Waterson, 1983). Kaposi's sarcoma also has an increased incidence in renal and cardiac transplant patients (Harwood et al., 1979; Lanza et al., 1983). One may surmise that late steps in the oncogenesis of CMV-infected cells are normally suppressed by a competent T-cell system. AIDS patients, being deficient in OKT4$^+$ T-cells, may be regarded as the human equivalent of nude mice, and it is possible though unlikely that the Kaposi's sarcoma results from the transfer of occult malignant cells from donor to immunodeficient recipient. The development of tumours of donor origin following organ transplantation has been documented (Kinlen, 1982), and Kaposi's sarcoma in AIDS might conceivably represent an analogous cellular transmission between

sexual partners, for which canine venereal sarcoma is an apt model
(Hayes *et al.*, 1983; Weiss, 1983).

PROSPECTS OF PREVENTING AND CONTROLLING VIRAL MALIGNANCIES

Wherever viruses prove to be the primary cause of cancer, prevention by immunization against viral infection must be seriously entertained. None of the viruses implicated in human cancer are endogenous in the sense of being genetically transmitted, although CMV and probably HTLV may sometimes be congenitally transmitted. Immunization against post-natal infection should be feasible with several of the oncogenic viruses, for example EBV (North *et al.*, 1982). Immunization against HBV will be a specially important task in view of the overall morbidity including liver cancer caused by this virus (Blumberg & London, 1981; WHO Scientific Group, 1983). With the prospects of chemically synthesized antigenic peptides and of recombinant DNA methods for producing antigens free of live virus in bacterial systems, large scale vaccine production should not be a problem.

Where primary immunization is not practical, intervention at the level of a rate-limiting cofactor could effectively prevent the great majority of viral cancers developing to a clinical presentation. For example, prevention of *Plasmodium* infection, which is so desirable owing to the direct morbidity of malaria, would incidentally eliminate all but the small number of sporadic cases of Burkitt's lymphoma.

Prevention of iatrogenic infection is an important consideration. The necessity for detection of HBV in donated blood is well recognized. There is evidence that both the AIDS agent (Ragni *et al.*, 1983) and HTLV (Miyoshi *et al.*, 1982; Saxinger & Gallo, 1982) are transmitted by the transfer of whole blood or crude blood products. It is of great importance to devise rapid screening methods if the risk of infection via blood transfusion is to be eliminated.

Early diagnosis of cancer in infected patients may become practicable by monitoring viral antibodies or antigens as markers of incipient disease. For example, mass screening for EBV-specific salivary IgA to detect early, treatable nasopharyngeal carcinoma is being developed in China (Zeng *et al.*, 1980). Individuals at particularly high risk of viral cancers (e.g., immunosuppressed patients) could be monitored for elevation of virus- or tumour-associated antigens.

Finally, serious thought will be given to the development of strategies for intervening in the course of viral malignant disease by aiming therapeutic agents at viral targets. The expression of viral antigens may lend itself to immunotherapy, as shown experimentally for human liver cancer xenografts (Shouval et al., 1982). Where the malignant phenotype itself is dependent on the expression and function of viral proteins, agents that specifically interfere with those functions may arrest the tumour while sparing normal tissues.

REFERENCES

ABO, W., TAKADA, K., KAMADA, M., IMAMURA, M., MOTOYA, T., IWANGA, M., AYA, T., YANO, S., NAKAO, T. & OSATO, T. (1982). Evolution of infectious mononucleosis into Epstein-Barr virus carrying monoclonal malignant lymphoma. *Lancet*, **i**, 1272–5.

BEASLEY, R. P. (1982). Hepatitis B virus as the etiologic agent in hepatocellular carcinoma. Epidemiologic considerations. *Hepatology*, **2**, 215–65.

BEASLEY, R. P., HWANG, L.-Y., LIN, C.-C. & CHIEN, C.-S. (1981). Hepatocellular carcinoma and hepatitis B virus: a prospective study of 22 707 men in Taiwan. *Lancet*, **ii**, 1129–33.

BLATTNER, W. A., KALYANARAMAN, V. S., ROBERT-GUROFF, M., LISTER, T. A., GALTON, D. A. G., SARIN, P., CRAWFORD, M. H., CATOVSKY, D., GREAVES, M. & GALLO, R. C. (1982). The human type-C retrovirus, HTLV, in blacks from the Caribbean, and relationship to adult T-cell leukemia/lymphoma. *International Journal of Cancer*, **30**, 257–64.

BLUMBERG, B. S. & LONDON, W. T. (1981). Hepatitis B virus and the prevention of primary hepatocellular carcinoma. *New England Journal of Medicine*, **304**, 782–4.

BOLDOGH, I., BASKAR, J. F., MAR, E.-C. & HUANG, E.-S. (1983). Human cytomegalovirus and herpes simplex type 2 virus in normal and adenocarcinomatous prostate glands. *Journal of the National Cancer Institute*, **70**, 819–26.

BOLDOGH, I., BETH, E., HUANG, E.-S., KYALWAZI, S. K. & GIRALDO, G. (1981). Kaposi's sarcoma. IV. Detection of CMV DNA, CMV RNA and CMNA in tumor biopsies. *International Journal of Cancer*, **28**, 469–74.

BURKITT, D. (1962). A children's cancer dependent on climatic factors. *Nature*, **194**, 232–4.

BURNET, F. M. (1969). *Self and Not-self*. Cambridge University Press.

CAIRNS, J. (1978). *Cancer: Science and Society*. San Francisco. W. H. Freeman & Co., 142 pp.

CAMACHO, A. & SPEAR, P. C. (1978). Transformation of hamster embryo fibroblasts by a specific fragment of the herpes simplex virus genome. *Cell*, **15**, 993–8.

CATOVSKY, D., GREAVES, M. F., ROSE, M., GALTON, D. A. G., GOOLDEN, A. W. G., McCLUSKEY, D. R., WHITE, J. M., LAMPERT, I., BOURIKAS, G., IRELAND, R., BROWNELL, A., BRIDGES, J. M., BLATTNER, W. A. & GALLO, R. C. (1982). Adult T-cell lymphoma-leukaemia in blacks from the West Indies. *Lancet*, **i**, 639–43.

CLAPHAM, P., NAGY, K. & WEISS, R. A. (1984). Pseudotypes of human T-cell leukemia virus: host range and neutralization by antisera. *Virology* (in press).

COSTA, J. & RABSON, A. S. (1983). Generalised Kaposi's sarcoma is not a neoplasm. *Lancet*, **i**, 58.

CRAWFORD, D. H., EDWARDS, J. M. B., SWENY, P., HOFFBRAND, A. V. & JANOSSY, G. (1981). Studies on long-term cell mediated immunity to Epstein-Barr virus in immunosuppressed renal allograft recipients. *International Journal of Cancer*, **28**, 705–9.

DE THÉ, G. (1980). Role of Epstein-Barr virus in human diseases: infectious mononucleosis, Burkitt's lymphoma, and nasopharyngeal carcinoma. In *Viral Oncology*, ed. G. Klein, pp. 769–97. New York: Raven Press.

DE THÉ, G., LAVOUÉ, M. F. & MUENZ, L. (1978). Differences in EBV antibody titres of patients with nasopharyngeal carcinoma originating from high, intermediate and low incidence areas. In *Nasopharyngeal Carcinoma: Etiology and Control*, ed. G. de Thé, & Y. Ito, International Agency for Research on Cancer Scientific Publication, **20**, 471–81.

DREW, W. L., CONANT, M. A., MINER, R. C., HUANG, E.-S., ZIEGLER, J. L., GROUNDWATER, J. R., GULLETT, J. H., VOLBERDING, P., ABRAMS, D. I. & MINTZ, L. (1982). Cytomegalovirus and Kaposi's sarcoma in young homosexual men. *Lancet*, **ii**, 125–7.

DUFF, R., KREIDER, J. W., LEVY, B. M., KATZ, M. & RAPP, F. (1974). Comparative pathology of cells transformed by herpes simplex virus type 1 or type 2. *Journal of the National Cancer Institute*, **53**, 1159–64.

EGLIN, R. P., SHARP, F., MACLEAN, A. B., MACNAB, J. C. M., CLEMENTS, J. B. & WILKIE, N. M. (1981). Detection of RNA complementary to herpes simplex virus DNA in human cervical squamous cell neoplasms. *Cancer Research*, **41**, 3597–603.

EPSTEIN, M. A. & ACHONG, B. G. (eds.) (1979). *The Epstein-Barr virus*. New York: Springer-Verlag.

EPSTEIN, M. A., ACHONG, B. G. & BARR, Y. M. (1964). Virus particles in cultural lymphoblasts from Burkitt's lymphoma. *Lancet*, **i**, 702.

ESSEX, M., MCLANE, M., LEE, T. H., FALK, L., HOWE, C. W. S., MULLINS, J. I., CABRADILLA, C. & FRANCIS, D. P. (1983). Antibodies to cell membrane antigens associated with human T-cell leukemia virus in patients with AIDS. *Science*, **220**, 859–62.

FLINT, S. J. (1980). Cell transformation induced by adenoviruses. In *DNA Tumour Viruses*, ed. J. Tooze, pp. 547–75. Cold Spring Harbor: Cold Spring Harbor Laboratory.

FRANCESCHI, S., DOLL, R., GALLWEY, J., LA VECCHIA, C., PETO, R. & SPRIGGS, A. I. (1983). Genital warts and cervical neoplasia: an epidemiological study. *British Journal of Cancer* (in press).

FRIEDMAN-KIEN, A., LAUBENSTEIN, L. J., RUBINSTEIN, P., BUIMOVICI-KEIN, E., MARMOR, M., STAHL, R., SPIGLAND, I., KIM, K. S., ZOLLA-PAZNER, S. (1982). Disseminated Kaposi's sarcoma in homosexual men. *Annals of Internal Medicine*, **96**, 693–700.

GALLO, R. C., SARIN, P. S., GELMANN, E. P., ROBERT-GUROFF, M., RICHARDSON, E., KALYANARAMAN, V. S., MANN, D., SIDHU, G. D., STAHL, R. E., ZOLLA-PAZNER, S., LEIBOWITCH, J. & POPOVIC, M. (1983). Isolation of human T-cell leukemia virus in acquired immune deficiency syndrome (AIDS). *Science*, **220**, 865–7.

GALLOWAY, D. A. & MCDOUGALL, J. K. (1983). The oncogenic potential of herpes simplex viruses: evidence for a 'hit-and-run' mechanism. *Nature*, **302**, 21–4.

GARDNER, S. D., FIELD, A. M., COLEMAN, D. V. & JULME, B. (1971). New human papovavirus (BK) isolated from urine after renal transplantation. *Lancet*, **i**, 1253–7.

GEDER, L., KNEIDER, J. & RAPP, F. (1977). Human cells transformed *in vitro* by human cytomegalovirus: tumorigenicity in athymic nude mice. *Journal of the National Cancer Institute*, **58**, 1003–9.

GEISSLER, E., SCHERNECK, S., THEILE, M., HEROLD, H.-J., STANECZEK, W., ZIMMERMAN, W., KRAUSE, H., PROKOPH, H., VOGEL, F. & PLATZER, H. (1980). Studies on the mutagenic and transforming activities of SV40-like viruses. In *Leukaemias, lymphomas and Papillomas: Comparative Aspects*, ed. P. A. Bachmann, Munich Symposia on Microbiology, pp. 43–55. London: Taylor & Francis Ltd.

GIRALDO, G., BETH, E., KOURILSKY, F. M., HENLE, W., MIKE, V., HURAUX, J. M., ANDERSEN, H. K., GHARBI, M. R., KYALWAZI, S. K. & PUISSANT, A. (1975). Antibody patterns of herpes virus in Kaposi's sarcoma: serological association of European Kaposi's sarcoma with cytomegalovirus. *International Journal of Cancer*, **15**, 839–48.

GISSMANN, L., WOLNIK, L., IKENBERG, H., KOLDOVSKY, U., SCHNÜRCH, H. G. & ZUR HAUSEN, H. (1983). Human papillomavirus types 6 and 11 DNA sequences in genital and laryngeal papillomas and in some cervical cancer biopsies. *Proceedings of the National Academy of Sciences (USA)*, **80**, 560–3.

GOUBIN, G., GOLDMAN, D. S., LUCE, J., NEIMAN, P. E. & COOPER, G. M. (1983). Molecular cloning and nucleotide sequence of a transforming gene detected by transfection of chicken B-cell lymphoma DNA. *Nature*, **302**, 114–19.

GREEN, M., BRACKMANN, K. J., SANDERS, P. R., LÖWENSTEIN, P. M., FREEL, J. H., EISENGER, M. & SWITLYK, S. A. (1982). Isolation of a human papillomavirus from a patient with epidermodysplasia verruciformis: presence of related viral DNA genomes in human urogenital tumors. *Proceedings of the National Academy of Sciences, (USA)*, **79**, 4437–41.

GREENE, M. H., YOUNG, T. I. & CLARK, W. H. (1981). Malignant melanoma in transplant recipients. *Lancet*, **i**, 1196–9.

HANAOKA, M., TAKATSUKI, K. & SHIMOYAMA, M. (eds.) (1982). Adult T-cell leukemia and related diseases. *Gann Monograph on Cancer Research*, **28**, 1–237.

HANTO, D. W., FRIZZERA, G., GAJL-PECZALSKA, K. J., SAKAMOTO, K., PURTILO, D. T., BALFOUR, H. H., SIMMONS, R. L. & NAJARIAN, J. S. (1982). Epstein-Barr-virus-induced B-cell lymphoma after renal transplantation: acyclovir therapy and transition from polyclonal to monoclonal B-cell proliferation. *New England Journal of Medicine*, **306**, 913–18.

HARWOOD, A. R., OSOBA, D., HOFTADED, S. L., GOLDSTEIN, M. B., CARDELLA, C. J., HOLECEK, M. J., KUNYNETZ, R. & GIAMMARCO, R. A. (1979). Kaposi's sarcoma in recipient of renal transplants. *American Journal of Medicine*, **67**, 759–65.

HAYES, H. M., BIGGAR, R. J., PICKLE, L. W., HOOVER, R. & TOFT, J. D. (1983). Canine transmissible venereal tumour: a model for Kaposi's sarcoma? *American Journal of Epidemiology*, **117**, 108–9.

HAYWARD, W. S., NEAL, B. G. & ASTRIN, S. M. (1981). Activation of a cellular *onc* gene by promoter insertion in ALV-induced lymphoid leukosis. *Nature*, **290**, 475–80.

HENDERSON, B. E., LOUIE, E., JING, J. S., BUELL, P. & GARDNER, M. B. (1976). Risk factors associated with nasopharyngeal carcinoma. *New England Journal of Medicine*, **295**, 1101–6.

HENLE, G. & HENLE, W. (1976). Epstein-Barr-virus-specific IgA serum antibodies as an outstanding feature of nasopharyngeal carcinoma. *International Journal of Cancer*, **17**, 1–7.

HEYWARD, W., BENDER, T. R., LANIER, A. P., FRANCIS, D. P., MCMAHON, B. J. & MAYNARD, J. E. (1982). Serological markers of hepatitis B virus and alpha-fetoprotein levels preceding primary hepatocellular carcinoma in Alaskan Eskimos. *Lancet*, **ii**, 889–91.

HINUMA, Y., KOMODA, H., CHOSA, T., KONDO, T., KOHAKURA, M., TAKENARA, T., KIKUCHI, M., ICHIMURA, M., YUNOKI, K., SATO, I., MATSUO, R., TAKIUCHI, Y., UCHINO, H., & HANAOKA, M. (1982). Antibodies to adult T-cell leukemia-virus-associated antigen (ATLA) in sera from patients with ATL and controls in Japan: a nation-wide sero-epidemiologic study. *International Journal of Cancer*, **29**, 631–5.

HOWLEY, P. M. (1980). Molecular biology of SV40 and the human polyomaviruses BK and JC. In *Viral Oncology*, ed. G. Klein, pp. 489–550. New York: Raven Press.

ITO, Y. (1983). Possible roles of Epstein-Barr virus, normal flora microbes and promoter plant diterpene esters in etiology of Burkitt's lymphoma. *Leukemia Reviews International*, **1**, 9–10.

JABLONSKA, S., DABROWSKI, J. & JAKUBOWICZ, K. (1972). Epidermodysplasia verruciformis as a model in studies on the role of papova viruses in oncogenesis. *Cancer Research*, **32**, 583–9.

JARRETT, W. F., MCNEILL, P., GRIMSHAW, W., SELMAN, I. & MCINTYRE, W. (1978). High incidence area of cattle cancer with a possible interaction between an environmental carcinogen and a papilloma virus. *Nature*, **274**, 215–17.

KALYANARAMAN, V. S., SARNGADHARAN, M. G., ROBERT-GUROFF, M., MIYOSHI, I., BLAYNEY, D., GOLDE, D. & GALLO, R. C. (1982). A new subtype of human T-cell leukemia virus (HTLV-II) associated with a T-cell variant of hairy cell leukemia. *Science*, **218**, 571–3.

KESSLER, I. I. (1976). Human cervical cancer as a venereal disease. *Cancer Research*, **36**, 783–91.

KINLEN, L. (1982). Immunosuppressive therapy and cancer. *Cancer Surveys*, **1**, 565–83.

KLEIN, G. (1983). Specific chromosomal translocations and the genesis of B-cell-derived tumors in mice and men. *Cell*, **32**, 311–15.

KLEIN, G., GIOVANELLA, B. C., LINDAHL, T., FIALKOW, P. J., SINGH, S. & STEHELIN, J. (1974). Direct evidence for the presence of Epstein-Barr virus DNA and nuclear antigen in malignant epithelial cells from patients with anaplastic carcinoma of the nasopharynx. *Proceedings of the National Academy of Sciences (USA)*, **71**, 4737–41.

LANZA, R. P., COOPER, D. K. C., CASSIDY, M. J. D. & BARNARD, C. N. (1983). Malignant neoplasms occurring after cardiac transplantation. *Journal of the American Medical Association*, **249**, 1746–8.

LUTWICK, L. I. (1979). Relation between aflatoxin, hepatitis B virus, and hepato-cellular carcinoma. *Lancet*, **i**, 755–7.

MACALLISTER, R. M., GOODHEART, C. R., MIRABAI, V. Q. & HUEBNER, R. J. (1966). Human adenoviruses: Tumor production in hamsters by types 12 and 18 grown from single plaques. *Proceedings of the Society for Experimental Biology and Medicine*, **122**, 454–6.

MCCANCE, D. J., WALKER, P. G., DYSON, J. L., COLEMAN, D. V. & SINGER, A. (1983). Presence of human papillomavirus DNA in cervical intraepithelial neoplasia. *British Medical Journal*, (in press).

MEISELS, A., ROY, M., FORTIER, M., MORIN, C., CASAS-CORDERO, M., SHAH, K. V. & TURGEON, H. (1981). Human papillomavirus infection of the cervix. *Acta Cytologica*, **25**, 7–16.

MILLER, G. (1980). Biology of Epstein-Barr Virus. In *Viral Oncology*, ed. G. Klein, pp. 713–38. New York: Raven Press.

MIYOSHI, I., FUJISHITA, M., TAGUCHI, H., OHTSUKI, Y., AKAGI, T., MORIMOTO, Y. M. & NAGASAKI, A. (1982). Caution against blood transfusion from donors seropositive to adult T-cell leukemia-associated antigens. *Lancet*, **i**, 683–4.

MIYOSHI, I., KUBONISHI, I., YOSHIMOTO, S., AKAGI, T., OHTSUKI, Y., SHIRAISHI, Y., NAGATA, K. & HINUMA, Y. (1981). Detection of type-C particles in cord leukocytes and human leukemic T-cells. *Nature*, **294**, 770–1.

MORTIMER, E. A., LEPOW, M. L., GOLD, E., ROBBINS, F. C., BURTON, G. J. & FRAUMENI, J. F. (1981). Long-term follow up of persons inadvertently inoculated with SV40 as neonates. *New England Journal of Medicine*, **305**, 1517–19.

MOSS, D. J., RICKINSON, A. B. & POPE, J. H. (1978). Long-term T-cell-mediated immunity to Epstein-Barr virus in man. I. Complete regression of virus-induced transformation in cultures of seropositive donor leukocytes. *International Journal of Cancer*, **22**, 662–8.

NAGY, K., CLAPHAM, P., CHEINGSONG-POPOV, R. & WEISS, R. A. (1983). Human T-cell leukemia virus type I: induction of syncytia and inhibition by patients' sera. *International Journal of Cancer*, **32**, 321–8.

NEIMAN, P. E., JORDAN, L., WEISS, R. A. & PAYNE, L. N. (1980). Malignant lymphoma of the Bursa of Fabricius: analysis of early transformation. *Cold Spring Harbor Conference of Cell Proliferation*, **8**, 519–28.

NILSSON, K. (1982). Phenotypic and cytogenetic characteristics of human B-lymphoid cell lines and their relevance for the etiology of Burkitt's lymphoma. *Advances in Cancer Research*, **37**, 319–80.

NORTH, J. R., MORGAN, A. J., THOMPSON, J. L. & EPSTEIN, M. A. (1982). Purified Epstein-Barr virus mw 340 000 glycoprotein induces potent virus-neutralizing antibodies when incorporated in liposomes. *Proceedings of the National Academy of Sciences, (USA)*, **79**, 7504–8.

OBATA, H., HAYASHI, N., MOTOIKE, Y., HISUMITSU, T., OKUDA, H., KOBAYASHI, S. & NISHIOKA, K. (1980). A prospective study on the development of hepatocellular carcinoma from liver cirrhosis persistent hepatitis B virus infection. *International Journal of Cancer*, **25**, 741–7.

OHNISHI, K., IIDA, S., IWAMA, S., GOTO, N., NOMURA, F., TAKASHI, M., MISHIMA, A., KONO, K., KIMURA, K., MUSHA, H., KOTOTA, K. & OKUDA, K. (1982). The effect of chronic habitual alcohol intake on the development of liver cirrhosis and hepatocellular carcinoma: relation to hepatitis B surface antigen carriage. *Cancer*, **49**, 672–7.

OKUDA, K., NAKASHIMA, T., SAKAMOTO, K., IKARI, T., HIDAKA, H., KUBO, Y., SAKUMA, K., MOTOIKE, Y., OKUDA, H. & OBATA, H. (1982). Hepatocellular carcinoma arising in non-cirrhotic and highly cirrhotic livers. *Cancer*, **49**, 450–5.

OLD, L. J., BOYSE, E. A., OETTGEN, H. F., DE HARVEN, E., GEERING, G., WILLIAMSON, E. & CLIFFORD, P. (1966). Precipitation antibody in human sera to an antigen present in cultured Burkitt's lymphoma cells. *Proceedings of the National Academy of Sciences (USA)*, **56**, 1699–704.

ORTH, G., JABLONSKA, J., FAVRE, M., CROISSANT, O., JARSABEK-CHORZELSKA, M. & RZESA, G. (1978). Characterization of two new types of human papilloma viruses in lesions of epidermodysplasia verruciformis. *Proceedings of the National Academy of Sciences (USA)*, **75**, 1537–41.

OSTROW, R. S., BENDER, M., NIIMURA, M., SEKI, T., KAWASKIMA, M., PASS, F. & FARAS, A. J. (1982). Human papillomavirus DNA in cutaneous primary and metastasized squamous carcinomas from patients with epidermodysplasia verruciformis. *Proceedings of the National Academy of Sciences (USA)*, **79**, 1634–8.

PADGETT, B. (1980). Human papovaviruses. In *DNA Tumor Viruses*, ed. J. Tooze, pp. 339–70. Cold Spring Harbor: Cold Spring Harbor Laboratory.

PADGETT, B. L., WALKER, D. L., ZU RHEIN, G. M., ECKROADE, R. J. & DESSELL, B. H. (1971). Cultivation of a papova-like virus from human brain with progressive multifocal leukoencephalopathy. *Lancet*, **i**, 1257–60.

PENN, I. (1982). The occurrence of cancer in immune deficiencies. *Current Problems in Cancer*, **6**, 2–64.

PETO, R. (1978). Epidemiology, multistage models, and short-term mutagenicity tests. *Cold Spring Harbor Conference on Cell Proliferation*, **4**, 1403–9.

POIESZ, B. J., RUSCETTI, F. W., GAZDAR, A. F., BUNN, P. A., MINNA, J. D. & GALLO, R. C. (1980). Detection and isolation of type-C retrovirus particles from fresh and cultured lymphocytes of a patient with cutaneous T-cell lymphoma. *Proceedings of the National Academy of Sciences (USA)*, **77**, 7415–19.

POIESZ, B. J., RUSCETTI, F. W., REITZ, M. S., KALYANARAMAN, V. S., & GALLO, R. C. (1981). Isolation of a new type-C retrovirus (HTLV) in primary uncultured cells of a patient with Sézary T-cell leukaemia. *Nature*, **294**, 268–71.

POPOVIC, M., KALYANARAMAN, V. S., SARNGADHARAN, M. G., ROBERT-GUROFF, M., NAKAO, Y., REITZ, M. S., MIYOSHI, Y., ITO, Y., MINOWADA, J. & GALLO, R. C. (1982). The virus of Japanese adult T-cell leukemia is a member of the human T-cell leukemia virus group. *Nature*, **300**, 63–6.

POPOVIC, M., SARIN, P. S., ROBERT-GUROFF, M., KALYANARAMAN, V. S., MANN, D., MINOWADA, J. & GALLO, R. C. (1983). Isolation and transmission of human retrovirus (human T-cell leukemia virus). *Science*, **219**, 856–9.

PRAWER, S. E., PASS, F., VANCE, J. C., GREENBERG, L. J., YUNIS, J. J. & ZELICKSON, A. S. (1977). Depressed immune functions in epidermodysplasia verruciformis. *Archivae Dermatologica*, **113**, 495–9.

PURTILO, D. T. (1981). Immune deficiency predisposing to Epstein-Barr virus-induced lymphoproliferative diseases: the X-linked lymphoproliferative syndrome as a model. *Advances in Cancer Research*, **34**, 279–312.

PURTILO, D. T. (1982). Viruses, tumours, and immune deficiency. *Lancet*, **i**, 684.

RAGNI, M. V., LEWIS, J. H., SPERO, J. A. & BONTEMPO, F. A. (1983). Acquired immunodeficiency-like syndrome in two haemophiliacs. *Lancet*, **i**, 213–14.

REID, R., STANHOPE, C. R., HERSCHMAN, B. R., BOOTH, E., PHIBBS, G. D. & SMITH, J. P. (1982). Genital warts and cervical cancer. I. Evidence of an association between subclinical infection and cervical malignancy. *Cancer*, **50**, 377–87.

REITZ, M. S., POIESZ, B. J., RUSCETTI, F. W. & GALLO, R. C. (1981). Characterization and distribution of nucleic acid sequences of a novel type-C retrovirus isolated from neoplastic lymphocytes. *Proceedings of the National Academy of Sciences (USA)*, **78**, 1887–91.

SAXINGER, W. C. & GALLO, R. C. (1982). Possible risk to recipients of blood from donors carrying serum markers of human T-cell virus. *Lancet*, **i**, 1074.

SCHRÖTER, G. P. J., WEIL, R., PENN, I., SPEARS, W. C. & WADDELL, W. R. (1982). Hepatocellular carcinoma associated with chronic hepatitis B virus infection after kidney transplantation. *Lancet*, **ii**, 381–2.

SHAFRITZ, D. A., SHOUVAL, D., SHERMAN, H. I., HADZIYANNIS, S. J. & KEW, M. C. (1981). Integration of hepatitis B virus DNA into the genome of liver cells in chronic liver disease and hepatocellular carcinoma. *New England Journal of Medicine*, **305**, 1067–73.

SHOKI-TABIBZADEH, S., KOSS, L. G., MOLNAR, J. & ROMNEY, S. (1981). Association of human papillomavirus with neoplastic processes in the genital tract of four women with impaired immunity. *Gynecology and Obstetrics*, **12**, 5129–40.

SHOUVAL, D., SHAFRITZ, D. A., ZURAWSKI, V. R., ISSELBACHER, K. J. & WANDS, J.

R. (1982). Immunotherapy in nude mice of human hepatoma using monoclonal antibodies against hepatitis B virus. *Nature*, **298**, 567–9.

SIMONS, M. J., CHAN, S. H., WEE, G. B., SHANMUGARATNAM, K., GOH, E. H., HO. J. H. C., CHAN, J. C. W., DARMALINGAM, S., PRAGAD, V., BETEUD, G., DAY, N. E. & DE THÉ, G. (1978). Nasopharyngeal carcinoma and histocompatibility antigens. In *Nasopharyngeal Carcinoma: Etiology and Control*, ed. G. de Thé & Y. Ito. *International Agency for Research on Cancer Scientific Publication*, **20**, 271–82.

SKEGG, D. C. G., CORWIN, P. A., PAUL, C. & DOLL, R. (1982). Importance of the male factor in cancer of the cervix. *Lancet*, **ii**, 581–3.

SLAVIN, G., CAMERON, H. M. & SINGH, H. (1969). Kaposi's sarcoma in mainland Tanzania: a report of 117 cases. *British Journal of Cancer*, **23**, 349–57.

SMEDILE, A., FARCI, P., VERME, G., CAREDDA, F., CARGNEL, A., CAPORASSO, N., DENTICO, P., TREPO, C., OPOLON, P., GIMSON, A., VERGANI, D., WILLIAMS, R. & RIZZETTO, M. (1982). Influence of Delta infection on severity of hepatitis B. *Lancet*, **ii**, 945–7.

SPRADBROW, P. B., BEARDMORE, G. L. & FRANCIS, J. L. (1983). Virions resembling papillomaviruses in hyperkaryotic lesions from sun-damaged skin. *Lancet*, **i**, 189.

STAGNO, S., PASS, R. F., MEYER, E. D., HENDERSON, R. E., MOORE, E. G., WALTON, P. D. & ALFORD, C. A. (1982). Congenital cytomegalovirus infection: the relative importance of primary and recurrent maternal infection. *New England Journal of Medicine*, **306**, 945–9.

SUMMERS, J. & MASON, W. S. (1982). Properties of the hepatitis-B-like viruses related to their taxonomic classification. *Hepatology*, **2**, 61–6.

SUMMERS, J., O'CONNELL, A., MAUPAS, P., GOUDEAU, A., COURSAGET, P. & DRUCKER, J. (1978). Hepatitis B virus DNA in primary hepatocellular carcinoma. *Journal of Medical Virology*, **2**, 207–14.

SZMUNESS, W. (1978). Hepatocellular carcinoma and hepatitis B virus: evidence for a causal association. *Progress in Medical Virology*, **24**, 40–69.

TAJIMA, K., TOMINAGA, S., SHIMIZU, H. & SUCHI, T. (1981). A hypothesis on the etiology of adult T-cell leukemia/lymphoma. *Gann*, **72**, 684–91.

TAJIMA, K., TOMINAGA, S., SUCHI, T., KAWAGOE, T., KOMODA, H., HINUMA, Y., ODA, T. & FUJITA, K. (1982). Epidemiological analysis of the distribution of antibody to adult T-cell leukemia-virus-associated antigen: possible horizontal transmission of adult T-cell leukemia virus. *Gann*, **73**, 893–901.

TRENTIN, J. J., YABE, Y. & TAYLOR, G. (1962). The quest for human cancer viruses. *Science*, **137**, 835–41.

TRICHOPOULOS, D. (1981). The causes of primary hepatocellular carcinoma in Greece. *Progress in Medical Virology*, **27**, 14–25.

UCHIYAMA, T., YODOI, J., SAGAWA, K., TAKATSUKI, K. & UCHINO, H. (1977). Adult T-cell leukemia in Japan: clinical and hematological features of 16 cases. *Blood*, **50**, 481–92.

VIERA, J., FRANK, E., SPIRA, T. J. & LANDESMAN, S. H. (1983). Acquired immune deficiency in Haitians: opportunistic infections in previously healthy Haitian immigrants. *New England Journal of Medicine*, **308**, 125–9.

WALKER, T. G., SINGER, L., DYSON, J. L., SHAH, K. V., TO, A. & COLEMAN, D. V. (1983). The prevalence of human papillomavirus antigen in patients with cervical intraepithelial neoplasia. *British Journal of Cancer*, **48**, 99–101.

WATERSON, A. P. (1983). Acquired immune deficiency syndrome. *British Medical Journal*, **286**, 743–6.

WEISS, R. A. (1982*a*). The search for human RNA tumor viruses. In *RNA Tumor Viruses*, ed. R. Weiss, N. Teich, H. Varmus & J. Coffin, pp. 1205–81. Cold Spring Harbor: Cold Spring Harbor Laboratory.

WEISS, R. A. (1982*b*). The persistence of retroviruses. In *Viral Persistence*, ed. B. W. J. Mahy, A. C. Minson & G. K. Darby, pp. 267–88. Cambridge University Press.

WEISS, R. A. (1983). Kaposi's sarcoma in AIDS: Are the tumour cells transmissible? *Lancet* (in press).

WHITTEMORE, A. S. (1978). Quantitative theories of oncogenesis. *Advances in Cancer Research*, **27**, 55–88.

WHO SCIENTIFIC GROUP (1983). Prevention of primary liver cancer. *Lancet*, **i**, 463–5.

WOLD, W. S. M., MACKEY, J. K., BRACKMANN, K. H., TAKEMORI, N., RIGDEN, P. & GREEN, M. (1978). Analysis of human tumors and human malignant cell lines for BK virus-specific DNA sequences. *Proceedings of the National Academy of Sciences (USA)*, **75**, 454–8.

WONG-STAAL, F., HAHN, B., MANZARI, V., COLOMBINI, S., FRANCHINI, G., GELMANN, E. P. & GALLO, R. C. (1983). A survey of human leukaemias for sequences of a human retrovirus. *Nature*, **302**, 626–8.

YOSHIDA, M., MIYOSHI, I. & HINUMA, Y. (1982). Isolation and characterization of retrovirus from cell lines of human adult T-cell leukemia and its implication in the disease. *Proceedings of the National Academy of Sciences (USA)*, **79**, 2031–5.

ZENG, Y., LIN, Y., LIU, C., CHEN, S., WEI, J., ZHU, J. & ZAI, H. (1980). Application of an immunoenzymatic method and an immunoautoradiographic method for a mass survey of nasopharyngeal carcinoma. *Intervirology*, **13**, 162–8.

ZUCKERMAN, A. J. (1982). Primary hepatocellular carcinoma and hepatitis B virus. *Transactions of the Royal Society of Tropical Medicine and Hygiene*, **76**, 711–18.

ZUR HAUSEN, H. (1982). Human genital cancer: synergism between two virus infections or synergism between a virus infection and initiating events? *Lancet*, **ii**, 1370–2.

ZUR HAUSEN, H. & GISSMANN, L. (1980). Papillomaviruses. In *Viral Oncology*, ed. G. Klein, pp. 433–45. New York: Raven Press.

NEW VIRUS DISEASE SYNDROMES

J. R. PATTISON*, F. BROWN† and A. A. BRUNT‡

*Department of Medical Microbiology, King's College Hospital Medical School, London SE5 8RX, UK
†Wellcome Research Laboratories, Beckenham, Kent BR3 3BS, UK
‡Glasshouse Crops Research Institute, Worthing Road, Rustington, Littlehampton, West Sussex BN16 3PU, UK

INTRODUCTION

The word virus was first used as a non-specific adjunct to the term 'ultra-microscopic, filterable' in order to define the agents of disease that were clearly infectious but for which no causative bacterium could be found. Since then, viruses have become well defined in terms of both physico-chemical and biological properties. Nevertheless, their disease-producing potential has remained of central interest; for almost a century now there has been much fruitful research defining the viral aetiology of diseases of humans, animals and plants. Advances continue to be made and even today there are diseases which are widely regarded as viral but for which the specific causative agent remains to be identified.

Such is the extent of recent advances in the aetiology of virus diseases that it is impossible to review the entire field in a single chapter. By and large we have omitted those diseases which have already been extensively reviewed in the literature. For example, we make no mention of the role of rotaviruses in gastroenteritis in humans and animals nor of the viruses causing the relatively recently described haemorrhagic fevers, viz., Lassa fever, Marburg disease and Ebola. Moreover the examples chosen inevitably reflect the special interests of the authors.

By and large the occurrence of clusters of cases or larger outbreaks of disease provide the stimulus for a search for the causative virus. Sometimes a new virus is found to be the cause of a long-established disease. Sometimes a new disease emerges and this may be shown to be due to a new virus or to a virus which has been known for some time in another context. In arranging the material for this paper we have attempted to give examples of each of these.

ANIMAL VIRUSES

Old disease – newly recognised virus

The discovery of the human parvovirus and its associated diseases provides a good example of the establishment of the viral aetiology of some long-established diseases of presumed infective origin.

In the early 1970s much screening for hepatitis B surface antigen (HBsAg) was performed by counter-immuno-electrophoresis (CIE) using human sera as a source of antibody to HBsAg (anti-HBs). Clearly these were polyvalent sera and contained many anti-viral antibodies other than anti-HBs. As alternative techniques for the detection of HBsAg were introduced, a group of sera giving anomalous reactions emerged. These sera gave a line of precipitate by CIE but were negative for HBsAg by the more sensitive techniques of reverse passive haemagglutination and radioimmunoassay (Cossart *et al.*, 1975). Electron microscopy of the antigen-containing sera revealed not the characteristic pleiomorphic particles of HBsAg but virus particles with a fairly uniform diameter averaging 23 nm and an appearance resembling parvoviruses. The virus banded in caesium chloride at a density of 1.36–1.40 but was not serologically related to any of the four adeno-associated virus types or rat virus and did not agglutinate a range of animal and avian erythrocytes (Cossart *et al.*, 1975).

Recently Summers, Jones & Anderson (1983) have determined the nature of the nucleic acid of a serologically identical virus found in the serum of an asymptomatic blood donor in 1982. The nucleic acid is single-stranded DNA, 5.5 kb in length. Thus, the virus can be classified as a member of the Parvoviridae. However the human virus packages complementary strands into separate virions (Summers *et al.*, 1983). To date, this property has been regarded as a feature of members of the *Dependovirus* and the *Densovirus* genera but not of the autonomously replicating vertebrate pathogens of the genus *Parvovirus* (Andrewes, Pereira & Wildy, 1978). However, at least one of the autonomous parvoviruses (LuIII) incorporates complementary single-stranded DNA separately and at equal efficiency (Muller & Siegl, 1983). Moreover, using hybridization techniques, the human parvovirus has been shown to have sequence homology with the genomes of LuIII, minute virus of mice and the porcine parvovirus but no homology with adeno-associated viruses (P. Tattersall, personal communication). At present, therefore, it seems justifiable to regard the human virus as an autonomous parvovirus although the lack of an *in vitro* culture system does not permit final exclusion of the requirement for a helper virus.

Early studies on the clinical disease associated with human parvovirus infection indicated that the infection was either asymptomatic or associated with a variety of mild, non-specific symptoms (Paver & Clarke, 1976; Shneerson, Mortimer & Vandervelde, 1980). However, in 1981, Pattison and co-workers provided the first evidence for an association between human parvovirus infection and aplastic crises in chronic haemolytic anaemia (Pattison *et al.*, 1981). Aplastic crises are a well recognized complication of sickle-cell anaemia (SCA: MacIver & Parker-Williams, 1961) and are characterized by a cessation of red cell production in the bone marrow for 5–7 days with consequent drop in haemoglobin and disappearance of reticulocytes (immature red blood cells) from the peripheral blood. Examination of the bone marrow reveals an almost complete absence of erythrocyte precursors. The crisis is self-limiting with repopulation of the bone marrow and eventual rise of peripheral blood haemoglobin and reticulocytes to steady state values. The condition has long been regarded as having an infectious (probably viral) aetiology since it is associated with vague virus-like antecedent symptoms, cases cluster in time and within families, children are most commonly infected and it is rare for an individual to have more than one attack in a lifetime. Diagnosis of parvovirus infection can be based on finding virus in acute phase sera, or specific IgM antibody in convalescent sera taken up to 3 months after the episode of infection (Anderson *et al.*, 1982*a*; Cohen, Mortimer & Pereira, 1983). Using these criteria, all 14 cases of aplastic crisis occurring in children with SCA in the UK in 1979 and 1980, 23 of 28 cases occurring in Kingston, Jamaica in the same years, and all 7 adult cases presenting to Cook County Hospital, Chicago during 1980–2 have been shown to be associated with primary parvovirus infection (Serjeant *et al.*, 1981; Anderson, 1982; Anderson *et al.*, 1982*a, b*; Rao *et al.*, 1983).

The first patients with aplastic crises to be investigated all had homozygous SCA. However, one of the Chicago patients had thalassaemia intermedia; aplastic crises due to parvovirus infection have also been found in pyruvate kinase deficiency (Duncan *et al.*, 1983) and hereditary spherocytosis (Kelleher, *et al.*, 1982; Mortimer, 1983). All these diseases are characterized by the presence of a chronic haemolytic anaemia and the likely pathogenesis of parvovirus-induced aplastic crises is related to this underlying condition. The erythroid series in the bone marrow is hyperactive in these conditions and is thus a prime target for infection with a parvovirus since such viruses have a predilection for rapidly dividing cells (see next section). Mortimer *et al.* (1983) have shown that the human parvovirus inhibits the formation of erythroid colony forming units when human bone marrow is cultured *in vitro*, thus strengthening the suggestion that the parvovirus-induced aplastic crises are a result of a direct cytotoxic effect of the virus on erythroid progenitors in the bone marrow.

Infection with the human parvovirus is common. Early studies (Cossart *et al.*, 1975; Edwards *et al.*, 1981; Anderson *et al.*, 1982*b*), using the relatively insensitive technique of CIE, showed that 30–40% of adults have parvovirus antibody. In a more recent study (Cohen *et al.*, 1983) 43% of 310 blood donors were antibody positive by CIE, but 61% were positive by radioimmunoassay for IgG antibody. The assumption that most of these infections are asymptomatic or mild and non-specific has been challenged by recent studies indicating that the human parvovirus is the cause of a common disease of children. Erythema infectiosum (Fifth disease) is a mild, acute exanthematous disease, first described 70 years ago, which has always been assumed to be due to a virus (Balfour, 1976; Gershon, 1979). The erythematous maculopapular rash often begins on the face, giving the so-called 'slapped cheek' appearance, while on the extremities the rash has a characteristic reticular or lacy appearance. There are few, if any, constitutional symptoms though arthralgia is not uncommon, especially in adults. Epidemics tend to occur in late winter and spring. Recently 41 cases from a large epidemic in London have been studied for evidence of recent human parvovirus infection (Anderson *et al.*, 1983). Parvovirus-specific IgM was detected in all sera from 31 cases in children, 2 in adolescents and 6 of the 8 adult cases. Other cases associated with primary parvovirus infection are now being found and, on the basis of this preliminary evidence, it is proposed that the human parvovirus is the hitherto elusive agent of erythema infectiosum.

New disease – new virus?

The sudden occurrence of epidemic disease due to canine parvovirus (CPV) infection certainly represented the emergence of a new disease and at first seemed to suggest the emergence of a new virus.

Subsequent work, however, has shown that the virus may have been derived from a pre-existing feline virus although its origin is far from certain.

Eugster & Nairn (1977) demonstrated parvovirus-like particles in the faeces of a litter of puppies with diarrhoea in Texas; in the following year there were a number of severe outbreaks of canine enteritis associated with the same virus (Eugster, Bendele & Jones, 1978). Almost immediately similar outbreaks were described in other parts of the USA (Appel, Scott & Carmichael, 1979), Canada (Gagnon & Povey, 1979), Australia (Kelly, 1978) and the UK (McCandlish et al., 1979). The causative role of CPV in this disease has been demonstrated by finding parvovirus-like particles in the faeces of dogs (e.g., Eugster & Nairn, 1977) by demonstrating CPV antigen in intestinal crypt epithelial cells using immunofluorescence (Miller, Evermann & Ott, 1980) and by demonstrating seroconversion for CPV antibody in association with the disease (Gagnon & Povey, 1979; Eugster, 1980).

At the same time as enteritis due to CPV emerged there were a number of outbreaks of sudden death in puppies due to myocarditis (Jeffries & Blakemore, 1979; Thompson et al., 1979; Wilkinson, 1979). In some outbreaks there were no premonitory signs but in others, especially those involving older puppies, there was severe diarrhoea, rapid dehydration and neutropaenia. CPV was shown to be the cause of these outbreaks of myocarditis (e.g., McCandlish et al., 1979) and the viruses isolated from cases of myocarditis and enteritis are serologically indistinguishable (Lenghaus & Studdert, 1980).

The occurrence of two distinct clinical syndromes due to the same virus requires explanation, especially since some infected litters exhibit one syndrome or the other and some a mixture of the two. The explanation lies in the dependence of parvovirus replication on events occurring late in the S-phase of the cell cycle (Siegl, 1976). As a consequence, extensive multiplication *in vivo* of parvoviruses is restricted to tissues providing a large number of dividing cells. In the very young pup there is a rapid growth of heart muscle and the mitotic index is high. At this time there is a relatively low turnover of intestinal cells. In weaning and weaned pups there is an increase in intestinal epithelial cell turnover with alteration in diet and establishment of gut flora while cardiac muscle cell division occurs at a diminished rate (McCandlish et al., 1979). Thus, the disease syndrome produced by CPV infection will depend upon the age of the animal and the related physiological state of its myocardium and intestinal epithelium.

One of the most intriguing questions about CPV concerns its origin and the reason why it was able to cause disease almost simultaneously in at least three continents during 1977–9. Soon after the first outbreaks of enteritis and myocarditis, the causative virus was isolated in cell culture of both feline and canine origin and shown to be a typical parvovirus (Johnson & Spradbrow, 1978). These authors also showed that CPV was not related to rat, mouse or porcine parvoviruses but showed serological cross-reactivity with feline panleucopaenia virus (FPV). This led to the proposition that the canine virus developed by mutation from FPV and was then disseminated in a biological product such as a vaccine (Johnson & Spradbrow, 1978; Lenghaus & Studdert, 1980). No CPV infection in dogs can be identified prior to 1976 (Johnson & Spradbrow, 1978; McCandlish et al., 1980), whereas disease which can be presumed to be due to FPV infection has been known for almost a century.

Filtration experiments (e.g., Hindle & Findlay, 1932) initially established the viral aetiology of feline enteritis and the virus was eventually characterized as a parvovirus (Johnson & Cruikshank, 1966; Johnson, Siegl & Gautschi, 1974). At first it was thought that members of the family Felidae were the only natural hosts of FPV. However severe epizootics of enteritis very similar to infectious feline enteritis were observed in ranch mink in Canada in 1947 and 1950–2 (e.g., Schoffield, 1949). A close serological relationship between FPV and mink enteritis virus (MEV) was established by cross-protection tests and the serological identity of the two viruses was subsequently demonstrated (Johnson, 1967). As with FPV and MEV, cross-protection studies indicate a serological relationship between FPV and CPV but these latter two viruses are clearly not serologically identical. Differences can be shown by serum neutralization (Lenghaus & Studdert, 1980), precipitation-in-agar tests (Flower, Wilcox & Robinson, 1980) and by variations in reactivity with monoclonal antibodies (Parrish, Carmichael & Antczak, 1982).

Most recently the relationship between CPV, FPV and MEV has been investigated by genome analysis. MEV and CPV can be readily distinguished by restriction enzyme analysis of their double-stranded replicative form DNA, the cleavage pattern of the two viruses differing in 11 of the 79 mapped restriction sites (McMaster, Tratschin & Siegl, 1981). Strains of MEV and wild-type FPV, however, differ by only one or two restriction sites (Tratschin et al.,

1982). These authors also examined five vaccine strains of FPV. Three of the five had identical genomes to MEV/FPV. However, one lacked seven of the restriction sites characteristically found at the 5' end of the MEV/FPV genome. This indicates that vaccine strains can differ substantially from prototype strains but the differences found with this particular vaccine strain were in no way characteristic of CPV. On the other hand another of the vaccine strains had three additional *Hinf*I sites compared to the MEV/FPV genome and all three of these sites are characteristic of the CPV genome. At least five additional mutations would be required to generate the typical CPV genome, but, nevertheless, it is difficult to ignore the theory that FPV is the ancestor of CPV. The corollary of this is that CPV was disseminated in the form of a feline vaccine and this seems more likely than the simultaneous mutation of FPV to a virus virulent for dogs in three continents.

New diseases – old viruses

During the last decade or so, three virus diseases of animals have emerged for which the agent responsible was already known as a cause of disease in another species. The diseases are encephalomyocarditis of pigs, swine vesicular disease and a virus infection of sea lions causing abortion. Encephalomyocarditis (EMC) virus was better known as an agent infecting rodents, whereas swine vesicular disease virus is better known as Coxsackie B5 virus, an agent causing illness in man. The third disease agent, that isolated from aborting sea lions, had been responsible for causing vesicular exanthema in pigs in the USA during 1932–56. This disease is clinically indistinguishable from foot-and-mouth disease and it assumed considerable economic importance because of this similarity. All three examples focus attention on the pig as a reservoir of infection.

Encephalomyocarditis of pigs
EMC virus was first isolated from cotton rats by Jungeblut & Sanders in 1940 (see Jungeblut, 1958). The virus has subsequently been recovered from a variety of mammals and birds, and also from blood-sucking arthropods in many different parts of the world. Although the virus is generally thought to be a rodent virus, in which group it causes both overt disease and inapparent infection, it

has been associated with occasional cases of central nervous system disease in humans and with several outbreaks of fatal myocarditis in pigs in geographically separated parts of the world. Outbreaks have been reported in the USA, Panama, Australia, New Zealand and South Africa. In a survey of sera, Tesh & Wallace (1978) found evidence for the presence of EMC virus neutralizing antibody in a wide range of species in Hawaii. They concluded that the virus infects many species, probably by the oral route. Oral infection of rats and mice did not often produce an intestinal carrier state and contact transmission among these rodents is rare. In fact, rats appeared to be dead-end hosts for the virus and unlikely to be the source of infection for other animal species. The popular view that rodents are the natural reservoir of the virus could not be substantiated; Tesh & Wallace were unable to identify the natural reservoir.

The fatal disease in pigs has been reported on several occasions from different parts of the world. The observations by Tesh & Wallace (1978) and by Sangar, Rowlands & Brown (1977) that more than 20% of pigs have antibody in their sera in the absence of any reports of disease point strongly to the presence of viruses with extremely variable pathogenic activities. Only a superficial comparison has been made of the physico-chemical properties of the viruses causing the outbreaks in pigs with the laboratory strains of the virus. D. J. Rowlands and his colleagues (unpublished) found that the pattern of separation of the capsid proteins in polyacrylamide gel electrophoresis (PAGE) was the same for several isolates. However, the length of the polycytidylic acid tract of the RNAs differed markedly, ranging from 50 to 500 nucleotides (Black et al., 1979). Similar ribonuclease T1 maps have shown many differences (K. J. H. Robson, unpublished) but an ordered comparison of virulent and avirulent strains has not been made. A comparison of the physico-chemical properties of isolates causing disease in pigs with those of isolates from other animals in the same geographical location at the same time would be illuminating.

Swine vesicular disease

Swine vesicular disease (SVD) was first described in Italy in 1966. The outbreaks, which were clinically indistinguishable from foot-and-mouth disease, occurred in pigs housed indoors (Nardelli et al., 1968). About 10 000 animals were involved. When the causal agent could not be allotted by serological methods to one of the seven known serotypes of foot-and-mouth disease virus, recourse was made to physico-chemical methods for its identification. It was shown to be an enterovirus on the basis of its appearance in the electron microscope, stability in acid conditions, buoyant density in caesium chloride and the base composition of its RNA (Nardelli *et al.*, 1968; Newman, Rowlands & Brown, 1973).

A crucial observation in the identification of the agent was made by Graves (1973) who showed that it was neutralized by antiserum to Coxsackie B5 virus. Subsequent work on the serological and physico-chemical properties of swine vesicular disease virus (SVDV) and Coxsackie B5 virus confirmed this relationship but drew attention to the variation between isolates of both viruses. The differences between individual Coxsackie B5 virus isolates by RNA homology tests was as high as 50% (as much as between these viruses and SVDV). The variation between different isolates of SVDV was much smaller, rarely exceeding 10%. Comparable differences were also found by a variety of serological tests (Harris *et al.*, 1979).

These observations suggest that the environmental pressure on Coxsackie B5, whether due to the antibody present in the sera of most individuals or other factors, is greater than that on SVDV. This would be expected in view of the much shorter life span of the pig. Nevertheless ribonuclease T1 maps of the RNAs of SVDV isolates in the UK over the period 1972–80 show the gradual emergence of new strains of the virus. The maps of recent isolates are readily distinguished from those obtained in 1972.

Graves (1973) observation was of considerable importance because it suggested the intriguing possibility that SVD was caused by a virus that normally affects humans. Indeed, at least four cases of Coxsackie-like illness were described in which the causative agent was SVDV (Brown, Goodridge & Burrows, 1976). However, there is still no convincing evidence that Coxsackie B5 virus will cause vesicular disease in pigs although both viruses produce brain and spinal cord lesions when administered by intravenous inoculation. Despite these observations, the emergence of SVD points to the vulnerability of a host species to viruses not normally associated with that species.

Vesicular exanthema of pigs
This disease was first described in pigs in the USA in 1932. Its close clinical similarity to foot-and-mouth disease resulted in extreme

precautionary measures being taken regarding the control of the disease and it was not until some years later that it was realized that a different agent was involved. Serological tests were then used to distinguish between the two viruses. The last outbreak of the disease, which has been confined to the USA, occurred in 1956.

Characterization of the agent was not pursued at that time. It was considered to be closely related to the picornaviruses but it was not until the occurrence in 1966 of SVD in Italy (referred to above) that serious efforts were made to assign it to one of the existing families of viruses. It soon became apparent that the virus differed from the picornaviruses in possessing a unique morphology and in containing only one major protein (Bachrach & Hess, 1973; Brown & Hull, 1973; Burroughs & Brown, 1974). This work led directly to a reassessment of the nature of the so-called feline picornavirus, an agent causing upper respiratory infection in cats (see review by Gillespie & Scott, 1973) which has a morphology indistinguishable from that of vesicular exanthema virus (VEV). Physico-chemical analysis showed that the feline virus was similar to the pig virus and they are both now recognized as belonging to a new family, the Caliciviridae (Matthews, 1982). The status of the viruses, isolated from the faeces of humans and pigs, which have a similar morphology is still uncertain because of the lack of information on their physico-chemical properties.

In 1972, several isolates of a calicivirus were made by Smith *et al.* (1973) from sea lions on San Miguel Island off the coast of California during a study of abortions in that species. The virus, which was called San Miguel sea lion virus, was found to be indistinguishable from VEV in its morphology and biochemical properties. Further studies of marine mammal populations have provided additional isolates from sea lions, Alaskan fur seals and elephant seals. The virus has also been isolated from the opal eye perch, which is a food source of the sea lion and from a liver fluke, which is a parasite of the sea lion (Smith, Skilling & Ritchie, 1978). Moreover, serological evidence has been obtained that the agent has been prevalent in other pinniped species and whales in the same area since at least 1961. Since neutralizing antibodies have also been found in several species of wild terrestrial mammals, it seems apparent that the agent is readily transmitted between marine and terrestrial species and lends support to the suggestion that the original outbreaks of VEV in pigs were caused by the feeding of these animals on the sea shore.

Detailed physico-chemical studies, including RNA hybridization experiments, have shown that the viruses isolated from the marine animals are very closely related to VEV (Burroughs, Doel & Brown, 1978). However, the feline virus appears to be only distantly related to these viruses. There is no clear indication of the relationship of the calicivirus-like particles isolated from human faeces to the pig and feline viruses.

PLANT VIRUS DISEASES

Old diseases – newly recognized viruses

Geminiviruses

Some diseases of tropical and sub-tropical crops that have caused concern since the early years of this century have recently been shown to be induced by newly recognized geminiviruses. Such diseases include maize streak (Fuller, 1901), tobacco leaf curl (Peters & Schwartz, 1912), sugar beet curly top (Ball, 1906) and possibly also cassava mosaic (Warburg, 1894). Some other important plant diseases which were first described 40–50 years ago (including chloris striate, euphorbia mosaic, tomato golden mosaic, tobacco yellow dwarf and bean summer death) are also now known to be induced by geminiviruses.

Leafhopper-borne geminiviruses such as maize streak virus (MSV) and chloris striate mosaic virus (CSMV) characteristically induce chlorotic leaf striping in susceptible cereal crops. Other geminiviruses transmitted by whiteflies (such as tobacco leaf curl virus, TLCV) typically cause severe leaf curling and distortion whereas those such as bean golden mosaic virus (BGMV) cause almost complete leaf chlorosis. Most geminiviruses severely reduce the growth and yield of infected plants and so cause severe crop losses. Because of their economic importance some geminiviruses, such as MSV and beet curly top virus (BCTV), have been investigated intensively for several decades. Thus the relationship between BCTV and the vector (*Circulifer tenellus*) in relation to epidemiology was known over 70 years ago, and disease control by use of resistant cultivars, insecticides and crop management has been practised since about 1950 (Bennett, 1971). Similarly, studies since 1920 on maize streak disease and its vector (*Cicadulina mbila*) by H. H. Storey and his colleagues have provided much information about its pathology and control (Bock, Guthrie & Woods, 1974). However, the aetiology of these and similar diseases has been elucidated only during the past decade, when improvements in purification methods permitted the viruses to be isolated and characterized.

The geminivirus group, first recognized by the International Commission for Taxonomy of Viruses (ICTV) in 1978 (Matthews, 1979), now contains five members and nine probable members (Matthews, 1982). The viruses are spread naturally by either whiteflies or leafhoppers, but only one or two are transmitted experimentally by mechanical methods of inoculation. The geminiviruses that have been most studied are BGMV, CSMV and Cassava latent virus (CLV). They have individual particles which measure about 18×20 nm but which often occur in pairs with overall dimensions of 20×30 nm. The particles contain a single coat protein (polypeptide molecular weight of $28–34 \times 10^3$) and about 20% positive-sense single-stranded DNA (molecular weight 0.7–0.8×10^6) which occurs predominantly as covalently closed circular molecules. The single-stranded DNA isolated from BGMV particles is infective when inoculated into bean protoplasts (Goodman, 1981). The single-stranded DNA of BGMV and other geminiviruses contains about 2510 nucleotides; they are thus the smallest nucleic acid molecules found in autonomously replicating viruses. Analyses of BGMV-specific double-stranded DNA from infected plants after treatment with restriction endonuclease, however, indicate that the fragment sizes contain approximately 5000 nucleotides, suggesting that the BGMV has a divided genome consisting of two circular DNA molecules indistinguishable in size but differing in nucleotide sequence and genetic content (Haber et al., 1981).

Electron microscope investigations of infected plant sections has shown that some geminiviruses occur only within phloem cells, although both those transmitted in cereal crops by leafhoppers (MSV and CSMV) are also readily detected in other tissues. All the geminiviruses that have been investigated have been found to accumulate within nuclei where replication is assumed to occur.

Serological investigations indicate that geminiviruses are immunologically distinct. Thus although tobacco yellow dwarf virus is related to BCTV (Thomas & Bowyer, 1980) and CLV to BGMV (Sequeira & Harrison, 1982), MSV and CLV are apparently unrelated (Bock et al., 1974; Bock et al.,1977) as are MSV, CSMV and wheat dwarf virus (Francki et al., 1979; Lindsten et al., 1980).

Luteoviruses

Because of technical difficulties, much less is known about luteo-viruses than most other virus groups. They are not sap-transmissible, are confined to host phloem tissues and are purified with difficulty (often less than 100 μg of pure virus is recovered kg^{-1} plant tissue); moreover, all are transmitted in the circulative manner by aphids, and some only from plants also containing a second 'assistor' virus. Nevertheless, intensive and careful studies on barley yellow dwarf (BYDV), beet western yellows (BWYV) and potato leaf roll (PLRV) viruses facilitated the subsequent recognition that other luteoviruses also cause some well known and important diseases (Rochow & Brakke, 1964; Duffus, 1972).

The luteovirus group was first recognized by ICTV in 1975 (Shepherd *et al.*, 1976), and it now contains 14 members and 19 possible members because they have been incompletely charac-terized and, as luteoviruses are often serologically interrelated and some have extensive natural host ranges, further study might show some to be strains of previously described viruses. Luteoviruses have isometric particles about 25 nm in diameter which sediment at approximately 115–127 S and contain positive sense single-stranded RNA (molecular weight 1.9–2.0 × 10^6) and a single structural polypeptide (molecular weight 24 × 10^3). Although the viruses are recovered in low yields from infected plants, they are moderately stable *in vitro*.

Plants infected with luteoviruses can be severely stunted, and when chronically infected, have mature leaves which are yellow and/or reddened and brittle. Such symptoms result, at least partially, from the collapse of conducting tissues in infected plants. The severity of symptoms is dependent upon the tolerance of the plant species and cultivar, and on the age and physiological condition of infected plants. Environmental conditions also greatly affect the development and severity of symptoms; thus, those of some viruses such as BWYV are intensified in plants grown in high light intensity and those of others (like BYDV) at lower ambient temperatures (Rochow & Duffus, 1980).

Many luteoviruses, although recognized only recently, are now known to induce many economically important diseases which have caused concern for two decades or more. Thus, individual luteo-viruses have been shown during the past few years to cause diseases such as malva yellows, pea leaf roll, milk-vetch dwarf, subterranean clover stunt, banana bunchy top, beet yellow net, carrot red leaf, celery yellow spot, cotton anthocyanosis, raspberry leaf curl, strawberry mild yellow edge, tobacco vein distortion, tomato yellow net and tomato yellow top (Rochow & Duffus, 1980).

The characteristics of some better known members of the luteovirus group have also facilitated the recognition that similar, hitherto undescribed viruses, cause other more recently observed diseases such as beet mild yellowing (Duffus & Russell, 1975), legume yellows (Duffus, 1979), solanum yellows (Milbrath & Duffus, 1978), soybean dwarf (Tamada & Kojima, 1977), tobacco necrotic dwarf (Kubo & Takanami, 1979) and filaree red leaf (Sylvester & Osler, 1977).

New diseases – newly recognized pathogens

Satellite pathogenic RNA in cucumoviruses

Cucumber mosaic virus (CMV), the type member of the cucumovirus group (Matthews, 1982), first caused concern almost 70 years ago to cucurbit growers in the USA (Doolittle, 1916; Jagger, 1916). It is now known to occur naturally throughout temperate and sub-tropical regions in more than 470 species in 67 plant families (Horvath, 1979), and new hosts are continually being reported (Kaper & Waterworth, 1981). Numerous aphid species are able to transmit the virus efficiently in the non-persistent manner.

CMV particles are isometric, about 30 nm in diameter, sedimenting at about 100 S and containing a single polypeptide species (molecular weight 24.5×10^3) and four species of single-stranded RNA (molecular weight 1.27, 1.13, 0.82 and 0.35×10^6). The three larger species form the viral genome and the smallest contains the coat protein gene; the two larger species are encapsidated separately, but the remaining two occur together. Thus although the three types of particles have similar sedimentation properties, they contain different RNA species. The three larger RNA molecules are required for infectivity (Peden & Symons, 1973; Lot et al., 1974); apart from a common sequence of about 200 nucleotides at their 3′ termini, the three species are not otherwise homologous. The nucleotide sequence of the smallest species is present also in the next largest.

It has been demonstrated experimentally that the corresponding molecules of different virus strains can be exchanged to produce stable hybrids or pseudo-recombinants (Habili & Francki, 1974). This possibly explains the origin of the very wide range of naturally occurring strains of CMV. In many hosts, strains of CMV often induce conspicuous symptoms in leaves and/or flowers. In recent years, however, it has been shown that symptoms are greatly modified when the virus occurs together with a satellite pathogenic RNA.

Although several satellite RNAs are now known, the first described and best known has been designated CARNA 5 (*CMV-associated RNA*) by Kaper & Waterworth (1977) and as (n) CARNA 5 by Kaper, Tousignant & Thompson (1981). CARNA 5 is a single-stranded RNA molecule with a molecular weight of about 0.1×10^6 (Kaper, Tousignant & Lot, 1976; Mossop & Francki, 1977; Gould & Symons, 1978) which, because of its secondary structure, is more stable than any of the four CMV RNAs (Mossop & Francki, 1979*a*). Its sequence of 335 nucleotides is known, and its capped 5' terminus suggests that it acts *in vivo* as a mRNA (Richards *et al.*, 1978). CARNA 5 shows no sequence homology with any of the four CMV RNAs (Gould & Symons, 1978) and, as it is not a subgenomic viral RNA and is totally dependent upon CMV for its replication, it is recognized as a satellite RNA (Diaz-Ruiz & Kaper, 1977; Lot, Jonard & Richards, 1977; Mossop & Francki, 1978). A double-stranded replicative form of CARNA 5 (molecular weight 0.22×10^6) has been isolated from infected tobacco plants and protoplasts (Kaper & Diaz-Ruiz, 1977; Takanami, Kubo & Imaizumi, 1977). Using the wheat germ system, translation studies *in vitro* indicate that CARNA 5 codes for the formation of two polypeptides (with molecular weights of 5.2×10^3 and 3.8×10^3) and has the following genetic map (Owens & Kaper, 1977):

Symptom severity in plants containing both CMV and CARNA 5 is dependent upon the host species. Thus, when present together in tomato, these viruses induce a lethal disease described as tomato necrosis, whereas CMV alone induces only leaf chlorosis and some leaf distortion. The severe necrotic disease was first observed in 1972, in France, where infection caused total loss of some field crops in Alsace (Putz *et al.*, 1974). It has also since occurred sporadically in Japan (Takanami, 1981) and the USA (Kaper & Waterworth, 1981).

When CARNA 5 and CMV are present together in tobacco (*Nicotiana tabacum*), pepper (*Capsicum frutescens*) and maize (*Zea mays*), however, symptoms are milder than in comparable plants containing CMV only (Takanami, 1981; Waterworth, Kaper & Tousignant, 1979). In these hosts, large amounts of satellite RNA are produced and the concentration of viral RNA is correspondingly lower; these results indicate that the remission of symptoms is possibly attributable to the reduction in virus replication due to the competition with that of the satellite RNA (Kaper & Tousignant, 1977).

Another satellite RNA (designated CMV satellite RNA), has the same basic sequence homology as CARNA 5, is dependent upon the two larger CMV RNAs for replication, and ameliorates symptoms of CMV in some hosts (Mossop & Francki, 1979*b*). However, unlike CARNA 5, when present with CMV this satellite induces mild symptoms only in tomato (Mossop & Francki, 1978). A satellite RNA which has yet to be further characterized has been shown to cause the so-called white leaf disease of tomato in New York State, and to occur with CMV without causing conspicuous symptoms in tobacco (Gonsalves, Provvidenti & Edwards, 1982). It is interesting that a satellite RNA associated with peanut stunt virus (PARNA 5), a cucumovirus serologically related to CMV, has no sequence homology with CARNA 5 (Kaper & Tousignant, 1978).

Newly recognized viruses

Intensification of research and modern methods of analysis have facilitated the rapid recognition of hitherto undescribed viruses as the causes of inadequately studied or newly recognized diseases. Because innumerable diseases have been shown in recent years to be virus-induced, we cite here only a few recently described examples.

Labile tobamoviruses. These are viruses which, in some respects, resemble very stable and well characterized tobamoviruses, such as tobacco mosaic and cucumber green mottle mosaic viruses, and are possible members of the tobamovirus group (Matthews, 1982). They differ markedly, however, in that they occur within infected plants in low concentration, have a low specific infectivity and are unstable *in vitro*; some have fungal vectors.

Despite considerable technical difficulties in studying such viruses, wheat soil-borne mosaic (WSBMV) and potato mop-top (PMTV) viruses were partially characterized during the late 1960s and early 1970s, and were then shown to be serologically distantly related to tobacco mosaic virus (Harrison & Jones, 1970; Kassanis, Woods & White, 1972; Tsuchizaki, Hibino & Saito, 1973). The development of techniques for investigating these two viruses, and defective strains of tobacco mosaic virus (Kassanis & Woods, 1969), undoubtedly facilitated the subsequent recognition that similar viruses induced diseases such as beet necrotic yellow vein (BNYVV: Tamada, Abe & Baba, 1975; Putz & Vuittenez, 1974), *Nicotiana velutina* mosaic (Randles, Harrison & Roberts, 1976), peanut clump (Thouvenel, Dollet & Fauquet, 1976), Indian peanut clump (Reddy *et al.*, 1983) and hypochoeris mosaic (Brunt & Stace-Smith, 1978). These viruses characteristically cause leaf chlorosis and stunting of infected plants.

The recognition that such viruses occur in all five continents and that some induce severe diseases in important crop plants stimulated studies on their natural mode of spread. These have resulted in the identification of *Polymyxa graminis*, *P. betae* and *Spongospora subterranea* as the vectors of WSBMV, BNYVV and PMTV, respectively (Calvert & Harrison, 1966; Estes & Brakke, 1966; Tamada *et al.*, 1975).

Labile tobamoviruses have fragile rod-shaped particles with a loosely coiled helix (helical pitch 2.6–2.8 nm) which characteristically uncoils and/or fractures. The particles are 18–21 nm in diameter and, presumably due to fragmentation, have a very wide length distribution; unlike those of stable tobamoviruses, very few particles are 300 nm long, but they characteristically have two predominant lengths of about 100–140 and 200–240 nm. The particles contain a single structural polypeptide of molecular weight 21×10^3 and four species (or fragments) of single-stranded RNA with molecular weights of 0.6, 0.7, 1.8 and 2.3×10^6.

Closteroviruses. Members of the closterovirus group occur throughout the world in various plant species, and some cause severe crop losses. Typical closteroviruses are transmitted in the semi-persistent manner by aphids, and have very flexuous filamentous particles 1200–2000 nm long which contain about 5% positive-sense single-stranded RNA (molecular weight $2.2–4.7 \times 10^6$) and one coat polypeptide $23–27 \times 10^3$). The particles often form cross-banded aggregates in phloem cells (Matthews, 1982). A few possible members of the group have very flexuous filamentous particles measuring only 600–800 nm.

Because of the technical difficulties in purifying closteroviruses, they have been less well characterized than viruses of other groups. Nevertheless, experience gained in purifying and characterizing sugarbeet yellows and apple chlorotic leafspot viruses (Bar-Joseph & Hull, 1974; Bar-Joseph, Hull & Lane, 1974) facilitated subsequent studies on the pathology of similar viruses such as burdock yellows, carnation necrotic fleck, citrus tristeza, carrot yellow leaf, clover yellows, lilac chlorotic leafspot, potato T and Heracleum latent (Bar-Joseph, Garnsey & Gonsalves, 1979).

Cryptic viruses. During the past few years, the introduction of sensitive assay procedures has permitted the detection and identification of so-called cryptic viruses in a wide range of plant species including beet (*Beta vulgaris*), broad bean (*Vicia faba*), carnation (*Dianthus caryophyllus*), ryegrass (*Lolium perenne*), poinsettia (*Euphorbia pulcherrima*), alfalfa (*Medicago sativa*) and clovers (*Trifolium* spp.). The viruses investigated so far are serologically distinct but are alike in having isometric particles about 30 nm in diameter which occur within symptomlessly infected plants in low concentration and, although not transmitted mechanically, are extensively seed borne. Those from carnation and ryegrass have been shown to contain segmented double-stranded RNA, and so resemble some viruses which infect fungi (Milne *et al.*, 1982).

Club-shaped virus-like particles in mushrooms. A novel type of virus-like particle has been associated with a lethal disease of mushrooms (*Agaricus bisporus*) occurring in the Federal Republic of Germany and South Africa (Lesemann & Koenig, 1977; Atkey & Barton, 1979). Fluid expressed from diseased mushrooms contains particles that, although variable in shape, are often clavate and about 150 nm long and 25–50 nm in diameter. The particles are sometimes found attached to, or apparently budding from, membrane fragments. The particles are easily disrupted by osmotic shock and by mechanical shearing forces. They have, however, been partially purified and concentrated by ion exchange and molecular permeation chromatography followed by diafiltration. The particles contain three major (molecular weights 49.5, 42.5 and 30×10^3) and two minor (molecular weights 24 and 19.5×10^3) polypeptides, but have yet to be further characterized.

PROSPECTS FOR THE FUTURE

We have cited in this chapter some examples of the continued recognition of viruses as the aetiological agents of diseases of humans, animals and plants. This has been a continuous process for the whole of this century and there is no reason to believe that we have reached the end of the line. However the pace of such discoveries may slow down partly because so much has been achieved already and partly because the problems which remain are likely to be the more difficult ones. In conclusion, it is worth considering whether any general principles can be established and if so whether we can make any predictions about the future.

If the occurrence of a disease itself acts as a stimulus for the search for a viral aetiology then the emergence of new disease syndromes will ensure that the activity continues. This occurs in a number of ways. First, there are the wholly new disease syndromes and these continue to emerge. In this category we recounted the example of canine parvovirus infection but the recently emerged disease which is currently of intense interest to virologists is the acquired immune deficiency syndrome (AIDS) in humans. The known epidemiological aspects of this disease strongly suggest that a virus is at least an essential component of the aetiology. Second, it is notable that there is a limited range of disease syndromes manifested by a particular host and the establishment of the viral aetiology of some cases allow subsets to emerge which are not related to known viruses. It is clear that on clinical grounds both swine vesicular disease and vesicular exanthema of pigs were confused with foot-and-mouth disease and yet it became clear that all three were due to different viruses. Human hepatitis is another good example, where the definition of the viruses of hepatitis A and hepatitis B has resulted in the emergence of the unsolved problem of non-A, non-B hepatitis. Third, and more subtly, host and environmental factors may play a role in determining the pattern of viral disease; this is exemplified by the various manifestations of Epstein-Barr virus infection. These range from the universal sub-clinical infection or associated infectious mononucleosis in adolescents and young adults, to the geographically limited Burkitt lymphoma and nasopharyngeal carcinoma and finally to the fatal lymphoproliferative syndrome which is a clear-cut X-linked recessive disorder.

We have noted, particularly in relation to plant virus diseases, some examples of unusual pathogenic agents (such as satellite pathogenic RNA). By contrast, the recently described viruses causing disease in animals tend to be members of already established groups. In addition to the examples we have given, hepatitis A virus is likely to be classified as an enterovirus; the virus of

Korean haemorrhagic fever has properties in common with orbi-viruses. More intriguing is the recognition of the way (not previously appreciated) in which well characterized viruses cross species barriers. It seems likely that virology still has some surprises lying in wait for us in this respect.

Finally it is worth considering how developments in virological techniques contribute to the establishment of the causative role of viruses in disease. In the case of diseases of plants there seems little doubt that the intensification of research, improvements in techniques and the application of modern analytical methods have led to the dramatic increase in knowledge of the aetiological role of viruses. Again, this contrasts with the situation in animal virus diseases in which at least the preliminary evidence for an aetiological role of a virus has been obtained using the microscopic, serologic or culture methods of classical virology. However, this presupposes that a productive viral infection occurs during the stage of disease being investigated. This will be true of acute virus infections but may not be true of persistent virus infections that may be associated with sub-acute and chronic diseases. Perhaps nucleic acid probes will help to open up this aspect as a new chapter of virus/disease associations.

REFERENCES

ANDERSON, M. J. (1982). The emerging story of a human parvovirus-like agent. *Journal of Hygiene*, **89**, 1–8.

ANDERSON, M. J., DAVIS, L. R., HODGSON, J., JONES, S. E., MURTAZA, L., PATTISON, J. R., STROUD, C. E. & WHITE, J. M. (1982b). Occurrence of infection with a parvovirus-like agent in children with sickle cell anaemia during a two-year period. *Journal of Clinical Pathology*, **35**, 744–9.

ANDERSON, M. J., DAVIS, L. R., JONES, S. E., PATTISON, J. R. & SERJEANT, G. R. (1982a). The development and use of an antibody capture radioimmunoassay for specific IgM to a human parvovirus-like agent. *Journal of Hygiene*, **88**, 309–24.

ANDERSON, M. J., JONES, S. E., FISHER-HOCH, S. P., LEWIS, E., HALL, S. M., BARTLETT, C. L. R., COHEN, B. J., MORTIMER, P. P. & PEREIRA, M. S. (1983). Human parvovirus, the cause of erythema infectiosum (Fifth disease)? *Lancet*, **i**, 1378.

ANDREWES, C. M., PEREIRA, H. G. & WILDY, P. (1978). *Viruses of Vertebrates*, 4th edn, 255 pp. London: Ballière Tindall.

APPEL, M. J. G., SCOTT, F. W. & CARMICHAEL, L. E. (1979). Isolation and immunisation studies of a canine parvo-like virus from dogs with haemorrhagic enteritis. *Veterinary Record*, **105**, 156–9.

ATKEY, P. T. & BARTON, R. J. (1979). Club-shaped virus-like particles. *Report of the Glasshouse Crops Research Institute for 1978*, 147 pp.

BACHRACH, H. L. & HESS, W. R. (1973). Animal picornaviruses with a single major species of capsid protein. *Biochemical and Biophysical Research Communications*, **55**, 141–9.

BALFOUR, H. H. (1976). Fifth disease: full fathom five. *American Journal of Diseases of Children*, **130**, 239–40.

BALL, E. D. (1906). Sugar beet curly top. *Report of the Utah Agricultural Experiment Station for 1904–5*.

BAR-JOSEPH, M., GARNSEY, S. M. & GONSALVES, D. (1979). The closteroviruses: a distinct group of elongated plant viruses. *Advances in Virus Research*, **25**, 93–168.

BAR-JOSEPH, M. & HULL, R. (1974) Purification and partial characterisation of sugar beet yellows virus. *Virology*, **62**, 552–62.

BAR-JOSEPH, M., HULL, R. & LANE, L. C. (1974). Biophysical and biochemical characterisation of apple chlorotic leafspot virus. *Virology*, **62**, 563–6.

BENNETT, C. W. (1971). The curly top disease of sugarbeet and other plants. *Monograph* 7. St Paul (Minn.): American Phytopathological Society. 82 pp.

BLACK, D. N., STEPHENSON, P., ROWLANDS, D. J. & BROWN, F. (1979). Sequence and location of the poly C tract in aphtho- and cardiovirus RNA. *Nucleic Acids Research*, **6**, 2381–90.

BOCK, K. R., GUTHRIE, E. J., MEREDITH, G. & BARKER, H. (1977). RNA and protein components of maize streak and cassava latent viruses. *Annals of Applied Biology*, **85**, 305–8.

BOCK, K. R., GUTHRIE, E. J. & WOODS, R. D. (1974). Purification of maize streak virus and its relationship with streak diseases of sugar cane and *Panicum maximum*. *Annals of Applied Biology*, **77**, 289–96.

BROWN, F., GOODRIDGE, D. & BURROWS, R. (1976). Infection of man by swine vesicular disease virus. *Journal of Comparative Pathology and Therapeutics*, **86**, 409–14.

BROWN, F. & HULL, R. (1973). Comparative virology of the small RNA viruses. *Journal of General Virology*, **20**, 43–60.

BRUNT, A. A. & STACE-SMITH, R. (1978). Some hosts, properties and possible affinities of a labile virus from *Hypochoeris radicata* (Compositae). *Annals of Applied Biology*, **90**, 205–14.

BURROUGHS, J. N. & BROWN, F. (1974). Physico-chemical evidence for reclassification of the caliciviruses. *Journal of General Virology*, **22**, 281–6.

BURROUGHS, J. N., DOEL, T. R. & BROWN, F. (1978). Relationship of San Miguel sea lion virus to other members of the calicivirus group. *Intervirology*, **10**, 51–9.

CALVERT, E. L. & HARRISON, B. D. (1966). Potato mop-top, a soil-borne virus. *Plant Pathology*, **15**, 134–9.

COHEN, B., MORTIMER, P. P. & PEREIRA, M. S. (1983). Diagnostic assays with monoclonal antibodies for the human serum parvovirus-like virus (SPLV). *Journal of Hygiene*, **91**, 113–30.

COSSART, Y. E., FIELD, A. M., CANT, B. & WIDDOWS, D. (1975). Parvovirus-like particles in human sera. *Lancet*, i, 71–3.

DIAZ-RUIZ, J. R. & KAPER, J. M. (1977). Cucumber mosaic virus-associated RNA 5. III. Little nucleotide sequence homology between CARNA 5 and helper RNA. *Virology*, **80**, 204–13.

DOOLITTLE, S. P. (1916). A new infectious mosaic disease of cucumber. *Phytopathology*, 6, 145–7.

DUFFUS, J. E. (1972). Beet western yellows virus. *Commonwealth Mycological Institute/Association of Applied Biologists Descriptions of Plant Viruses*, No. 89, 4 pp.

DUFFUS, J. E. (1979). Legume yellows virus, a new persistent aphid-transmitted virus of legumes in California. *Phytopathology*, **69**, 217–21.

DUFFUS, J. E. & RUSSELL, G. E. (1975). Serological relationship between beet western yellows and beet mild yellowing viruses. *Phytopathology*, **65**, 811–15.

DUNCAN, J. R., POTTER, C. G., CAPPELLINI, M. D., KURTZ, J. B., ANDERSON, M. J. & WEATHERALL, D. J. (1983). Aplastic crisis due to parvovirus infection in pyruvate kinase deficiency. *Lancet*, **ii**, 14–16.

EDWARDS, J. M. B., KESSEL, I., GARDNER, S. K., EATON, B. R., POLLOCK, T. M., GIBSON, P., WOODROOF, M. & PORTER, A. D. (1981). A study of antibodies to five viruses in sera from 210 acute admissions to a paediatric unit: EB virus; hepatitis A virus; parvovirus-like B19 and polyoma virus BK and JC. *Abstracts of European Association Against Virus Diseases*, p. 14.

ESTES, A. P. & BRAKKE, M. K. (1966). Correlation of *Polymyxa graminis* with transmission of soil-borne wheat mosaic virus. *Virology*, **28**, 772–4.

EUGSTER, A. K. (1980). Studies on canine parvovirus infections: development of an inactivated vaccine. *American Journal of Veterinary Research*, **41**, 2020–4.

EUGSTER, A. K., BENDELE, R. A. & JONES, L. P. (1978). Parvovirus infection in dogs. *Journal of the American Veterinary Medical Association*, **173**, 1340–1.

EUGSTER, A. K. & NAIRN, C. (1977). Diarrhoea in puppies: parvovirus-like particles demonstrated in their faeces. *Southwestern Veterinarian*, **30**, 50–60.

FLOWER, R. L. P., WILCOX, G. E. & ROBINSON, W. F. (1980). Antigenic differences between canine parvovirus and feline panleucopaenia virus. *Veterinary Record*, **107**, 254–6.

FRANCKI, R. I. B., HATTA, T., GRYLLS, N. E. & GRIVELL, C. J. (1979). The particle morphology and some other properties of chloris striate mosaic virus. *Annals of Applied Biology*, **91**, 51–9.

FULLER, C. (1901). Mealie variegation. *Report of the Government Entomologist for 1899–1900*, pp. 17–19. Natal: Natal Government Printer.

GAGNON, A. N. & POVEY, R. C. (1979). A possible parvovirus associated with an epidemic gastroenteritis of dogs in Canada. *Veterinary Record*, **104**, 263–4.

GERSHON, A. A. (1979). Erythema infectiosum (Fifth disease). In *Principles and Practice of Infectious Diseases*, ed. G. L. Mandel, R. H. Douglas & J. E. Bennett, pp. 1453–6. New York: Wiley.

GILLESPIE, J. H. & SCOTT, F. W. (1973). Feline viral infections. *Advances in Veterinary Science and Comparative Medicine*, **17**, 163–200.

GONSALVES, D., PROVVIDENTI, R. & EDWARDS, M. C. (1982). Tomato white leaf: the relation of an apparent satellite RNA and cucumber mosaic virus. *Phytopathology*, **72**, 1533–8.

GOODMAN, R. M. (1981). Geminiviruses. *Journal of General Virology*, **54**, 9–21.

GOULD, A. R. & SYMONS, R. H. (1978). Alfalfa mosaic virus RNA: determination of the sequence homology between the four RNA species and a comparison with the four RNA species of cucumber mosaic virus. *European Journal of Biochemistry*, **91**, 269–78.

GRAVES, J. H. (1973). Serological relationship of swine vesicular disease virus and Coxsackie B5 virus. *Nature*, **245**, 314–15.

HABER, S., IKEGAMI, M., BAJET, N. B. & GOODMAN, R. M. (1981). Evidence for a divided genome in bean golden mosaic virus, a geminivirus. *Nature*, **289**, 324–6.

HABILI, N. & FRANCKI, R. I. B. (1974). Comparative studies on tomato aspermy and cucumber mosaic viruses. *Virology*, **57**, 392–401.

HARRIS, T. J., UNDERWOOD, B. O., KNOWLES, N. J., CROWTHER, J. R. & BROWN, F. (1979). Molecular approach to the epidemiology of swine vesicular disease: correlation of variation in the virus structural polypeptides with serological properties. *Infection and Immunity*, **24**, 593–9.

HARRISON, B. D. & JONES, R. A. C. (1970). Host range and some properties of potato mop-top virus. *Annals of Applied Biology*, **65**, 393–402.

HINDLE, E. & FINDLAY, G. M. (1932). Studies on feline distemper. *Journal of Comparative Pathology*, **45**, 11–22.

HORVATH, J. (1979). New artificial hosts and non-hosts of plant viruses and their role in the identification and separation of viruses. X. Cucumovirus group: cucumber mosaic virus. *Acta Phytopathologica Academiae Scientiarum, Hungariae*, **14**, 285–95.

JAGGER, I. E. (1916). Experiments with the cucumber mosaic disease. *Phytopathology*, **6**, 148–51.

JEFFRIES, A. R. & BLAKEMORE, W. F. (1979). Myocarditis and enteritis in puppies associated with parvovirus. *Veterinary Record*, **104**, 221.

JOHNSON, R. H. (1967). Feline panleucopaenia virus – *in vitro* comparison of strains with a mink enteritis virus. *Journal of Small Animal Practice*, **8**, 319–23.

JOHNSON, R. H. & CRUIKSHANK, J. G. (1966). Problems in classification of feline panleucopaenia virus. *Nature*, **212**, 622–3.

JOHNSON, R. H., SIEGL, G. & GAUTSCHI, M. (1974). Characteristics of feline panleucopaenia virus strains enabling definitive classification as parvoviruses. *Archives für die Gesamte Virusforschung*, **46**, 315–24.

JOHNSON, R. H. & SPRADBROW, P. B. (1978). Isolation from dogs with severe enteritis of a parvovirus related to feline panleucopaenia virus. *The Australian Veterinary Journal*, **55**, 151.

JUNGEBLUT, C. W. (1958). Columbia SK group of viruses. In *Handbuch der Virusforschung*, ed. C. Hallauer & K. F. Meyer. Band III, 459–500. Vienna: Springer-Verlag.

KAPER, J. M. & DIAZ-RUIZ, J. R. (1977). Molecular weights of the double-stranded RNAs of cucumber mosaic virus strain S and its associated RNA5. *Virology*, **80**, 214–17.

KAPER, J. M. & TOUSIGNANT, M. E. (1977). Cucumber mosaic virus-associated RNA 5. I. Role of host plant and helper strain in determining amount of associated RNA 5 with virions. *Virology*, **80**, 186–95.

KAPER, J. M. & TOUSIGNANT, M. E. (1978). Cucumber mosaic virus-associated RNA 5. V. Extensive nucleotide sequence homology among CARNA 5 preparations and different CMV strains. *Virology*, **85**, 323–7.

KAPER, J. M., TOUSIGNANT, M. E. & LOT, H. (1976). A low molecular weight replicating RNA associated with a divided genome virus: defective or satellite RNA? *Biochemical and Biophysical Research Communications*, **72**, 1237–43.

KAPER, J. M., TOUSIGNANT, M. E. & THOMPSON, S. M. (1981). Cucumber mosaic virus-associated RNA 5. VIII. Identification and partial characterisation of a CARNA 5 incapable of inducing tomato necrosis. *Virology*, **114**, 526–33.

KAPER, J. M. & WATERWORTH, H. E. (1977). Cucumber mosaic virus associated RNA 5: causal agent for tomato necrosis. *Science*, **196**, 429–31.

KAPER, J. M. & WATERWORTH, H. E. (1981). Cucumoviruses. In *Handbook of Plant Virus Infections, Comparative Diagnosis*, ed. E. Kurstak, pp. 257–332. Amsterdam: Elsevier/North Holland Biomedical Press.

KASSANIS, B. & WOODS, R. D. (1969). Properties of some defective strains of tobacco mosaic virus and their behaviour as affected by inhibitors during storage in sap. *Annals of Applied Biology*, **64**, 213–24.

KASSANIS, B., WOODS, R. D. & WHITE, R. F. (1972). Some properties of potato mop-top virus and its serological relationship to tobacco mosaic virus. *Journal of General Virology*, **14**, 123–32.

KELLEHER, J. F., LUBAN, N. L. C., MORTIMER, P. P. & KAMIMURA, T. (1982). Parvovirus: a specific cause of aplastic crisis in hereditary spherocytosis. *Blood*, **60**, Suppl. 1, p. 36a.

KELLY, W. R. (1978). An enteric disease of dogs resembling feline panleucopaenia. *Australian Veterinary Journal*, **54**, 593.

KUBO, S. & TAKANAMI, Y. (1979). Infection of tobacco mesophyll protoplasts with tobacco necrotic dwarf virus, a phloem-limited virus. *Journal of General Virology*, **42**, 387–98.

LENGHAUS, C. & STUDDERT, M. J. (1980). Relationships of canine panleucopaenia (enteritis) and myocarditis parvoviruses to feline panleucopaenia virus. *Australian Veterinary Journal*, **56**, 152–3.

LESEMANN, D. E. & KOENIG, R. (1977). Association of club-shaped virus-like particles with a severe disease of *Agaricus bisporus*. *Phytopathologische Zeitschrift*, **89**, 161–9.

LINDSTEN, K., LINDSTEN, B., ABDELMOETI, M. & JUNTTI, N. (1980). Purification and some properties of wheat dwarf virus. In *Proceedings of the Third Conference in Virus Diseases of Gramineae in Europe*, Rothamsted, UK.

LOT, H., JONARD, G. & RICHARDS, K. E. (1977). Cucumber mosaic virus RNA 5. Partial characterisation and evidence for no sequence homologies with genomic RNAs. *FEBS Letters*, **80**, 395–400.

LOT, H., MARCHOUX, G., MARROU, J., KAPER, J. M., WEST, C. K., VAN VLOTEN-DOTING, L. & HULL, R. (1974). Evidence for three functional RNA species in several strains of cucumber mosaic virus. *Journal of General Virology*, **22**, 81–93.

McCANDLISH, I. A. P., THOMPSON, H., CORNWELL, H. J. C., LAIRD, H. & WRIGHT, N. G. (1979). Isolation of a parvovirus from dogs in Britain. *Veterinary Record*, **105**, 167–8.

McCANDLISH, I. A. P., THOMPSON, H., CORNWELL, H. J. AND McCARTNEY, L. (1980). Canine parvovirus infection. *Veterinary Record*, **107**, 204.

MacIVER, J. E. & PARKER-WILLIAMS, E. J. (1961). The aplastic crisis in sickle cell anaemia. *Lancet*, i, 1086–9.

McMASTER, G. K., TRATSCHIN, J.-D. & SIEGL, G. (1981). Comparison of canine parvovirus with mink enteritis virus by restriction site mapping. *Journal of Virology*, **38**, 368–71.

MATTHEWS, R. E. F. (1979). Classification and nomenclature of viruses. *Intervirology*, **12**, 129–296.

MATTHEWS, R. E. F. (1982). Classification and nomenclature of viruses. *Intervirology*, **17**, 1–199.

MILBRATH, G. M. & DUFFUS, J. E. (1978). Solanum yellows virus. *Phytopathology News*, **12**, 170.

MILLER, J., EVERMANN, J. & OTT, R. (1980). Immunofluorescence test for canine coronovirus and parvovirus. *Western Veterinarian*, **18**, 14–19.

MILNE, R. G., BOCCARDO, G., LISA, V. & LUISONI, E. (1982). Cryptic plant viruses. *IV International Conference on Comparative Virology, Banff, Canada, October 17–22, 1982*. Abstract W14–17.

MORTIMER, P. P. (1983). Hypothesis: the aplastic crisis of hereditary spherocytosis is due to a single transmissible agent. *Journal of Clinical Pathology*, **36**, 445–8.

MORTIMER, P. P., HUMPHRIES, R. K., MOORE, J. G., PURCELL, R. H. & YOUNG, N. S. (1983). A human parvovirus-like virus inhibits haematopoietic colony formation *in vitro*. *Nature*, **305**, 426–9.

MOSSOP, D. W. & FRANCKI, R. I. B. (1977). Association of RNA 3 with aphid transmission of cucumber mosaic virus. *Virology*, **81**, 177–81.

MOSSOP, D. W. & FRANCKI, R. I. B. (1978). Survival of satellite RNA *in vivo* and its dependence on cucumber mosaic virus for replication. *Virology*, **86**, 562–6.

MOSSOP, D. W. & FRANCKI, R. I. B. (1979a). The stability of satellite viral RNAs *in vivo* and *in vitro*. *Virology*, **94**, 243–53.

Mossop, D. W. & Francki, R. I. B. (1979b). Comparative studies on two satellite RNAs of cucumber mosaic virus. *Virology*, **95**, 395–404.

Muller, D.-E. & Siegl, G. (1983). Maturation of parvovirus LuIII in a subcellular system. I. Optimal conditions for *in vitro* synthesis and encapsidation of viral DNA. *Journal of General Virology*, **64**, 1043–54.

Nardelli, L., Lodetti, E., Gualandi, G. L., Burrows, R., Goodridge, D., Brown, F. & Cartwright, B. (1968). A foot-and-mouth disease syndrome in pigs caused by an enterovirus. *Nature*, **219**, 1275–6.

Newman, J. F., Rowlands, D. J. & Brown, F. (1973). A physico-chemical subgrouping of mammalian picornaviruses. *Journal of General Virology*, **18**, 171–80.

Owens, R. A. & Kaper, J. M. (1977). Cucumber mosaic virus-associated RNA 5. II. *In vitro* translation in a wheat germ protein synthesis system. *Virology*, **80**, 196–203.

Parrish, C. R., Carmichael, L. E. & Antczak, D. F. (1982). Antigenic relationships between canine parvovirus type 2, feline panleucopaenia virus and mink enteritis virus using conventional antisera and monoclonal antibodies. *Archives of Virology*, **72**, 267–78.

Pattison, J. R., Jones, S. E., Hodgson, J., Davis, L. R., White, J. M., Stroud, C. E. & Murtaza, L. (1981). Parvovirus infections and hypoplastic crises in sickle-cell anaemia. *Lancet*, i, 664–5.

Paver, W. K. & Clarke, S. K. R. (1976). Comparison of human fecal and serum parvo-like viruses. *Journal of Clinical Microbiology*, **4**, 67–70.

Peden, K. W. C. & Symons, R. H. (1973). Cucumber mosaic virus contains a functionally divided genome. *Virology*, **53**, 487–92.

Peters, L. & Schwartz, H. (1912). Krankheiten und Beschadingen des Tabaks. *Mitteilungen Biologische Bundesanstalt Land und Forstwirtschaft*, **13**, 58–64, Berlin-Dahlem.

Putz, C., Kuszala, J., Kuszala, M. & Spindler, C. (1974). Variation du pouvoir pathogène des isolats du virus de la mosaique du concombre associée à la nécrose de la tomate. *Annales de Phytopathologie*, **6**, 139–54.

Putz, C. & Vuittenez, A. (1974). Observations de particules virales chez des betteraves presentent, en Alsace, des symptomes de 'rhizomanie'. *Annales de Phytopathologie*, **6**, 129–38.

Randles, J. W., Harrison, B. D. & Roberts, I. M. (1976). Nicotiana velutina mosaic virus: purification, properties and affinities with other rod-shaped viruses. *Annals of Applied Biology*, **84**, 193–204.

Rao, K. R. P., Patel, A. R., Anderson, M. J., Hodgson, J., Jones, S. E. & Pattison, J. R. (1983). Infection with parvovirus-like virus and aplastic crisis in chronic haemolytic anaemia. *Annals of Internal Medicine*, in press.

Reddy, D. V. R., Rajeshwari, R., Iizuka, N., Lesemann, D. E., Nolt, B. L. & Goto, T. (1983). The occurrence of Indian peanut clump, a soil-borne virus disease of groundnuts (*Arachis hypogaea*) in India. *Annals of Applied Biology*, **102**, 305–10.

Richards, K. E., Jonard, G., Jacquemond, M. & Lot, H. (1978). Nucleotide sequence of cucumber mosaic virus-associated RNA 5. *Virology*, **89**, 394–408.

Rochow, W. F. & Brakke, M. K. (1964). Purification of barley yellow dwarf virus. *Virology*, **24**, 310–22.

Rochow, W. F. & Duffus, J. E. (1980). Luteoviruses and yellows diseases. In *Handbook of Plant Virus Infections, Comparative Diagnosis*, ed. E. Kurstak, pp. 147–70. Amsterdam: Elsevier/North Holland Biomedical Press.

Sanger, D. V., Rowlands, D. J. & Brown, F. (1977). Encephalomyocarditis virus antibodies in sera from apparently normal pigs. *Veterinary Record*, **100**, 240–1.

SCHOFFIELD, F. W. (1949). Virus enteritis in mink. *North American Veterinarian*, **30**, 651–4.

SEQUEIRA, J. C. & HARRISON, B. D. (1982). Serological studies on cassava latent virus. *Annals of Applied Biology*, **101**, 33–42.

SERJEANT, G. R., MASON, K., TOPLEY, J. M., SERJEANT, B. M., PATTISON, J. R., JONES, S. E. & MOHAMED, R. (1981). Outbreak of aplastic crises in sickle cell anaemia associated with parvovirus-like agent. *Lancet*, ii, 595–7.

SHEPHERD, R. J., FRANCKI, R. I. B., HIRTH, L., HOLLINGS, M., INOUYE, T., MACLEOD, R., PURCIFULL, D. E., SINHA, R. C., TREMAINE, J. H., VALENTA, V. & WETTER, C. (1976). New groups of plant viruses approved by the International Committee on Taxonomy of Viruses, September 1975. *Intervirology*, **6**, 181–4.

SHNEERSON, J. M., MORTIMER, P. P. & VANDERVELDE, E. M. (1980). Febrile illness due to a parvovirus. *British Medical Journal*, **2**, 1580.

SIEGL, G. (1976). *The Parvoviruses*. Vienna: Springer-Verlag.

SMITH, A. W., AKERS, T. G., MADIN, S. H. & VEDROS, N. A. (1973). San Miguel sea lion virus isolation, preliminary characterisation and relationship to vesicular exanthema of swine virus. *Nature*, **244**, 108–10.

SMITH, A. W., SKILLING, D. E. & RITCHIE, A. E. (1978). Immuno-electron microscopic comparison of caliciviruses. *American Journal of Veterinary Research*, **39**, 1531–3.

SUMMERS, J., JONES, S. E. & ANDERSON, M. J. (1983). Characterisation of the agent of erythrocyte aplasia as a human parvovirus. *Journal of General Virology*, **64**, 2527–32.

SYLVESTER, E. S. & OSLER, R. (1977). Further studies on the transmission of the filaree red-leaf virus by the aphid *Acyrthosiphon perlargonii zerosalphum*. *Environmental Entomology*, **6**, 39–42.

TAKANAMI, Y. (1981). A striking change in symptoms in cucumber mosaic virus-infected tobacco plants induced by a satellite RNA. *Virology*, **109**, 120–6.

TAKANAMI, Y., KUBO, S. & IMAIZUMI, S. (1977). Synthesis of single- and double-stranded cucumber mosaic virus RNAs in tobacco mesophyll protoplasts. *Virology*, **80**, 376–89.

TAMADA, T., ABE, H. & BABA, T. (1975). Beet necrotic yellow vein virus and its relation to the fungus *Polymyxa betae*. *Proceedings of the 1st International Congress of the International Association of Microbiological Societies*, **3**, 313–20.

TAMADA, T. & KOJIMA, M. (1977). Soybean dwarf virus. *Commonwealth Mycological Institute/Association of Applied Biologists Descriptions of Plant Viruses*, No. 179, 4 pp.

TESH, R. B. & WALLACE, G. D. (1978). Observations on the natural history of encephalomyocarditis virus. *American Journal of Tropical Medicine and Hygiene*, **27**, 133–45.

THOMAS, J. E. & BOWYER, J. W. (1980). Properties of tobacco yellow dwarf and bean summer death viruses. *Phytopathology*, **70**, 214–17.

THOMPSON, H., MCCANDLISH, I. A. P., CORNWELL, H. J. C., WRIGHT, N. A. & ROGERSON, P. (1979). Myocarditis in puppies. *Veterinary Record*, **104**, 107–8.

THOUVENEL, J.-C., DOLLET, M. & FAUQUET, C. (1976). Some properties of peanut clump, a newly discovered virus. *Annals of Applied Biology*, **84**, 311–20.

TRATSCHIN, J.-D., MCMASTER, G. K., KRONAUER, G. & SIEGL, G. (1982). Canine parvovirus: relationship to wild-type and vaccine strains of feline panleucopaenia virus and mink enteritis virus. *Journal of General Virology*, **61**, 33–41.

TSUCHIZAKI, T., HIBINO, H. & SAITO, Y. (1973). Comparisons of soil-borne wheat mosaic virus isolates from Japan and the United States. *Phytopathology*, **63**, 634–9.

WARBURG, O. (1894). Cassava mosaic. *Mitteilungen des Deutschen Schutzgebiet*, **7**, 131–99.

WATERWORTH, H. E., KAPER, J. M. & TOUSIGNANT, M. E. (1979). Cucumber mosaic virus satellite RNA. *Science*, **204**, 845–7.

WILKINSON, G. T. (1979). Myocarditis in puppies, unidentified feline illness and gingivitis in cats. *Veterinary Record*, **104**, 149.

THE ERADICATION OF VIRUS INFECTIONS

DAVID A. J. TYRRELL

Medical Research Council Common Cold Unit, Harvard Hospital, Salisbury SP2 8BW, UK

It is a bold and attractive idea that we should eliminate virus infections. Some might say it is impossible but now that the object has been achieved for smallpox it is natural to suggest that we should do the same for other viruses, certainly those which cause any serious or frequent disease. The following is a personal view of what eradication means, how it might be achieved and whether it is practical or not.

First of all, it is important to clarify what is meant by eradication. The word implies taking out the roots of something and means that the disease (and more importantly the organism that produces it) is no longer found and no longer being transmitted in a geographical area so that no further measures are needed to prevent infection there. Global eradication has a particularly profound effect since, once it is achieved, countermeasures can be discontinued; for instance, the trouble and expense of vaccination against smallpox and of certification are over, and no child need be exposed to the small but real risk of an adverse reaction to the vaccine. Eradication of a virus may succeed in a smaller area, for instance foot-and-mouth disease and rabies have been eradicated from the British Isles. In order to maintain that state, however, a co-ordinated set of expensive measures are needed: restrictions on the import of food, animals and animal products; quarantine facilities; diagnostic laboratories and so on. Terminology can be confusing, and a group considering the problems of human infections suggested that this localized eradication be called 'elimination' (International Conference, 1982). If it is achieved, precautions may be discontinued within a cleared area – for instance we never vaccinate against rabies if a patient receives an animal bite in Britain – but precautions are usually needed to maintain the sanitary cordon around the area. This may nevertheless be less expensive than the alternative strategy of the *control* of a disease, i.e., substantial reduction in the incidence though the virus continues to be transmitted within the population. For instance, poliomyelitis is controlled in this country by ensuring that children receive oral poliovirus vaccine at an early age. So the price of freedom from this disease will be continued universal vaccination, and we know that if the vaccination programme flags in any area, there is a significant chance that virulent virus and paralytic disease will return.

However, even if we limit ourselves to those which infect vertebrates, it is clear that viruses exist in enormous variety and very few generalizations are possible. If we wish to consider the possibility of eradication of specific disease it is necessary to look at important attributes which may make eradication possible or impossible. This depends first, on the natural history and mode of transmission of the causative virus and second, on whether some sort of antiviral measures are available.

THE NATURAL HISTORY

There are some viruses that are very unlikely candidates for eradication because they are carried life long and may be released in infectious form later in life. An example would be varicella/zoster virus. This may disappear from isolated communities for years but can recur, without being introduced from outside, because an elderly individual who had chickenpox earlier, perhaps as a child, develops shingles and sheds virus which then causes chickenpox in a non-immune child or adult. Secondary cases may then occur among other susceptible individuals.

Another group comprises those viruses with which subclinical or silent infections are common. It is easy to see why. An important part of achieving or maintaining eradication is to have a simple and effective means of detecting the presence of the virus and these are almost always based on recognizing cases of illness. Community wide virus tests are usually impractical. If an animal pathogen loses a little virulence, so that infected herds are not recognized to be ill, then it immediately becomes more difficult to detect them and therefore to stop the infection spreading. Thus we are very unlikely even to attempt to eradicate diseases due to many of the enteroviruses since they usually cause silent infections of children, although they do occasionally cause meningitis.

Yet other viruses have a reservoir in species other than that usually affected by disease and if it is impractical to eliminate infection from the reservoir then clearly eradication is impossible. An example of this would be a disease like African Swine Fever which is endemic in wild life in Africa and in the arthropods which parasitize them. In human disease, we have the example of yellow fever which can be eliminated from towns, but which continues to circulate in wild life in the jungles, and it is impossible to prevent it emerging from time to time.

ANTIVIRAL MEASURES

If a virus is to be eliminated, effective antiviral measures must be available. They may be very simple, e.g., the slaughter policy used to control foot-and-mouth disease in Britain, but they have to be used as part of a carefully thought out policy by a trained and prepared organization. Vaccination is often used in other parts of the world to control foot-and-mouth disease, but vaccinated animals may exhibit modified disease when infected and it may be more difficult to identify an outbreak quickly, so vaccination in addition to a detect-and-slaughter policy would make it less effective. Thus, the present simple policy of killing off infected stock very rapidly after the disease appears is regarded as more effective and less expensive; this is particularly true as we are trying to maintain the situation in the face of very few introductions, so that most of the vaccine administered would be to animals which would never be exposed to the infection. Provided rapid diagnosis of affected animals and herds is possible, slaughter is the only measure that is needed and would probably control African Swine Fever if it entered Britain even though no vaccine is available.

On the other hand, vaccination was a key measure in the smallpox eradication programme. At first, most emphasis was placed on vaccinating the whole population, but it was found that smallpox persisted in spite of this. It became clear that vaccination was not in fact being carried out uniformly in all sections of the population, so that virus was continuing to circulate in relatively unvaccinated subpopulations which were easily overlooked. The field methods were then altered so that cases were specifically sought out. When they were found, strenuous efforts were made to vaccinate all contacts. Thus, mass vaccination failed as an eradication policy and was replaced by one of rigorous case finding and containment by ring vaccination and quarantine. Of course, the vaccination had to be effective. The programme therefore included research to develop a potent freeze-dried (and therefore stable) vaccine, and measures to ensure that large amounts of high quality vaccine were made. The programme depended on the development and introduction of the bifurcated needle which made it possible for relatively untrained people to vaccinate effectively. Finally, it required a well organized, well funded and appropriate system of applying it in the field. When discussing vaccination, it is important to consider not only the product but also the method and regime of administration; nevertheless there is no doubt that proper vaccination was the one specific antiviral measure by which smallpox was eradicated (Henderson, 1977).

In theory, other measures might be used – strict isolation and quarantine, for example. This is the basis on which Lassa Fever virus has been contained when it has been introduced into the UK. It has apparently been effective but probably only because the disease is not as communicable and infectious as was at first thought. Quarantine apparently had some success in preventing the introduction of yellow fever into the USA before the epidemiology of the disease was really understood, but early workers could not believe it was infectious because case to case transmission was so rare. Now we realize that the transmission of a disease of that sort would have to be controlled by eliminating the arthropod vectors or by substantially reducing their numbers. Vaccination could only prevent disease and reduce transmission to a limited extent (Strode, 1951). Antibacterial treatment was the main measure used in the WHO programme for the eradication of yaws, but although antiviral drugs are now becoming available there is no immediate prospect of their being used for virus eradication, though they might be considered for this purpose in the future if very effective and inexpensive ones were developed.

SOME MODELS OF EPIDEMIC BEHAVIOUR

Over the years, a great many efforts have been made to construct mathematical models of the epidemic process. They have succeeded to a varying extent. They do not usually enable us to predict the occurrence of the course of epidemics except in general terms. Nevertheless, efficient models which emphasize different aspects of the process enable us to discuss epidemics in more rigorous and quantitative terms and show certain general properties of epidemics which may not be intuitively obvious (Fine, 1982). One is called the epidemic threshold, meaning that there must be a certain minimum combination of communicability of infection, contact between individuals and proportion of susceptible individuals for one case to lead to another. The expression 'basic reproduction rate' has been used in the past. If the rate is greater than one, then one case leads to more than one infection and rates build up, like the chain reaction in an atomic bomb. If the basic reproductive rate is lower than one then, although infection may continue for a time, it will in the end die out. It also turns out that towards the end of an epidemic, infections will die out before all the susceptible individuals have been infected and rendered immune. This characteristic is responsible for the phenomenon described as 'herd immunity' and from it can be deduced that it might be possible to eradicate an infection from a community by vaccination, even though not all were vaccinated or if the vaccine did not protect 100% of those to whom it was given. On the other hand, individuals susceptible to infection would remain and if the virus had only been eliminated from an area and not eradicated globally such individuals would be at risk if they moved into an endemic area – this can be seen when unvaccinated individuals remain well while they live in Britain but catch poliomyelitis when they travel abroad.

THE PRACTICAL POSSIBILITIES

There has been considerable debate worldwide on the possibility of eradication of infectious diseases. Indeed the construction of a reference list for this paper is made easier by the recent publication of two extensive and authoritative debates on the subject (Round Table, 1981, International Conference, 1982). My purpose is to take the biological and clinical principles which have just been outlined and to amplify and illustrate them using some of the material brought out in the above debates. I shall concentrate on the field I know best, namely human virus disease.

In the first place, which viruses might be eradicated? Some clearly cannot be eradicated. The common respiratory and enteric viruses are too numerous, and too often subclinical to be considered as candidates. Some have reservoirs outside the human species from which it is most unlikely they can be driven – rabies would be an example. As already mentioned, the herpes viruses will always be with us. The two infections most likely to be eradicated are measles and poliomyelitis.

Table 1. *Comparison of theoretical and technical factors affecting eradication of smallpox and measles*

	Smallpox	Measles
Non-human host	No	No
Long-term carrier host	No	No
Obvious illness	Yes	Yes
Immunity from disease	Life long	Life long
Immunity from vaccine	Long-term	Long-term
Effectiveness of vaccine	High	High
Stability of vaccine	Stable	Labile
Evidence of immunity	Visible	Not visible
Infectivity	Moderate	High
Universal vaccination	Not essential	Probably essential
Search-containment	Highly effective	Probably effective

After Hinshaw (1982).

Hinshaw (1982) has pointed out that in a number of ways measles resembles smallpox as a candidate for eradication. I reproduce as Table 1 his assessment of the relevant factors. There are, of course, significant differences. Measles spreads more effectively and more rapidly than smallpox. One cannot confirm by clinical examination whether a patient has been vaccinated or suffered from the disease. Some of the differences are diminishing – for instance, a more stable measles vaccine is now available. Some of the problems of eradication would be similar to those found in the smallpox campaign – for instance, in the USA, disease has been found to appear suddenly in unrecognized, undervaccinated subgroups of the general population. If an eradication campaign were partly successful, infection might be deferred to adult life, perhaps occurring in individuals who were not protected when they were vaccinated. Such people might then suffer from severe and possibly fatal disease as seen in the terrible community wide epidemics which occurred in remote communities such as those of Iceland and the Pacific islands. However, the vaccination programme has been reasonably success-ful in the USA. The disease is now rare and I have no doubt that with further intense surveillance and vaccination, transmission within the continental USA will cease. However that will leave the USA as a measle-free island in a worldwide sea of the virus. If vaccination is relaxed it can only be a few years before virus is reintroduced, perhaps by travelling families and, if enough suscepti-ble children have accumulated, transmission within the country will resume. However even this is quite similar to the situation of most developed countries at the time the smallpox eradication campaign was started. Indeed, it seems to me that since epidemiology and immunology are such inexact sciences we cannot safely predict from academic analysis whether a vaccination and surveillance programme can eliminate measles virus from a substantial and varied area of the world. We need an experiment or field-study to show that it is possible. It seems that the current experience in the USA may provide us with what we need. Of course, such an experiment should be repeated and it grieves me that we are so far from doing the same in the UK, where we have ample supplies of well made vaccines and well developed medical services. Control of the disease in another area would support the idea that eradication was worth aiming for.

Poliomyelitis seems at first sight to be a less promising target for eradication. Subclinical infections are common and oral poliovirus vaccine strains are shed and transmitted, while successful vaccination with inactivated virus does not prevent infection of the intestinal tract. Yet, contrary to many expectations, poliovirus seems to have been eliminated from several countries, particularly in Scandinavia where they have immunized their population with inactivated vaccine. In the UK and the USA polioviruses still circulate but are apparently vaccine derived. Most cases of poliomyelitis are due to the low incidence of vaccine induced disease in vaccinated individuals and their contacts, or to unvaccinated travellers who have picked up virulent virus in endemic areas. Virus may persist in the environment for a while but there is, as far as we know, no reservoir or alternative host.

THE APPLICATION IN DEVELOPING COUNTRIES

The advantages of virus eradication are much more obvious in a developing country than in a rich industrialized one. Firstly, the disease problems are greater. Measles exacts a heavy toll of deaths in infancy and childhood, either directly or by leading to pneumonia or by exacerbating problems arising from dietary deficiencies. It is known that there is much life long crippling due to poliomyelitis (Ofosu-Amaah, Kratzer & Nicholas, 1977) and that the prevalence is increasing in some areas; there is presumably mortality which has not been measured. However there are technical difficulties in vaccinating in these areas. Measles vaccine is thermolabile and even though a refrigerated transport system (the 'cold chain') is provided, it may not reach the clinic in a living state. However, now that better stabilized vaccines are being manufactured, this particular problem may be solved. Polio vaccine is said to give a low frequency of 'takes' when given to children in developing countries. This might be due to interference by a heavy load of enteric viruses, or to some other environmental factor. However, Sabin has recently argued strongly that if potent well transported vaccine is given in a properly organized programme, it is effective in these circumstances (Sabin, 1982). This is certainly supported by the dramatic fall in the frequency in Brazil once such measures were taken in areas where the disease was persisting in the face of what were said to be proper vaccination programmes.

SOCIOECONOMIC FACTORS

Some very important comments have been made by Yekutiel (1981), who was in close contact with many aspects of the WHO eradication programmes. He has listed what he regards as essential preconditions for a successful eradication programme which, stated shortly, run as follows:

(1) There should be a critical measure that is completely effective in breaking transmission, simple in application and relatively inexpensive.
(2) The disease should have epidemiological features facilitating effective case detection and surveillance in the advanced stages of the eradication programme.
(3) The disease must be of recognized socioeconomic importance, national or international.
(4) There should be a specific reason for eradication rather than control of the disease.
(5) There should be adequate financial, administrative, manpower and health service resources.
(6) There should be the necessary socioecological conditions.

It is interesting that – in addition to the preconditions (1), (2), and (4), which are equivalent to points we have already discussed – he indicates that there are three more points which can play an equally important role in determining whether the programme will succeed.

He illustrates these from a number of campaigns. He says, for instance, that some malaria campaigns failed because financial and administrative support flagged after the initial phase, and because people did not accept the idea of spraying their homes. Problems of insecticide resistance only developed later.

Politics, both local and national, play an important role. No large scale programme can be implemented unless it is positively supported by individuals or groups with political power which is maintained for a period of time. In developing countries it is often perceived that it is better to put limited resources into economic development projects than to try to affect health by more direct action. Nevertheless, the World Bank has in recent years accepted that in some conditions health-providing measures may be an effective way of bringing benefit to the community (Golliday & Liese, 1980). Every programme aimed at total eradication must succeed throughout an area, so not only must the organization maintain high standards throughout each region and area in which it operates, but it must also obtain the goodwill and co-operation of every local community or, as we have already seen, the campaign may fail. One of the great successes of the smallpox eradication programme was that the basic principles and techniques were applied in a great many different socioeconomic settings with due regard to how people would respond. A series of orders issued emphatically from on high will only have partial success. The organization had to be flexible and able to identify areas where the programme had not succeeded, analyse why this was, and take specific measures to correct these faults. These corrective measures included bringing in substantial international resources when, as often happened, there was no way in which the national resources of funds, staff or expertise would be sufficient. These were forthcoming because other countries had a national interest in eradication in the remaining areas, since if success was achieved they could discontinue control measures and start saving money and lives.

CONCLUSIONS

The only infectious disease to have been eradicated from the world is smallpox, and we have seen how there were special features in favour of this succeeding. Measles and poliomyelitis might be among the next targets, although the difficulties may be greater. It seems to me that the best line to take would be to proceed stepwise, in other words to prove that measles can be eliminated from an entire region and transmission interrupted by a thorough vaccination campaign – this might then be extended to developing countries. If the practicability and the advantages could be demonstrated, rather than presented as theoretical propositions, it might be possible to recruit the political, financial and technical backing that a global eradication programme would require. The same approach would be appropriate with slight variations for poliomyelitis, though in this case it might be desirable to develop a high potency, inexpensive, killed virus vaccine and show that it is effective in reducing transmission in a developing country before seriously considering the other factors. It might be possible to eradicate with a live virus, but it seems to me to be a serious risk that somewhere vaccine strains would revert to virulence. There are good reasons for trying to improve our methods of control and regional elimination of poliomyelitis; if they succeed they can be regarded as preparation for a programme of eradication, but such a programme would require also comprehensive international commitment. Eradication must be decades away.

Somewhat more fancifully, it might be possible to prevent future pandemics of influenza A by preventing animal influenza viruses, particularly those from pigs and birds, gaining access to the human population where they may mutate or recombine to produce a new virus pathogenic for man. However, unless it can be shown that there are certain particular circumstances in which this is likely to happen then there is little hope of keeping the species apart on a worldwide scale. Influenza A viruses have in the past 'died out', at least when a new serotype has appeared, but they might not do so again, so one would only eliminate a pandemic form of the disease and leave an endemic or mild epidemic form unaffected.

Eradication is thus a strategy which can be very satisfactory, but is only feasible for a very limited number of infections. We should remember, however, that other strategies may have a substantial effect on the impact of a disease. Oral rehydration can greatly reduce the impact of rotavirus diarrhoea, better nutrition and nursing can help the child with a respiratory virus infection, good antivirals against the herpes viruses are now available. So if eradication is impossible there are other things we can do – all is not lost!

REFERENCES

FINE, P. (1982). Application of mathematical models to the epidemiology of influenza: a critique in influenza models. In *Influenza Models – Prospects for Development and Use*, ed. P. Selby, pp. 18–85. Lancaster; Medical Technical Publishing Press Ltd.

GOLLIDAY, F. L. & LIESE, B. H. (1980). Issues in the institutionalization and management of rural health care: making technology appropriate. *Proceedings of the Royal Society of London. B*, **209**, 173–80.

HENDERSON, D. A. (1977). Smallpox eradication. *Proceedings of the Royal Society of London. B*, **199**, 83–97.

HINSHAW, A. R. (1982). World eradication of measles. *Reviews of Infectious Diseases*, **4**, 933–6.

INTERNATIONAL CONFERENCE (1982). Can Infectious Diseases be Eradicated? *Reviews of Infectious Diseases*, **4**, 912–84.

OFOSU-AMAAH, S., KRATZER, J. H. & NICHOLAS, P. D. (1977). Is poliomyelitis a serious problem in developing countries? Lameness in Ghanian schools. *British Medical Journal*, **1**, 1012-14

ROUND TABLE (1981). Lessons from the big eradication campaigns. *World Health Forum*, **2**, 465–90.

SABIN, A. B. (1982). Vaccine control of poliomyelitis in the 1980s. *Yale Journal of Biology and Medicine*, **55**, 383–9.

STRODE, G. K. (1951). *Yellow Fever*. New York: McGraw-Hill.

YEKUTIEL, P. (1981). Lessons from the big eradication campaigns. *World Health Forum*, **2**, 465–81.

MINIMAL INFECTIOUS AGENTS: THE VIROIDS

H. L. SÄNGER

Max-Planck-Institut für Biochemie, D–8033 Planegg-Martinsried bei München, FDR

INTRODUCTION

Viroids are the smallest well-characterized infectious agents presently known. Their discovery is an interesting example of serendipity in modern science, because they were found although not actually sought for. In fact, these originally presumed virus-like agents finally turned out to be completely different from what had been expected and extended our concept of infectious microbial disease agents into the realm of small RNA molecules.

In the late 1960s several research groups were trying to isolate and characterize the elusive agents of certain 'virus' diseases of plants which had defied isolation and identification. It was T. O. Diener (1971*b*) who first published evidence that the potato spindle tuber 'virus' (PSTV) is actually a replicating low molecular weight RNA. In order to differentiate this novel type of agent from the conventional viruses, with their comparatively large and encapsidated genomes, he proposed the term 'viroid' for it. At the same time the viroid aetiology of additional virus-like plant diseases was independently established in several other laboratories.

Rapid progress has been made in the elucidation of different biochemical and biophysical properties of PSTV and other viroids since this initial discovery so that, today, viroids are structurally the best studied RNA molecules next to tRNA. Much less is known on the mechanism of viroid replication and pathogenicity, on which viroid research will focus in future. It is the aim of this report to discuss the present concepts of these still enigmatic processes and some recent new developments in the viroid field.

VIROID DISEASES

Thus far, viroids have only been found in higher plants and a dozen viroid-incited plant diseases are presently known (Table 1). Their

Table 1. *The viroid diseases presently known and the abbreviations of the corresponding viroids*

Viroid disease	Viroid	References[a]
1. Potato spindle tuber	PSTV	1,2
2. Citrus exocortis	CEV	3,4
3. Chrysanthemum stunt	CSV	5,6
4. Chrysanthemum chlorotic mottle	CCMV	7,8
5. Cucumber pale fruit	CPFV	9,10
6. Coconut 'cadang-cadang'	CCCV	11,12
7. Hop stunt	HSV	13,14
8. Avocado sunblotch	ASBV	15,16
9. Tomato bunchy top	TBTV	17,18
10. Tomato apical stunt	TASV	18,19
11. Tomato 'planta macho'	TPMV	20
12. Burdock stunt	BSV	21

[a]References: 1. Diener, 1971b; 2. Singh & Clark, 1971; 3. Sänger, 1972; 4. Semancik & Weathers, 1972a; 5. Hollings & Stone, 1973; 6. Diener & Lawson, 1973; 7. Romaine & Horst, 1975; 8. Horst & Romaine, 1975; 9. van Dorst & Peters, 1974; 10. Sänger et al., 1976; 11. Randles, 1975; 12. Randles et al., 1976; 13. Sasaki & Shikata, 1977a: 14. Sasaki & Shikata, 1977b; 15. Thomas & Mohamed, 1979; 16. Dale & Allen, 1979; 17. McClean, 1931; 18. Walter, 1981; 19. Semancik & Weathers, 1972b; 20. Galindo, Smith & Diener, 1982; 21. Chen et al., 1982.

names reflect the characteristic disease symptoms caused by the corresponding viroids. There is little doubt that additional viroid diseases will be detected in future, so that their number will gradually increase with time.

In susceptible host plants, viroids may cause symptoms which range from malformations and discolourations of leaves (CEV, CCMV) and fruits (PSTV, ASBV, CPFV) to retardation of general plant growth (CSV, BSV, HSV); death of the whole plant may even occur (CCCV). (These abbreviated names are given in full in Table 1.) Therefore, most of the viroid diseases are of economic importance. The most dramatic effect is caused by CCCV which results in disastrous losses in coconut palms in the Philippines (Fig. 1). It has been estimated that more than 30 million trees have been killed by this viroid in the past and that every year another 500 000 palms will die, which makes CCCV a main threat to the coconut production in the Philippines (Randles, 1975). The chrysanthemum stunt disease caused a serious epidemic in cultivated cut- and pot-flower chrysanthemums in the USA during 1945–7 (Hollings & Stone, 1970). It had been spread inadvertently throughout the USA with infested

Fig. 1. Coconut plantation in the Philippines with palm trees killed by the viroid of the coconut 'cadang-cadang' disease (CCCV) and subsequently decapitated by a typhoon.

but symptomless cuttings, which were distributed mainly by one propagator (Keller, 1951, 1953). When it was realized that the disease was easily transmissible by contaminated tools and hands, pertinent control measures were rapidly adopted; only indexed plants, kept in isolation were used for further propagation. Considerable losses have also been reported for PSTV-infected potato (Singh, Finnie & Bagnall, 1971) and CEV-infected citrus species and varieties (Calavan, Weathers & Christiansen, 1968).

The viroid aetiology of all these diseases has been established by demonstrating the extra band of viroid RNA by polyacrylamide gel electrophoresis of nucleic acid preparations directly extracted from diseased plant tissue with the aid of phenol. This band is absent in RNA samples from healthy control plants. In addition, the viroid nature of the extra RNA band has been demonstrated by assaying its infectivity, wherever possible. For this purpose the gel is cut into slices and the buffer homogenates individually prepared from them are bioassayed on suitable indicator plants.

VIROID TRANSMISSION, HOST RANGE AND CONTROL

The viroid of the spindle tuber disease of potato (PSTV) and the stunt disease of chrysanthemum (CSV) are known to spread rapidly to neighbouring plants by the combination of wounding and contact of foliage during cultivation and handling in the field and in greenhouses.

Vertical transmission through seed and pollen has been reported for PSTV in certain solanaceous host species, which causes serious problems in obtaining healthy plant material by breeding. Early work indicated that, under field conditions, the potato spindle tuber disease could also be accidentally transmitted at low frequencies via biting insects and by aphids (Schultz & Folsom, 1923 ab). Recent investigations have confirmed that the two aphid species *Macrosiphum euphorbiae* (De Bokx & Piron, 1981) and *Myzus persicae* (Lee & Sänger, 1982) are capable of transmitting true PSTV from infected to healthy potato and tomato plants respectively at frequencies below 6%.

Most viroid diseases, however, are inadvertently spread by humans via the vegetative propagation of viroid-infested but symptomless plants through tubers (potato), cuttings (chrysanthemum) or grafting (citrus). Thus, CEV has been disseminated inadvertently into all citrus-growing areas of the world by distributing and grafting citrus varieties which were symptomless carriers of the agent. In addition, CEV may also be transmitted to neighbouring trees within a citrus plantation, by contaminated tools, during pruning and cutting. The mode of natural transmission of the cadang-cadang disease of the coconut palm, however, which is propagated only by seeds, is still unknown.

Studies of experimental host range have shown that viroids are often latent in wild plant species, whereas they are generally pathogenic in cultivated plants (Singh, 1973). In fact, the wild potato species *Solanum phureja* and *S. stenotonum* were found to harbour viroids which are capable of inducing potato spindle tuber disease in cultivated potato. Many observations lead to the suspicion that viroid diseases of cultivated plants are of recent origin and that their world-wide spread by human agricultural and horticultural activities is strongly favoured by the monoculture of crop plants and ornamentals.

The experimental host range of different viroids differs considerably. PSTV is able to infect systemically numerous plant species of several families (Singh, 1973), whereas CCMV seems to infect only chrysanthemum varieties. CCCV has, so far, only been transmitted successfully to a few other palm species and ASBV seems to infect avocado (*Persea americana*) and cinnamon (*Cinnamonum zelanicum*), which are both members of the family Lauracea. The host range of the other viroids is limited to certain species of a few specific plant families. Thus CEV can be transmitted from citrus to the herbaceous composite *Gynura aurantiaca* and to different solanaceous hosts including tomato, potato, petunia and scopolia. CPFV infects many species of the cucurbit family, but it may also be transmitted to certain solanaceous hosts. The host plant has no pronounced influence on the primary sequence of different viroids propagated in them (Dickson, Diener & Robertson, 1978).

All viroids described so far have been experimentally transmitted by mechanical means using crude tissue homogenates. The success of transmission is considerably increased if measures are taken to eliminate the action of nucleases, which inactivate viroid infectivity quite rapidly by degrading the nuclease-sensitive viroid RNA molecule. Correspondingly, tissue homogenates are usually freshly prepared with cold and slightly alkaline buffers, in the presence of the nuclease-absorbing clay mineral bentonite, and immediately used for inoculation. Under these conditions standard inoculation techniques, as used with conventional plant viruses or their RNAs, may be applied. Thus, rubbing of the inoculum onto carborundum-dusted leaves is used experimentally in many host–viroid combinations (tomato–PSTV, –CEV, –CPFV). In certain host plants, stem-puncturing or stem-slashing with razor blades or scalpels (gynura–CEV, chrysanthemum–CSV, cucumber–CPFV) is more efficient in viroid transmission. Grafting has played an important role in the early experimental transmission of the citrus exocortis disease (Garnsey & Whidden, 1973).

Quantitative experimental work, based on the infectivity of viroids as determined by bioassays, is somewhat complicated, because infectivity is generally measured by dilution end point titrations using the all-or-nothing type response of a generalized (systemic) viroid infection. Since most experiments with viroids result in large effects, the differences in titres can be determined with fair accuracy by this type of bioassay.

The dissemination of viroid diseases may be controlled efficiently by prompt removal of all apparently diseased plants and by the selection of healthy plant material for further propagation. In all cases where the disease agent is readily transmitted mechanically or by contaminated tools, their sterilization with flaming or with chemicals (like sodium hypochlorite or sodium hydroxide plus formalin) has proved successful in preventing further spread of the disease (Roistacher, Calavan & Blue, 1969). In addition, indexed viroid-free plant material to be used as propagating stock should be maintained in isolation and be treated with great caution to prevent recontamination. If the pertinent control measures are strictly followed, the incidence and spread of viroid diseases can be efficiently reduced.

In the case of PSTV-infected potatoes (Stace-Smith & Mellor, 1970) and CSV-infected chrysanthemums (Hollings & Stone, 1970) a combination of heat treatment and meristem-tip culture or axillary-bud culture has resulted in the recovery of 2–4% of viroid-free regenerated plants. In view of the preferentially enhanced viroid replication at elevated temperatures, meristem-tip cultures of potatoes after 'cold-treatment' has been successfully applied and about half of the regenerated plants were found to be viroid-free (Lizaraga *et al.*, 1980).

Thus far, no chemical treatment has resulted in the cure of a viroid disease. The antibiotics tetracycline and penicillin, which are effective against prokaryotic pathogens, failed to cure the cadang-cadang disease of coconut (Randles, Rillo & Diener, 1976). An emulsion containing 1% piperonyl butoxide, 0.1% Triton X100 and water acts only as a protectant against PSTV infection in potato and *Scopolia sinensis* if viroid inoculation of the sprayed leaves is carried out within 4 days. A similar treatment kills tomato plants (Singh, Michniewicz & Narang, 1975; Singh, 1977).

VIROID DIAGNOSIS

Because of the great impact of viroids on agriculture and horticulture and their potential threat to crop plants, improved methods for viroid detection are needed so that infested plants can be eliminated from further propagation. Polyacrylamide gel analysis may be used for routinely diagnosing the absence or presence of viroids in crop plants (Morris & Wright, 1975; Morris & Smith, 1977; Pfannenstiel, Slack & Lane, 1980; MacQuair *et al.*, 1981). However, the absence of a visible viroid band cannot be considered as a reliable proof for the absence of the agent. Since viroid RNA may represent only a minute fraction (usually less than 0.001%) of the total nucleic acids directly extracted from diseased plants, its concentration could be beyond the level of detection after staining the gel with methylene blue, toluidine blue, or ethidium bromide. Although the recently developed and highly sensitive silver-staining method (Sammons, Adams & Nishizawa, 1981) will detect as little as 1 ng

$(0.03\,\mathrm{ng\,mm^{-2}})$ of nucleic acids in polyacrylamide gels (Sommerville & Wang, 1981; Goldmann & Merril, 1982) gel analysis needs to be complemented by appropriate bioassays if critical decisions have to be made.

A completely new approach for assaying for viroids, even under practical conditions, has recently been developed by applying recombinant DNA techniques and molecular hybridization. The required viroid-complementary DNA (cDNA) can be synthesized with the aid of reverse transcriptase using viroid RNA as template (Owens, 1978; Palukaitis & Symons, 1978; Palukaitis et al., 1981). The use of tailor-made, sequence-specific primers has produced viroid-specific cDNA which corresponds to the full length of the corresponding viroid (Rohde, Schnölzer & Sänger, 1981b; Rohde et al., 1981a). In the case of PSTV, the corresponding cDNA has already been subjected to molecular cloning in order to obtain PSTV-specific hybridization probes (Owens & Cress, 1980). This new technique allows practically unlimited amplification and a very efficient radioactive labelling of the viroid-specific cDNA so that simplified techniques of molecular hybridization will be used routinely in future to screen for the presence of viroids. One of these highly specific and sensitive techniques, the 'dot-spot test' has been shown to be applicable for the diagnosis of PSTV in large numbers of samples (Owens & Diener, 1981) as required under practical conditions in institutions involved in plant propagation, breeding and introduction.

There are two new and promising alternatives to the use of hazardous and unstable radioisotopes for the detection of viroids. The first is the so-called bio-blot test which makes use of biotin-labelled DNA probes which are prepared by nick-translation in the presence of biotinylated nucleotide analogues and hybridized to DNA or RNA immobilized on nitrocellulose filters. After removal of the residual probe, the filters are briefly incubated with a preformed complex made with avidin or streptavidin and biotiny-lated polymers of intestinal alkaline phosphatase. The filters are then incubated with a mixture of 5-bromo-4-chloro-3-indolyl phosphate and nitro blue tetrazolium, which results in a purple blue colour reaction at the sites of hybridization (Leary, Brigati & Ward, 1983). This nonradiographic method of probe detection will certainly be adapted to viroid diagnosis, provided that the promise holds true that it will detect target sequences in the 1–10 pg range also in crude plant extracts.

The second method called 'bidirectional gel analysis' (Schumacher, Randles & Riesner, 1983b) is much simpler and can therefore be easily performed even in the field laboratories. It combines the high sensitivity of the silver staining technique for nucleic acids (Sammons et al., 1981; Sommerville & Wang, 1981; Goldmann & Merril, 1982) and the unique structure and physicochemical properties of viroids. It is based on the observation that the circular viroid RNA molecules become unfolded under denaturing conditions and are therefore greatly retarded in their electrophoretic mobility and thus clearly separated from any other nucleic acid species which comigrate with them under non-denaturing conditions (Sänger et al., 1979).

For viroid detection, 0.1–0.5 g leaf tissue is first homogenized in buffer in the presence of phenol and sodium dodecylsulfate. The homogenate is then centrifuged for phase separation. The upper aqueous phase is recovered, phenol extracted and centrifuged again. The total nucleic acids are finally concentrated from the aqueous phase by ethanol precipitation and redissolved in 100–500 μl buffer. Aliquots from 20 such samples can be run together on a 5% polyacrylamide slab under native conditions where the nucleic acids are separated primarily on the basis of their size. After the run, the region of the presumed viroid RNAs is determined by briefly staining the gel with ethidium bromide which allows the localization of this area in a lane where a reference sample has been run. A gel strip (width c. 1 cm) is then cut across the gel in such a way that it comprises the viroid region of all lanes. The strip is set onto the bottom of a gel chamber and a 5% gel which contains 8M urea is polymerized on top of it (Fig. 2a,b). The subsequent gel run is carried out at 50 °C in order to provide denaturing conditions. After silver-staining of the gel, it becomes evident that because of their retarded migration the circular and the linear forms of viroid RNA are well separated from all other cellular nucleic acid species and therefore clearly detectable (Fig. 2c,d). Since a concentration of a 0.06 μg of viroid per gram infected tissue can be detected (Schumacher et al., 1983b) the sensitivity of this test approaches the sensitivity range of molecular hybridization on 'dot blots'.

VIROID STRUCTURE

The wealth of information which has accumulated on the structure of viroids has repeatedly been reviewed in great detail (see Riesner

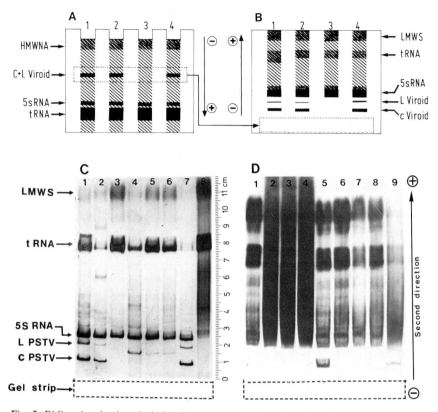

Fig. 2. Bidirectional gel analysis for the presence of viroids in ethanol-concentrated phenol extract of total nucleic acids from plant tissue on 5% polyacrylamide gels.

A. Schematic presentation of the first gel run of the total nucleic acids under native conditions.

B. Second run of the cut-out viroid-containing region from the first gel under denaturing conditions as provided by 8M urea in the top gel and a gel temperature during the run of 50 °C.

C. Analysis for the presence of viroids in different plants after silver-staining of the second fully denaturing gel. **lane 1:** PSTV in tomato, **lane 2:** CEV in Gynura aurantiaca, **lane 3:** healthy cucumber, **lanes 4,5,6:** CPFV in cucumber, **lane 7:** PSTV in tomato (note the difference in migration and the apparent absence of linear molecules.), **lane 8:** healthy Gynura.

D. Detection in different Chrysanthemum cultivars (lanes 1 to 7) of CSV (lane 5) and in Etrog citron (Citrus medica) (lanes 8 and 9) of CEV (lane 9). The original viroid-containing gel strip indicated at the bottom of gels C and D usually becomes separated during the staining-procedure from the gel which had been cast on top of it. The heavy blackening of several lanes shows that the nucleic acid extracts from certain plant species and cultivars contain silver-staining plant substances which may have interfered with the analysis as visible in gel D, lanes 5 and 9.

HMWNA: high molecular weight nucleic acids; LMWS: low molecular weight substances; C-PSTV: circular PSTV; L-PSTV: linear PSTV.

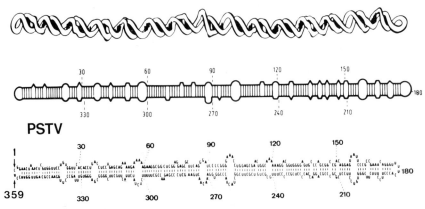

PSTV

Fig. 3. The nucleotide sequence (primary structure) and the resulting secondary structure of PSTV as shown in a two-dimensional base-pairing scheme (**bottom**), in the corresponding diagrammatic representation (**middle**) and in a three-dimensional manner (**top**). This structure is based on maximal intramolecular base-pairing taking into consideration the observed nuclease- and bisulphite-sensitivity of PSTV and the data obtained from detailed physiochemical investigations including binding studies with dyes and specific tRNAs.

et al., 1981; Sänger, 1982) so that only the features of presumed direct functional relevance for replication and pathogenicity need to be briefly summarized here. Biochemical, electronmicroscopical, thermodynamic, kinetic, hydrodynamic and sequence analysis on several viroid species have established that the structure of all viroids follows the same general principles. They are single-stranded, covalently closed circular RNA molecules which exist in their native state as highly base-paired rod-like structures (Sänger *et al.*, 1976) which are characterized by the serial arrangement of short double-stranded segments and unpaired regions forming small internal loops and a loop at both ends of the rod, as first demonstrated in the case of PSTV (Gross *et al.*, 1978), see Fig. 3. Different lines of evidence substantiate that there is no additional base-pairing between distant loops of the molecule, so that viroids are devoid of any tertiary structure folding. The sequencing work on additional viroids has shown that the unit length of their nucleotide chain ranges in size between 240 and 380 nucleotides (Table 2). Although considerable differences exist in their nucleotide sequences, their relative structural stiffness is reflected in a 'persistence length' of 30 nm which means that all viroids can bend in solution to only about a quarter of a circle. The specific features of the secondary structure of viroids have been strictly conserved (see Table 2 and Fig. 4) which must therefore be most essential for their replication and 'survival'.

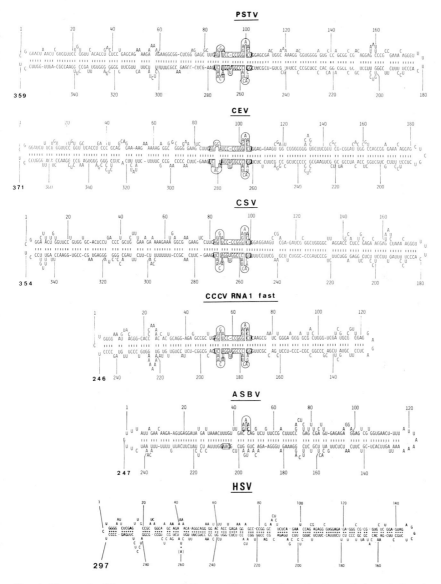

Fig. 4. The nucleotide sequence and the resulting secondary structure of the unit length molecules (monomers) of the six viroid 'species' PSTV, CEV, CSV, CCCV, ASBV and HSV. The boxed-in sequences of PSTV, CEV, CSV and CCCV are common to these four viroids whereas ASBV shares only a pentanucleotide with them. For references to the corresponding sequencing work see Table 3.

Table 2. *Chain lengths of different viroids and viroid isolates and their sequence homology with PSTV*

Viroids	Number of ribonucleotides	Sequence homology with PSTV (%)	References
PSTV			
Pathogenic prototype	359	(100)	Gross *et al.*, 1978
Mild isolate	359	99	Gross *et al.*, 1981
CEV			
Californian isolate	371	73	Gross *et al.*, 1982
Australian isolate	371	59	Visvader *et al.*, 1982
CSV			
English isolate	354	73	Gross *et al.*, 1982
Australian isolate	356	69	Haseloff & Symons 1981
CCCV			
RNA 1 fast[a]	246	11	Haseloff, Mohamed & Symons, 1982
RNA 1 slow[b]	287	11	Haseloff *et al.*, 1982
RNA 2 fast	492	11	Haseloff *et al.*, 1982
RNA 2 slow	574	11	Haseloff *et al.*, 1982
ASBV	247	18	Symons, 1981
HSV	297	55	Ohno *et al.*, 1983

[a]Certain isolates consisted of a mixture of two species of 246 and 247 nucleotides respectively.
[b]From nine isolates sequenced, one consisted of 287 nucleotides, one of 297 nucleotides, six consisted of 296 nucleotides and one of 301 nucleotides, so that the duplicated sequences of RNA 1 fast by which RNA 1 slow can be produced consist of 41, 50 and 55 nucleotides, respectively.

Of special interest is the structural complexity of CCCV which, in contrast to all other viroids, exists in four variant RNA species. They are generated from the monomer by sequence duplication and named according to their electrophoretic mobility. Two of these four RNAs, called CCCV RNA 1 fast (246 nucleotides) and CCCV RNA 2 (492 nucleotides) are present in infected palms at early stages of the disease (Imperial, Rodriguez & Randles, 1981). As infection progresses over a period of years, two additional RNAs, called CCCV RNA 1 slow (287 nucleotides) and CCCV RNA 2 slow (574 nucleotides) respectively, appear and predominate (Mohamed *et al.*, in press).

Sequence analysis has shown that four of the six isolates of CCCV RNA 1 fast consist of two populations of molecules 246 and 247 nucleotides long and differing in the presence or absence of a C at nucleotide 198. The relative proportions of the two CCCV RNA 1 fast subspecies vary between different isolates but the various sequence differences between the isolates do not appear to correlate with differences in geographic location. Since RNA 1 fast is essentially identical in different isolates, it can be considered as the unit length molecule of the CCCV RNA system. Its three other RNA species can be derived from RNA 1 fast by duplication of sequences (Haseloff, Mohamed & Symons, 1982) (Fig. 5).

Thus, the sequence analysis of nine isolates of RNA 1 slow from single palms from different localities in the Philippines revealed that they contain the entire sequence and structure of the smaller CCCV RNA 1 fast plus an additional duplicated sequence and structure of 41, 50, or 55 nucleotides, respectively. They are derived from the right hand region of RNA 1 fast between nucleotides 95 to 150 and added to the right hand end of the native molecule of CCCV RNA 1 fast between nucleotides 123 and 124 to produce the corresponding species of RNA 1 slow (Fig. 6) and corresponding molecules with chain lengths of 287, 296, 297 and 301 nucleotides have been found.

The CCCV RNAs 2 fast are 492 and 494 nucleotides long and represent perfect dimers of the RNA 1 fast with chain length of 246 and 247 nucleotides. The same applies for the sequenced RNA 2 slow of isolate Baao54 which proved to be a dimer of the corresponding RNA 1 slow with its 287 nucleotides. Fig. 6 shows that despite these sequence duplications the native structure of the RNAs remains rod-like. While the monomeric forms (RNA 1 fast and RNA 1 slow) exist as a single rod-like conformer, their corresponding duplexes (RNA 2 fast and RNA 2 slow) are potentially able to form two rod-like conformers as well as a large number of intermediate cruciform-shaped structures.

All these data would indicate that the CCCV RNA slow forms are generated from the RNA fast forms by single, rare sequence duplication events occurring separately in each infected palm. If so, all slow RNAs would originate from single parent molecules and accumulate in preference to the fast RNAs because of a competitive advantage in replication.

CONFORMATIONAL CHANGES

Since viroid molecules may exert their biological functions through conformational changes, the analysis of their structural transformation during thermal denaturation *in vitro* has been studied in detail

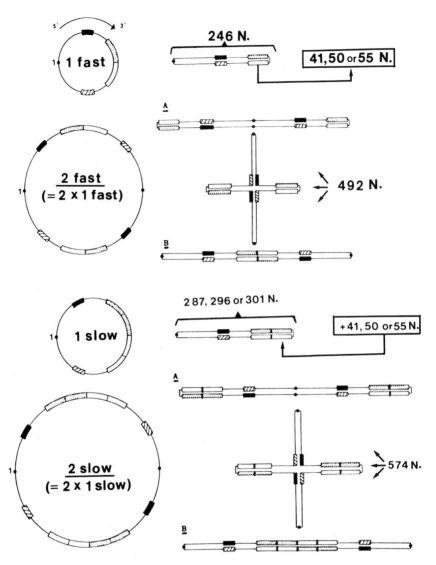

Fig. 5. Structural relationship between the four CCCV RNAs according to Haseloff *et al.*, 1982. RNA 1 fast with its 246 nucleotides represents the unit length molecule of CCCV from which RNA 2 fast is generated by duplication so that a molecule with 492 nucleotides is formed. Three of several structural alternatives are shown for RNA 2 fast. RNA 1 slow is derived from RNA 1 fast by duplication of a region of 41,50 or 55 nucleotides, respectively, which originate from the right hand end of 1 fast as shown in detail in Fig. 6. RNA 2 slow is again created by duplication of RNA 1 slow and 3 of several structural alternatives are shown.

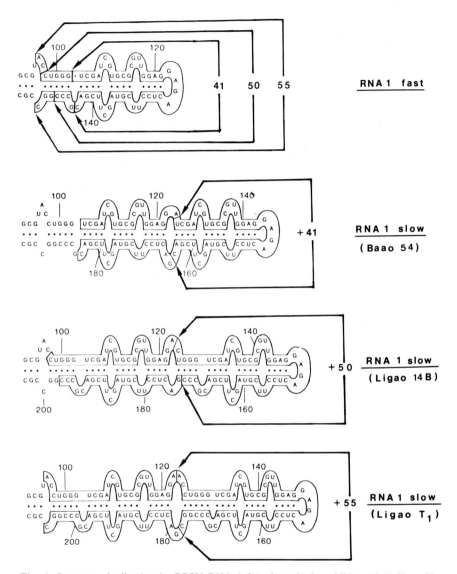

Fig. 6. Sequence duplication in CCCV RNA 1 fast through the addition of 41,50 or 55 nucleotides, respectively. They are derived from the right hand end of this molecule, as indicated on top by arrows and inserted between nucleotides 123 and 124 of RNA 1 fast to give three different types of RNA 1 slow forms. All sequenced RNA 1 slow isolates sequenced so far correspond to one of these forms (Haseloff *et al.*, 1982).

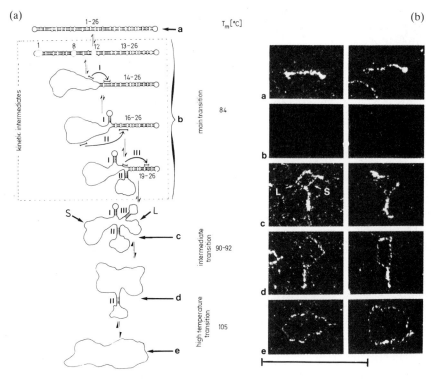

Fig. 7. Mechanism of denaturation and renaturation of the PSTV molecule with different structural intermediates.

   (a) Schematic representation of the denaturation mechanism. The double-stranded segments of the molecule are numbered 1–26 from left to right. The T_m values are given for 1 M NaCl. (b) Selected PSTV molecules from dark-field electron micrographs obtained after spreading of denatured PSTV RNA. The different structural intermediates a to e in the micrographs correspond to the intermediates a to e in the denaturation scheme. The conformers b are not found in electron micrographs because they represent kinetic intermediates. S and L denote the small and large single-stranded segment, respectively. Bar 100 nm.

(Riesner *et al.*, 1979). Thus the melting process of all the ones studied so far is not a mere dissociation of their 115–130 base-pairs but it follows a rather complicated and unique scheme as compared to all other nucleic acids. In the case of PSTV, the left half of the molecule opens in one cooperative step, leaving the right half still intact. This is dissociated only at higher temperatures via a series of intermediate stages so that finally the native rod-like structure is transformed into a completely unfolded circle (Fig. 7). All these structural states can be found using the electron microscope under appropriate conditions of preparation (Klotz & Sänger, 1981).

   The structural transitions of viroids have also been calculated at a

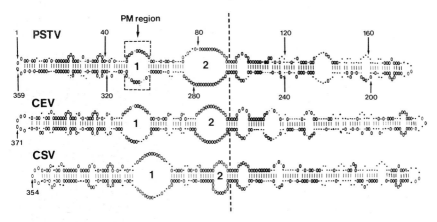

Fig. 8. Secondary structure of PSTV, CEV and CSV at a temperature 3 °C below the main transition. The nucleotides which are identical in the three viroids are represented by open circles, the varying ones by points. The dashed vertical line in the middle of the rod-like structures represents the border of stability. It divides the molecules of the three viroid species into a well conserved but unstable and hence flexible left half with the premelting loops 1 and 2 and into a stable right half with significantly less sequences being conserved. However, the region of and adjacent to the borderline of stability is strictly conserved in all three viroids. The boxed-in shadowed area indicates the pathogenicity-modulating (PM) region of PSTV. The open circles represent the nucleotides which are identical in the three viroid species whereas the variable ones are marked by dots.

temperature 3 °C below the main transition (Riesner *et al.*, 1983). At this temperature the regions of lowest stability undergo premelting transitions. The premelting regions have been compared for CEV, CSV and PSTV (which exhibit 60–70% sequence homology). Two premelting regions are found in each of the three viroids (Fig. 8); these are in strictly conserved parts of the molecule. The left premelting region contains the oligo A sequence which is suspected of acting as a promotor in viroid replication, although there is no experimental proof for this so far. The second premelting region is directly adjacent to one of the most stable helices. This borderline in the thermal stability is conserved in all of the three viroids in the form of a longer stretch of identical sequences. It divides the molecule into a less stable left half and a more stable right half. During evolution, the left part of these viroids has evidently been selected for particular sequences embedded in a structure which dissociates easily whereas the right part has evolved towards a stable secondary structure with significantly fewer sequences being conserved. As will be discussed in one of the following paragraphs, the modulation of the stability of the premelting region seems to influence the virulence of the PSTV molecule.

GENERAL PROBLEMS OF VIROID REPLICATION

The mechanism of viroid replication and pathogenesis is still largely a matter of conjecture. The lack of information in this area is mainly due to system-inherent experimental limitations. Thus, higher plants and the corresponding cell systems are less well suited to study the dynamic aspects of cell metabolism after virus infection than bacterial and animal cells.

Unlike certain viruses, viroids cannot carry their own replicase with them, because they are 'naked' RNA molecules and not encapsidated in a protein coat and envelope. Furthermore, it is obvious that the potential genetic information of the viroid molecule, with its 240–380 nucleotides, is insufficient to code for a complete viroid specific replicase, even if overlapping reading frames and three rounds of translation are assumed. Because of this limited coding capacity, one must assume that viroid replication depends on the nucleic-acid-synthesizing machinery pre-existing in the host cell (Fig. 9).

Although all attempts so far have failed to translate viroid RNA *in vitro* into proteins or peptides it is nevertheless conceivable that the plus viroid molecule or its complementary strand could be translated *in vivo* into a small polypeptide which is capable of converting a host enzyme into a viroid-specific replicase. This would resemble a strategy similar to the one used by the RNA bacteriophage Qβ, in which case a phage-coded 65 000 molecular-weight protein is combined with three host proteins (ribosomal protein S1 and the protein synthesis elongation factors EF-Tu and EF-Ts) to create the Qβ-specific replicase with its four subunits (see Kamen, 1975). Also overlapping reading frames could be used to code for different proteins of completely different functions, as realized by the DNA bacteriophage φ X174 (Sänger *et al.*, 1977). Finally, the possibility exists that the corresponding viroid molecule might require specific processing and/or modification before its translation proceeds.

Regarding the possible pathways of viroid replication by nucleic-acid-synthesizing enzymes pre-existing in the host, we have to consider that healthy plant cells are known to contain two classes of RNA-synthetising enzymes, namely the DNA-dependent RNA polymerases I, II and III (see Duda, 1976; Wollgiehn, 1982), and the RNA-dependent RNA polymerase (see Fraenkel-Conrat, 1979). Theoretically, viroids could therefore be replicated through a

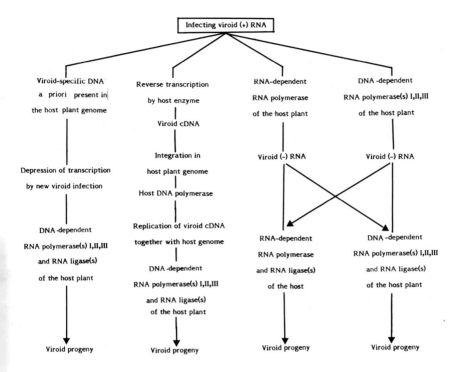

Fig. 9. Hypothetical pathways of viroid replication.

complementary RNA, as are the RNA plant viruses, in which case the RNA-dependent RNA polymerase of the host could be utilized. The corresponding transcription experiments *in vitro* showed that this enzyme from healthy tomato leaf tissue is capable of synthesizing PSTV copies of apparently full length when PSTV RNA is offered as template (Boege, Rohde & Sänger, 1982).

Viroid replication could also follow the strategy of the RNA tumour viruses, the genome of which is first transcribed by the reverse transcriptase into DNA, which then serves as template for the final synthesis of the retrovirus progeny RNA. Since no reverse transcriptase activity has been detected in higher plants so far, one would be forced to assume that upon infection the viroid-specific DNA template is newly synthesized in a way as yet unknown.

On the other hand, the viroid-specific DNA complement could already be present in the host cell as a part of the host genome, but in a repressed state. In this case one could visualize that the infecting viroid RNA acts as a regulatory molecule by triggering the de-repression of the transcription of the presumed viroid complementary sequences present in the host genome. If so, the viroid-specific DNA could serve as template for one of the DNA-dependent RNA polymerases from which the viroid progeny RNA could be synthesized. Unequivocal evidence for any of these routes would be the presence of viroid-specific DNA in viroid-infected tissue. The previously published experimental evidence for this has been refuted and there is now general agreement that viroids are replicated *in vivo* via oligomeric (−)RNA intermediates (Grill & Semancik, 1978; Branch, Robertson & Dickson, 1981; Rohde & Sänger, 1981) and that viroid specific DNA sequences do not exist (Branch & Dickson, 1980; Zaitlin *et al.*, 1980).

From the existence in the host cell of linear and circular viroid molecules, it may be concluded that the two characteristic steps in splicing of pre-mRNA (namely, nucleolytic cleavage and subsequent ligation) are also essential steps in viroid maturation. The insufficient genetic information content of viroids would indicate also that for this process the machinery pre-existing in the host cell is utilized. The recent finding that RNA ligase can be isolated from wheat germ, which is able to circularize linear RNA by the formation of a 2′-phosphomonoester, 3′,5′-phosphodiester linkage (Konarska *et al.*, 1981), strongly suggests that other higher plants (including viroid hosts) might also contain RNA ligases which could produce viroid circles from the corresponding linear precursor molecules. The wheat germ ligase, in fact, is capable of transforming linear viroid RNA into viroid circles *in vitro* (Branch *et al.*, 1982; Kikuchi *et al.*, 1982).

One of the unique features of viroid RNA is its similarity to a DNA molecule of the same size and G:C content as far as the overall shape, the thermal stability and the cooperativity of denaturation is concerned (Riesner *et al.*, 1979). It was conceivable therefore, that viroids might be efficient templates for one of the DNA-dependent RNA polymerases of the host. This notion gained additional support from the finding that RNA polymerase from *Escherichia coli* is able to select from a random mixture of copolymers a RNA molecule with a chain length between 100 and 150 nucleotides which is capable of multiplying indefinitely (Biebricher & Orgel, 1973). The results from corresponding experiments *in vivo* and *in vitro* with DNA-dependent RNA polymerase II from viroid host plants will be discussed below.

VIROID REPLICATION IN PROTOPLASTS AND CELL CULTURES

The first studies on viroid replication at the cellular level were carried out with protoplasts isolated from tomato leaf tissue (Mühlbach, Camacho-Henriquez & Sänger, 1977). The general metabolic capacity of these protoplasts was substantiated by their ability to regenerate new cell walls after 3 d, to divide within 1 wk and to produce complete tobacco mosaic virus (TMV) particles after infection with TMV RNA *in vitro*. Viroid replication can be followed in these protoplasts by incorporation of [^{3}H]uridine after infection *in vitro* (Mühlbach & Sänger 1977). To discriminate unambiguously between the different RNA polymerases possibly involved in viroid replication, the mushroom toxin α-amanitin was applied. In eukaryotic cells, α-amanitin is known to inhibit selectively the DNA-dependent RNA polymerases II and III at low and high concentrations, respectively, whereas RNA polymerase I is not inhibited at all (Lindell *et al.*, 1970; Zylber & Penman, 1971; Weinmann & Roeder, 1974; Wieland & Faulstich, 1978). Our experiments showed that at 50 μg ml^{-1} culture medium α-amanitin inhibits viroid replication to about 75% whereas the biosynthesis of the prominent cellular RNA species tRNA, 5S RNA, 7S RNA and ribosomal RNA is not appreciably affected (Fig. 10). Control experiments with tomato protoplasts inoculated with tobacco mosaic virus RNA showed that the replication and accumulation of TMV was not affected by α-amanitin under these conditions, where TMV replication is known to increase logarithmically. Therefore, the marked inhibition of viroid replication by α-amanitin is unlikely to be due to a secondary effect of this drug on cell metabolism in general.

In view of the concentration-dependent specificity of the α-amanitin inhibition, the interpretation of such experiments *in vivo* depends on knowing the intracellular concentration of this inhibitor; this was therefore determined by using the ^{3}H-labelled toxin. It was found that, under the conditions where viroid replication is specifically inhibited, the intracellular concentration of the inhibitor was in the range of 10^{-8}M (where RNA polymerase II is specifically inhibited). Since inhibition of RNA polymerase III requires a 1000-fold higher concentration (Weinmann & Roeder, 1974) and since RNA polymerase I is not affected at all by α-amanitin, it was postulated that the RNA polymerase II system is involved in viroid replication (Mühlbach & Sänger, 1979).

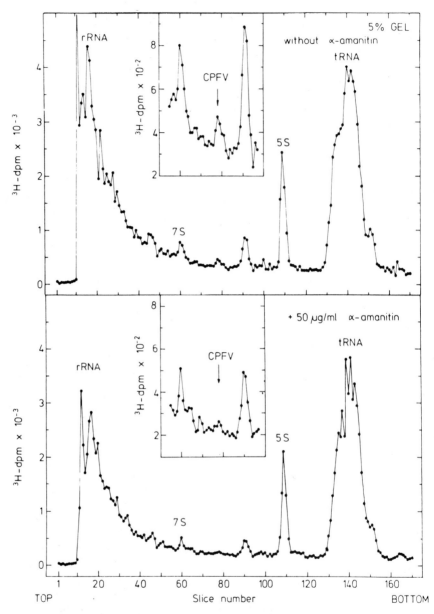

Fig. 10. Analysis of 2 M LiCl-soluble RNA from viroid (CPFV)-infected tomato protoplasts in 5% polyacrylamide gels. RNA pattern obtained in the absence (top) and in the presence of 50 μg ml^{-1} α-amanitin (bottom) which was found to correspond to an intracellular concentration of 10^{-8} M. For better comparison the viroid region of the gels is shown in an enlarged scale in the inserts.

It should be emphasized, however, that the observed α-amanitin inhibition cannot be considered as definitive proof of direct replication of viroid RNA by the nuclear DNA-dependent RNA polymerase II. There are several possible mechanisms, by which this enzyme system could be indirectly involved as a result of its central role in the formation of mRNAs in the host cell. There is also the remote possibility that viroid replication may proceed through a yet unknown pathway which is α-amanitin-sensitive.

Further progress in the study of viroid replication at the cellular level can be expected from permanent plant cell cultures which are continuously synthesizing viroid RNA. In callus cultures from a PSTV-infected wild-type potato (*Solanum demissum* L.) and tomato (*Lycopersicon peruvianum* L. Mill.) and in cell suspension cultures derived from potato (*Solanum tuberosum* L.) protoplasts inoculated *in vitro* with PSTV, viroid replication was found to persist for now over 3 years through more than 40 subcultures. The continuous *de novo* viroid synthesis was substantiated by the incorporation of ^{3}H-uridine and of ^{32}P-orthophosphate into PSTV RNA and its subsequent analysis by gel electrophoresis and autoradiography (Mühlbach & Sänger, 1981). Similar observations of sustained long-term viroid replication were reported for cell suspensions prepared from tomato (Zelcer *et al.*, 1981) and potato (Mühlbach, Faustmann & Sänger, 1983). The maximum rates of ^{32}P incorporation into PSTV RNA varied between 0.4% of that into soluble RNA (Zelcer *et al.*, 1981) and 2–3% of that into tRNA (Mühlbach & Sänger, 1981) and to 10% of the total cellular RNA (Mühlbach *et al.*, 1983). These differences are most probably due to the different systems used in these experiments.

It is interesting to note that the continuous viroid (PSTV) replication in plant tissue culture through numerous cell divisions seems to differ from the experience with conventional plant viruses which are often eliminated from continuously dividing meristematic tissue and cells (Hollings, 1965; Ingram, 1976). This would provide additional evidence of fundamental differences between the mechanisms of replication of plant viruses and of viroids. The reader interested in the use and specific problems of protoplasts and cell cultures in viroid and plant virus research is referred to a recent review by Mühlbach (1982).

TRANSCRIPTION *IN VITRO* OF VIROID RNA BY DNA-DEPENDENT RNA POLYMERASE II OF PLANT ORIGIN

The assumption that the DNA-dependent RNA polymerase II of the host cell might be involved in viroid replication was tested in experiments *in vitro* (Rackwitz, Rohde & Sänger, 1981) for which this enzyme had been purified from tomato tissue and wheatgerm. It was found with both enzymes that standard single-stranded DNA templates are transcribed less efficiently than double-stranded DNA templates and that the transcription of single-stranded DNA templates is considerably improved in the presence of oligo-RNA primers.

Synthetic and natural RNAs also served as templates for transcription, thus confirming previous results on the same enzyme from pea and cauliflower (Sasaki *et al.*, 1974*ab*, 1976). TMV RNA and soluble RNA (5S RNA and tRNA) from yeast were relatively poor templates, whereas the synthetic template/primer poly(rA)/oligo(rU) was transcribed with high efficiency. Most surprisingly, however, of all natural RNA templates tested, viroid RNA was transcribed with the highest efficiency by DNA-dependent RNA polymerase II from tomato and wheatgerm. Although overall transcription was about the same when circular and linear viroid PSTV RNA molecules were used as templates, the pattern of product formation varied slightly between different template preparations. When the transcription products were analyzed under denaturing conditions where circular and linear viroid molecules are clearly separated, it became evident that the largest transcripts from both polymerases had the same migrating properties and hence the same size as the linear viroid molecule (Fig. 11). Several control experiments demonstrated that this RNA is a genuine transcript and not due to terminal labelling of linear viroid template molecules. In addition to the full-length copies, several smaller viroid transcription products were found and further experiments are needed to decide whether they are caused by site-specific nicking or by run-off transcription on randomly nicked template molecules.

The highly specific interaction between viroid RNA and polymerase II was further substantiated by the inhibition of transcription at 10^{-8}M α-amanitin, by binding studies between ^{125}I-labelled PSTV RNA and tomato polymerase II using the cellulose filter technique, and by northern blot analysis. Viroid RNA was found to form a binary complex with polymerase II, to compete with DNA for the template binding site on the enzyme and to strongly inhibit DNA-directed RNA-synthesis.

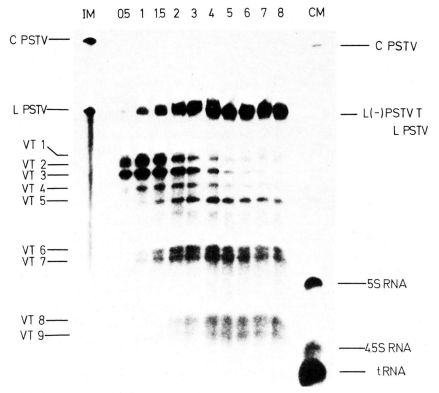

Fig. 11. Electrophoretic analysis of the kinetics of viroid (PSTV)-directed RNA transcription by DNA-dependent RNA polymerase II from tomato. The transcription products after different times of incubation were visualized by fluorography of dried gels. The numbers on top of the slots give the time of incubation in hours. C PSTV: ^{125}I-labelled circular PSTV; L PSTV: ^{125}I-labelled linear PSTV;L PSTV T: linear ^{32}P-labelled PSTV full-length transcript. VT1–VT9 represent viroid specific transcription products. VT1 migrates close to VT2 and due to overexposure the two bands cannot be discriminated in the autoradiogram.

Recent investigations have shown that polymerase II initiates specifically at nucleotides 168, 300 and 49 of PSTV, CSV and CEV, respectively. It was found, in addition, that polymerase III is also capable of specific initiation and transcribes viroid RNA into complete copies whereas, so far, polymerase I produced only smaller transcription products at low yield (unpublished results from the author's laboratory).

TRANSCRIPTION *IN VITRO* OF VIROID RNA BY RNA-DEPENDENT RNA POLYMERASE PURIFIED FROM HEALTHY HOST TISSUE

Higher plants are characterized by the presence of RNA-dependent RNA polymerase which, therefore, is another candidate for viroid replication. Such enzyme activities have been detected in several healthy plant species (Duda, Zaitlin & Siegel, 1973; Duda, 1976; Ikegami & Fraenkel-Conrat, 1978, 1979b; Romaine & Zaitlin, 1978; Lazar *et al.*, 1979; Takanami & Fraenkel-Conrat, 1982), including tomato, a well-known host plant for many viroids (Boege & Sänger 1980). An RNA-dependent RNA-polymerase activity is increased about two-to-six-fold after infection by various RNA plant viruses (Fraenkel-Conrat, 1979) and especially cucumber mosaic virus has been reported to elicit particularly high enzyme activity in cucumber cotyledons (Gilliland & Symons, 1968; May, Gilliland & Symons, 1969; May & Symons, 1971). It thus appears that upon virus infection these enzymes are taken over for the replication of the viral RNA, a conjecture which is still controversial. In analogy to the well characterized RNA replicase from *E. coli* infected with RNA bacteriophage Qβ (Kamen, 1975), the RNA-dependent RNA polymerase in virus-infected plants has therefore also been designated 'RNA-replicase'. Although the catalytic and biochemical properties of the RNA replicase seem to be very similar to those of the RNA-dependent RNA polymerase pre-existing in the uninfected plant cell, the actual relationship between these two enzyme activities is still unresolved. Therefore, the enzyme from healthy plants should be called RNA-dependent RNA polymerase. Its biological role in plant cell metabolism is still unknown. It has been suggested that it may serve to synthesize double-stranded RNAs, possibly with regulatory functions (Ikegami & Fraenkel-Conrat, 1979b).

Regarding viroid replication, it was found that RNA-dependent RNA polymerase from virus-free healthy tomato leaf tissue accepts PSTV RNA as template and produces full-length PSTV copies *in vitro*, as judged from their electrophoretic mobility on denaturing 5% polyacrylamide gels and from molecular hybridization on northern blots under stringent conditions (Boege *et al.*, 1982). Viroid transcription requires the presence of Mn^{2+} and/or Mg^{2+} and is not inhibited by a concentration of 10^{-5}M α-amanitin. The percentage of viroid template copying under optimal conditions was found to vary around 0.1%. Of special interest is the appearance of high molecular weight products which barely enter the 2.5% stacking gel or the 5% resolving gel. With other RNA templates, these enzymes were found to produce extremely heterodisperse transcription products of low molecular weight, irrespective of the nature and size of the template and the source from which they were isolated (Duda, 1976;

Romaine & Zaitlin, 1978; Chifflot *et al.*, 1980). Therefore the full-length viroid copy is the first well-defined and homogeneous product reported to be synthesized *in vitro* by an RNA-dependent RNA polymerase from healthy plants. A comparable synthesis of distinct transcription products has so far only been reported for the RNA-dependent RNA replicase from barley leaves infected with brome mosaic virus (Hardy *et al.*, 1979).

Regarding the relations between the structure of the RNA templates and their transcription efficiency, it was previously found that RNA-dependent RNA polymerase transcribes single-stranded RNA more efficiently than double-stranded RNA (Fraenkel-Conrat, 1979; Ikegami & Fraenkel-Conrat, 1978, 1979*ab*; Boege & Sänger, 1980). In view of these results, the observed viroid transcription into copies of full length is not too surprising, because viroid RNA displays features of both double-stranded and single-stranded RNAs. The elucidation of the details of the transcription process and of the nature of the transcripts larger than unit length requires further biochemical analysis.

INTERMEDIATES AND POSSIBLE MECHANISM OF VIROID REPLICATION

In viroid-infected tissue infections, circular viroid RNA molecules accumulate predominantly. They are called arbitrarily viroid (+)RNA. The complementary RNA copies of these molecules representing intermediates of viroid replication are denoted viroid (−)RNA. Such (−)RNA intermediates have been detected by molecular hybridization in viroid-infected plant tissue (Grill & Semancik, 1978, 1980; Owens & Cress, 1980; Rohde & Sänger, 1981; Branch *et al.*, 1981), whereas in extracts from healthy plants no such viroid complementary RNA was found. The characterization of these (−)RNA strands revealed that they are heterogenous in size, forming either four (Branch *et al.*, 1981) or six (Mühlbach *et al.*, 1983) discrete bands in polyacrylamide gels, thus representing multimers of the PSTV unit length RNA (359 nucleotides). They seem to be present in complexes containing extensive double-stranded regions, but after treatment with RNases, under conditions favouring digestion of single-stranded regions, the high molecular weight unit-length (−)strands appear (Branch *et al.*, 1981). Fingerprint analysis of hybridized ^{125}I-labelled PSTV following recovery from the hybrids demonstrated that all regions of the PSTV molecules are represented in the (−)RNA strands (Zelcer *et al.*, 1982). Further characterization of the double-stranded structures indicated that monomeric circular and linear (+)PSTV strands are complexed with multimeric (−)RNA strands and the synthesis of the double-stranded complexes increases simultaneously with the synthesis of the single-stranded PSTV-RNA (Owens & Diener, 1982).

In recent experiments with nuclei purified from a suspension culture of PSTV-infected potato cells and northern blot analysis, two oligomeric forms of PSTV(+)RNA and four forms of PSTV(−)RNA were found which represent 1–4 times the viroid unit length (Spiesmacher *et al.*, 1983).

From the different oligomeric forms which have been observed (Fig. 12); the following steps in viroid replication can be visualized: in the first step of the cycle the infectious circular viroid (+)RNA molecule is transcribed within the nucleus into viroid (−)RNA oligomers up to seven times viroid unit length. The mechanism most favoured for this step at present is the rolling-circle model previously proposed to explain the replication of certain viral RNAs (Brown & Martin, 1965), in particular the (+) strand of the DNA bacteriophage ϕX 174 (Gilbert & Dressler, 1968). This transcription could be catalyzed by the nuclear DNA-dependent RNA-polymerase II, because (*a*) viroid replication is specifically inhibited by α-amanitin (Mühlbach & Sänger, 1979) and (*b*) because this enzyme is capable of transcribing *in vitro* viroid RNA into products including copies of full-length and even longer transcripts (Rackwitz *et al.*, 1981, and unpublished results).

In a second step the (−)RNA oligomers may serve as template for the transcription into viroid (+)RNA oligomers which must be considered as the precursors of the PSTV (+)RNA monomers. The 'mature' circular viroid (+)RNA molecules (consisting of 359 nucleotides) are then generated in a third reaction step by specific endonucleolytic cleavage of the PSTV(+) oligomers into unit-length linear strands, which must become modified at their termini before they are finally ligated to covalently closed circles. It is conceivable that the cleavage and ligating steps could be carried out by enzymes normally involved in splicing of the host cell RNA. In fact, a novel type of RNA ligase, which is capable of ligating linear viroid RNA *in vitro* to circular molecules has been found in tissue of higher plants (wheat germ) (Branch *et al.*, 1982; Kikuchi *et al.*, 1982) and in

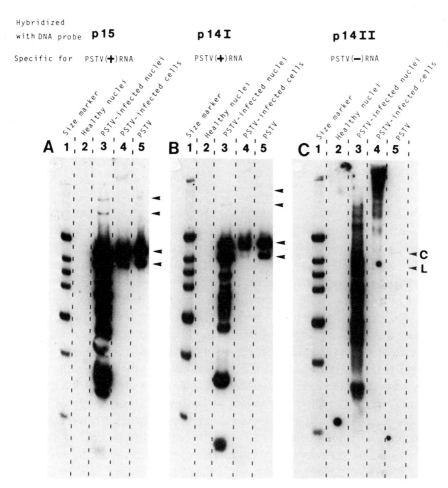

Hybridized
with DNA probe **p15**   **p14 I**   **p14 II**

Specific for   PSTV(+)RNA   PSTV(+)RNA   PSTV(−)RNA

Fig. 12. Northern blots of nucleic acids extracted from isolated potato cell nuclei hybridized to different ^{32}P-labelled PSTV-specific synthetic oligodeoxyribonucleotide probes to demonstrate the presence of intermediates of viroid replication in nuclei.
A. PSTV(+) RNA detection with probe p15 and B. with probe p14 I. C. Detection of PSTV(−) RNA with probe p14 II. In lane A3 and B 3 the bands of the PSTV(+) RNA monomers, of the slower migrating dimers and trimers and of the faster migrating smaller PSTV-specific uncompleted or specifically cleared molecules can be distinguished. Also PSTV(−) RNA molecules up to fourfold or sixfold unit-viroid-length can be detected in isolated nuclei (lane C2) or in intact cells (lane C4), respectively. However, smaller than unit-viroid-length molecules are only found in the nuclei (lane C2). The size markers are Hinf I fragments of pBR 322 with a length of 1631, 517/506, 396, 344, 298, 222/221, 154 and 75 nucleotides.

Chlamydomonas (Kikuchi *et al.*, 1982). Therefore one may anticipate that viroid host plants contain a similar type of RNA-ligating enzyme, which is involved in viroid circularization *in vivo*. Alternatively, a 'self-splicing' mechanism as previously proposed (Zaug,

Grabowski & Cech, 1983) could be responsible for these steps. Additional experiments *in vivo* and *in vitro* are needed to further substantiate the proposed precursor–product relationship between the different viroid-specific RNA species, from which a more complete picture of the replication mechanism of these unique plant pathogens will hopefully emerge.

SUBCELLULAR LOCATION OF VIROIDS AND OF THEIR REPLICATION

One of the problems related to the replication, pathogenicity and possible origin of viroids is their location in the infected host plant. Previous studies based on infectivity tests suggested that viroids were associated primarily with nuclei (Sänger, 1972; Takahashi & Diener, 1975) and/or membranes (Semancik *et al.*, 1976). At that time, however, there were neither sufficiently stringent purification procedures for subcellular and subnuclear components from plant tissue nor was direct quantitative analysis of viroid concentration possible. Consequently, no reliable conclusions about the site of viroid replication or the relative distribution and copy number of viroids could be drawn. Recent investigations have succeeded in clarifying some of the uncertainties in this area.

Previous attempts to follow the process of viroid replication in nuclei from infected leaves by incorporation of radioactive precursors of RNA biosynthesis did not detect discrete bands of radioactively labelled RNA species representing intermediates of viroid replication (Takahashi & Diener, 1975). Therefore, we searched for the presence of intermediates of viroid (PSTV) replication in highly purified nuclei isolated from potato cell suspension cultures infected with PSTV (Mühlbach *et al.*, 1983). We applied molecular hybridization on northern blots using as radioactive probes 'tailor-made' oligodeoxyribonucleotides with sequence complementarity for distinct regions only present in PSTV(+) and PSTV(−)RNA molecules, respectively. With this approach we were able to establish unequivocally that PSTV(+) and PSTV(−)RNA monomers and different oligomeric forms of PSTV(+) and PSTV(−)RNA are present in the cell nucleus (Fig. 12). In addition to the circular and linear PSTV(+)RNA monomers, two oligomeric forms of PSTV(+)RNA are found, which correspond in size to RNA strands of two and three times viroid RNA unit length. Longer-than-unit

Table 3. *Copy number of PSTV (+) RNA molecules*

Tomato leaf cells	200–10 000
Chloroplasts	1–2
Nuclei	200–10 000
Nucleoli	200–10 000

length PSTV(−)RNA molecules are detected as different bands in the range from monomeric up to tetrameric forms. These findings clearly show that not only longer-than-unit-length PSTV(−)RNA intermediates but also PSTV(+)RNA oligomers are involved in viroid replication. The demonstration that all these PSTV-specific RNA molecules are present in RNA preparations isolated from highly purified nuclei strongly favours the concept that the nucleus is, in fact, the intracellular site of viroid replication.

The intracellular localization of viroids within the cell is one of the prerequisites to understanding the interaction between viroid molecules and host components. Although the term 'viroid' was introduced originally to indicate that the infectious entity is a coat-protein-free nucleic acid, this was of course not meant to imply that viroids might be present in the cell as completely 'naked' nucleic acids. Under the assumption that viroids are, in fact, associated with nuclei one could expect them to be associated with different subnuclear components such as chromatin, nucleoli or ribonucleoprotein particles.

In order to clarify this problem, investigations with highly purified nuclei from viroid-infected tomato leaf tissue and fractionation studies on subnuclear components have recently been carried out (Schumacher *et al.* 1983*a*). It was found that viroids sediment nearly quantitatively together with the nucleolar fraction, from which their association with the nucleolus was inferred. This association can be abolished by raising the ionic strength of the buffer to 0.66м. Since protein–nucleic-acid complexes are dissociated and nucleic-acid–nucleic-acid complexes are stabilized by increasing ionic strength one can assume that viroids are complexed in the nucleolus via protein–nucleic-acid interactions. Regarding the number of viroid copies per nucleolus, the highly sensitive bidirectional gel electrophoretic study (Schumacher *et al.*, 1983*b*) gave an average number of 200–10 000 viroid molecules, depending upon the progress of the disease (Table 3). Although the majority of viroid molecules is present inside the nucleus, one cannot conclude that the cytoplasm or other cell organelles are completely devoid of viroids. However, the few viroid molecules which were found per chloroplast are most probably due to contamination, so that their complete absence in this organelle *in vivo* can be assumed. Microscopic studies show that the average diameter of tomato nuclei is 4–5 μm from which an intranuclear concentration of viroids of 10^{-7}M can be calculated. The dissociation constants of various protein–nucleic-acid complexes, such as the ones between nucleic acids and polymerases, would guarantee complex formation in this or even lower concentration ranges so that the preconditions for a viroid–protein complex in the nucleolus are fulfilled.

The finding that the bulk of viroid RNA is associated with the nucleoli (Schumacher *et al.*, 1983*a*) suggested the comparison between PSTV RNA and other RNA species present in the nucleolus. The most prominent nucleolar RNA of comparable size is RNA U3 which is capped and consists of 214 nucleotides. Although RNA U3 is always found to be bound to the 28S RNA present in the nucleolus, it is most probably transcribed outside the nucleolus by DNA-dependent RNA polymerase II (Busch *et al.*, 1982). Therefore, it becomes possible that a similar difference between the site of synthesis and accumulation may also hold for viroids. An additional feature of interest became apparent when the sequences of PSTV RNA (Gross *et al.*, 1978) and RNA U3 from Novikoff hepatoma (Reddy *et al.*, 1974) were compared. It appeared that these two RNAs share several areas of sequence homology. In addition, when the two sequences are aligned in a slightly different colinear manner, a different set of homologies becomes apparent. The implications of these sequence homologies with respect to viroid origin and function remain to be established.

POSSIBLE MECHANISMS OF VIROID PATHOGENESIS

Viroids are composed entirely of RNA, and their small genomes (240–380 nucleotides) are characterized by the absence of the protein synthesis initiation triplet AUG from both the plus and minus strand sequences. Moreover, PSTV and CEV have failed to stimulate or inhibit protein synthesis *in vitro* (Davies, Kaesberg & Diener, 1974; Hall *et al.*, 1974; Semancik, Conjero & Gerhart, 1977) and no viroid-specific protein could so far be detected in vivo

(Camacho-Henriquez & Sänger 1982*ab*). All these data suggest that viroids may not code for any protein products, that they are replicated by the nucleic acid synthesizing machinery pre-existing in the host cell and that viroid pathogenesis may be based on the direct interference of the viroid RNA molecules themselves with the nucleic acid metabolism of the host cell. In fact, it has been speculated by numerous authors (Semancik & Weathers, 1972*a*; Robertson & Dickson, 1974; Reanney, 1975; Dickson & Robertson, 1976; Diener, 1978; Roberts, 1978; Crick, 1979; Diener, 1979*ab*; Roberts, 1980; Dickson, 1981) that viroids might be some escaped regulatory RNA that became replicable. This notion has gained support from the comparison of the established viroid sequences with the sequence of the small stable nuclear U1a RNA (Diener, 1981*ab*, 1982; Gross *et al.*, 1982). U1a RNA is 165 nucleotides long (Reddy *et al.*, 1974), has been strictly conserved in its primary sequence during evolution (Bralant *et al.*, 1980) and is assumed to play a key role in the splicing of eukaryotic mRNA (Lerner *et al.*, 1980). Its proposed function in the splicing process is based on the considerable sequence complementarity of a large part of its 5′ end to the consensus sequence at the 5′- and 3′-proximal regions of the intervening sequences (introns) of mammalian genes. Therefore, it has been assumed that due to this complementarity a U1a RNA molecule could simultaneously form base pairs with both ends of the intron so that the two coding regions of the mRNA precursor are brought so close together that splicing could occur (Lerner *et al.*, 1980; Rogers & Wall, 1980).

When the sequences of U1 RNA and viroids were compared, it was found (Gross *et al.*, 1982) that the ACCUG sequences of U1a RNA (which might be involved in the recognition of pre-mRNA splice junctions) and similar sequences (e.g., ACCCG) occur in several locations of all viroids sequenced so far, with the exception of ASBV. Consequently viroids could bind to these recognition sites on the pre-mRNA and thus interfere with mRNA processing.

In addition, the viroid (−)RNA intermediates would also be able to form a corresponding palindrome-like structure which would be able to form base pairs with U1a RNA. Finally, viroids contain a conserved oligo(uridine) sequence which corresponds to an oligo(uridine) stretch in U1a RNA. Consequently, the viroid (−)RNAs contain an oligo(adenosine) sequence, also occurring frequently in introns, 50–80 nucleotides upstream of the splice site of eukaryotic pre-mRNA.

All these considerations reveal a possible interference between viroids and mRNA synthesis and processing which would explain the pathogenic action of viroids. First, the transcription experiments *in vitro* (Rackwitz *et al.*, 1981) have shown that plus (and minus) viroids' RNA molecules can act as potent competitors for the genomic DNA templates of the host by occupying the DNA-dependent RNA polymerase II and directing it towards viroid replication, so that mRNA synthesis is disturbed. Second, pathogenicity could also be exerted at the subsequent stages of mRNA processing. The viroid RNAs could act as competitors for U1a RNA and mRNA precursors, respectively, in the splicing process (i.e., the correct recognition and alignment of intron–exon junctions, the removal of introns and the ligation of the corresponding exons to give functional mRNA molecules). With these presumptions, one could postulate that viroids are replicated and exert their pathogenic functions by imitating both the genomic DNA and the splice sites of the pre-mRNA of the host.

Unfortunately, however, there are two serious shortcomings in this hypothesis. First, the differences in the pathogenicity of different viroid 'species' or different isolates of one and the same viroid are difficult to explain. One could argue that sequence variations which might result in affinity changes for different classes of mRNA precursors lead to the observed differences in symptom expression. Nevertheless, the understanding of the gradual modulation of symptoms, as brought about by only a few nucleotide exchanges in the case of PSTV isolates of different pathogenicity, poses a complex and intriguing problem. Because they exist only as a highly structured, non-translated 'silent' RNA their induction of different symptoms can only be related to a very specific interaction with a cellular component(s). One is forced to assume that the observed few nucleotide exchanges and the resulting minor structural variation of the viroid RNA molecule modulate this interaction in such a distinctive way that the expression of different symptoms is induced. Second, the possible involvement of U1 RNA in splicing was in itself, still a matter of conjecture (Trapnell, Tolstoshev & Crystal, 1980; Sharp, 1981; Zieve, 1981), until recently the U1 RNA–protein complex was found to bind a 5′ splice site *in vitro* (Mount *et al.*, 1983).

Approaching the problem from a more biological standpoint, it must be emphasized that pathogenicity is a complex biological property and that the expression of disease symptoms is not only specified by the genome of the viroid isolate but also by the genome of the host plant. Thus not only do lethal, severe, intermediate, and mild PSTV strains exist, but different tomato cultivars may respond with severe (cv. Rutgers), mild (cv. Rentita), or practically no symptoms (cv. Hilda 72) upon infection with one and the same PSTV isolate, although replication and accumulation is about the same in all these combinations (Sänger, unpublished). Moreover, symptom development is dependent on a threshold concentration of the viroid, as is demonstrated after inoculation with highly dilute inocula. In this case, pronounced symptoms may be recognized only after an extended period or in the newly growing axillary buds after decapitation, which can be directly related to the increased viroid concentration in the corresponding tissue. Finally, environmental factors such as the nature, intensity, and length of illumination and the temperature during plant growth may greatly influence the response of the host plant to viroid infection and also the intensity of viroid replication itself (Sänger & Ramm, 1975). It can be assumed, therefore, that the interaction between viroid and host cell and the mechanism of pathogenicity will turn out to be much more complex than it appears at first sight.

CORRELATION BETWEEN VIROID PATHOGENICITY AND STRUCTURE

Different PSTV field isolates exist which produce disease in tomato plants characterized by very mild, mild, intermediate and severe symptoms and even by the death of the infected plants, respectively (Fig. 13). The possibility of reversely transcribing this highly structured RNA into full length PSTV-specific cDNA (Rohde et al., 1981ab) initiated the sequence analysis of such PSTV isolates at the DNA level. PSTV-specific cDNA was produced with the aid of reverse transcriptase by extension of synthetic DNA primers which were synthesized by the phosphite method on a polymer support and ^{32}P-labelled at their 5' end with phage T4-induced 5'-hydroxylphosphokinase. Three DNA primer molecules were used which are complementary to different unique sequences located at different sites of the rod-like secondary structure of the PSTV RNA. Thus, the transcripts obtained with one primer comprised the regions which are covered during the transcription with the two other primers, respectively. From the resulting site-specifically initiated

Fig. 13. Tomato plants (cultivar Rutgers) 8 weeks after infection with PSTV field isolates which differ in their pathogenicity. From left to right plants infected with a lethal, severe, intermediate and mild PSTV isolated, respectively, and the uninfected control plant.

transcription products, molecules in the size range of the full-length cDNA were isolated by gel permeation chromatography and then sequenced according to the Maxam–Gilbert technique. The dideoxy-chain-terminating technique was applied in parallel experiments using the same 5' labelled primers.

The comparison of the established nucleotide sequences of various PSTV isolates of different virulence shows that in all isolates the total number of 359 nucleotides is strictly conserved which therefore seems to represent a characteristic and functionally essential feature of PSTV. Taking the isolate producing the mildest symptoms as standard, all other isolates differ in certain nucleotides which are either exchanged, inserted or deleted. All these changes are found in only three regions of the PSTV primary sequence i.e., at nucleotides 40–55, 115–25 and 305–20. In the secondary structure model, these regions are located in two distinct domains, one on the left and the other one on the right hand part of the rod-shaped molecule (Fig. 14).

The number of nucleotide changes in the left hand part of the various isolates was found to increase with their increasing

PSTV

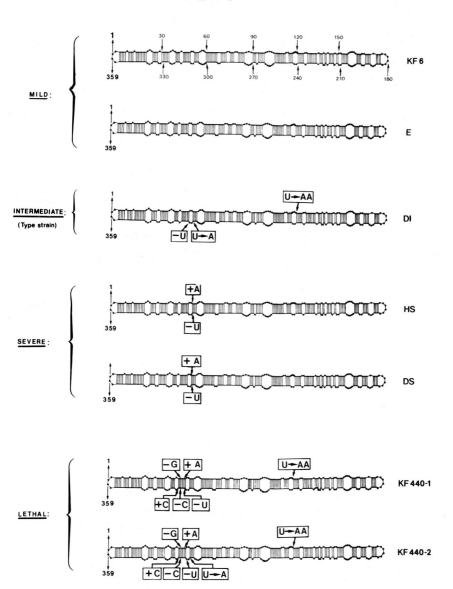

Fig. 14. Location of the nucleotide exchanges in various PSTV isolates causing mild, intermediate, severe and lethal symptoms in tomato as shown in Fig. 13. The exchanges are related to the mild isolate as standard and indicated by boxes. It should be noted that the resulting influence on the secondary structure is not shown.

Table 4. *Calculated melting temperature (T_m) of the pathogenicity-modulating region in the left hand part of various PSTV isolates*

PSTV isolate	$T_m(°C)$
KF 6 (mild)	80
DI (intermediate)	66
HS (severe)	53
KF 440 (lethal)	49

pathogenicity from two to six changes which result in a decrease of the number of base pairs, thus leading to an increasing structural instability of the corresponding region. The change observed in the right hand part of the PSTV structure always involved the same uridine which is replaced by two adenosines. It only occurred when, due to deletions in the left hand part, the total number of the 359 nucleotides had to be reestablished.

If the influence of these nucleotide changes on the thermodynamic properties of the corresponding structural domain is analysed, it becomes evident that the pathogenicity of the various PSTV isolates is in some way determined by the modulation of the stability of the first premelting region in the left half of the RNA molecule which contains an oligo A sequence, undergoes premelting transitions, and is highly conserved in the three viroid 'species' PSTV, CEV and CSV (see Fig. 8). With increasing pathogenicity this pathogenicity-modulating (PM) region becomes increasingly unstable (Table 4) and hence single-stranded and therefore more amenable to possible interactions with cellular components. These structural data and our recent findings that viroids accumulate in the nucleolus and that their association with this subcellular component is via a nucleic-acid–protein interaction (Schumacher *et al.*, 1983*a*) lead to the novel working hypothesis that viroids might form a complex with DNA-dependent RNA polymerase I which is also localized in the nucleolus. In addition, we found striking homologies between the two promotor sequences of the mouse rRNA gene (Grummt, 1982) and sequences of the PM-region of PSTV and its adjacent nucleotides. Therefore, one could postulate that PSTV might also exert its pathogenic functions by capturing the RNA polymerase I transcription factors. As a consequence, these protein(s) could not interact with the corresponding promotor region of the polymerase I gene so that its transcription would be disturbed. One of the preconditions for this working hypothesis would be that polymerase I and the factors of the PSTV host plant tomato are, of course, capable of recognizing these sequences. Polymerase I unlike polymerase II and III, has so far not been shown to be able to initiate specifically and to transcribe viroid RNA *in vitro* (unpublished results from the author's laboratory). Therefore, one could postulate that the mechanism of viroid pathogenicity would be based on binding of viroids to polymerase I and/or its transcription factors which would result in competitive inhibition rather than in its usurpation for viroid replication itself.

In contrast to all other hypotheses advanced so far, the present one has the great advantage that it is based on established pathogenicity-related structural data. Consequently, the modulation of viroid pathogenicity can be explained by differences in the accessibility of the nucleotides in the PM region which, due to the observed nucleotide exchanges, becomes increasingly single-stranded with increasing virulence and thus increasingly capable of interaction with polymerase I and probably some other proteinaceous nuclear components. Several corollaries of this hypothesis can be tested experimentally and are presently under investigation.

THE POSSIBLE ORIGIN OF VIROIDS

One question always arising during discussions on viroids concerns their possible origin. Before the unique structure of viroids had been established, the possibility was considered that viroids might have originated from conventional viruses by degeneration. They were alternatively regarded as a kind of 'primitive virus that had not yet developed the genetic sophistication to code for one or more capsid proteins capable of assembling into a protective capsid' (Diener, 1979a). The erroneous finding that sequences complementary to PSTV RNA exist in the DNA from several uninfected host species (Hadidi *et al.*, 1976) suggested that PSTV originated from host genetic material. Since the affinity between PSTV and the plant DNA seemed to be higher in several solanaceous species than in plants of more distant phylogenetic relationship, the idea arose that PSTV originated from genes normally present in certain solanaceous plant species (Hadidi *et al.*, 1976; Diener, 1978). Once the absence of any viroid-specific DNA in

Table 5. *The presumed origin of viroids*

Previous hypotheses

1. Degenerated viruses
2. Primitive viruses
3. Host genetic material

Present hypotheses

1. Mutated low molecular weight nuclear RNA
2. Escaped introns
3. Escaped 'virusoids'
4. Aberrant 'signal RNAs' or extensively deleted 'fusons'
5. Escaped extrachromosomal procaryotic elements

---

healthy or viroid-infected plants was unequivocally established (Branch & Dickson, 1980; Zaitlin *et al.*, 1980; Rohde & Sänger, 1981) this idea was, of course, no longer tenable. In the light of present knowledge, five additional hypotheses on viroid origin have been discussed.

1. Viroids might have originated from low molecular weight nuclear RNAs which are believed to perform normal regulatory functions within the cell nucleus. Mutation of these RNAs to allow their replication and/or chance introduction into a foreign plant species (especially into cultivated ones) could have changed these normal RNAs into pathogenic ones (Diener, 1974, 1979*b*).

2. Viroids may have originated during RNA splicing processes from spliced out and circularized introns, which acquired the ability to escape degradation and to become efficiently replicated by the nucleic acid synthesizing machinery of the host cell (Roberts, 1978; Crick, 1979; Diener, 1979*b*, 1981*b*; Gross *et al.*, 1982). This hypothesis gained some support from the finding that a surprising sequence homology between the small nuclear U1 RNA and PSTV RNA exists. These sequences may therefore be considered as relicts of the hypothetical origin of viroids as 'escaped introns'.

Although the 'escaped-intron' hypothesis gained some support from the finding that intron-derived circular RNA exists in mitochondria of yeast (Halbreich *et al.*, 1980) and in nuclei of *Tetrahymena thermophila* (Grabowski, Zang & Cech, 1981) a note of caution should be added here. None of the U1 RNA sequences is present in the plus or minus strand of ASBV (Symons, 1981) so that it must be assumed that this viroid is of different origin. Moreover, there is no published evidence as yet that U1 RNA exists in plants and the involvement of this RNA in splicing is still largely a matter of conjecture. However, intervening sequences with splice junctions similar to those of animal origin have been detected in plants in the genes for bean phaseolin (Sun, Slightom & Hall, 1981) and for soybean leghaemoglobin a and b (Jensen *et al.*, 1981). Thus, future work will have to show whether or not this very suggestive 'escaped-intron' hypothesis of the origin of viroids is applicable.

3. Viroids may have originated from the small circular RNA components of certain recently discovered plant viruses with multi-partite genomes. However, these viroid-like satellite RNAs (which are also called 'virusoids') have only their overall secondary struc-ture, size and circularity in common with the true viroids. The detailed analysis of several thermodynamic and kinetic properties has revealed that the 'virusoids' are clearly distinct from viroids with regard to the Gibbs free energy of structure formation, the mean T_m value cooperativity and the hydrodynamic stiffness. Since they are much more similar in all these parameters to calculated random sequences than the unique viroid RNA molecules (Randles, Steger & Riesner, 1982; Riesner, Kaper & Randles, 1982) this origin becomes less probable.

4. According to another, even more speculative, hypothesis, viroids and RNA viruses might form a novel class of RNA elements that exchange genetic information between eukaryotic cells (Zimmern, 1982). This theory is based is based on the previously discussed conjectures which connect the biosynthesis and structure of viroids with the splicing machinery of the eukaryotic cells and the existence of intron-derived circular RNAs in mitochondria of yeast and nuclei of *Tetrahymena*.

The hypothetical novel cellular elements would consist of two interacting classes of RNA molecules which are called 'signal RNAs' and 'antenna RNAs'. The 'signal RNAs' can be excised and/or inserted directly into another RNA and they can also be transported across membranes, between intracellular compartments and between cells. The 'antenna RNAs' are the target of the 'signal RNAs' and the fused product of both is called a 'fusion'. Based primarily on inductive arguments this unorthodox system has been advanced to provide a mechanism for cellular regulation by which the cellular genetic control via RNA splicing and the spatial arrangement of cells in an organism could be linked. This mechan-ism would have the advantage that, once the signal and antenna RNAs have been transcribed, the system could operate independently of genomic DNA. Moreover it could be a means by which the homogeneity of dispersed repeated genes is maintained.

According to this highly speculative hypothesis viroids are considered to represent aberrant 'signal RNAs' or small, extensively deleted, 'fusons' which have retained little more than an origin of replication and a splice junction and are no longer under the control of copy number.

5. Viroids might have evolved as RNA pathogens through the infection of higher plants by prokaryotes. This hypothesis of viroid origin is based on the finding that not only eukaryotic but also both prokaryotic DNA-dependent RNA polymerase and DNA-dependent DNA polymerase from *E. coli* are able to transcribe viroid RNA quite efficiently *in vitro* into viroid-complementary RNA and DNA copies, respectively (Rohde *et al.*, 1982). It is conceivable, therefore, that viroids may have originated in prokaryotes either as DNA or RNA and that viroid host plants originally acquired viroid-specific sequences through infection with prokaryotes, which are the common causal agents of plant diseases.

The transfer of nucleic acid sequences from a plant pathogen to its host is, for example, well established in the case of infections by virulent strains of *Agrobacterium tumefaciens*. These are known to contain Ti plasmids which induce so-called crown gall tumours in dicotyledonous plants; part of the tumorigenic T-DNA sequences become stably integrated into the plant cell genome (van Montagu & Schell, 1982) and their expression is under control of the DNA-dependent RNA-polymerase of the host (Willmitzer, Schmalenbach & Schell, 1981). Secondly, the prokaryotic plasmid pBR 322 contains 'TATA' box-like sequences that promote specific initiation of transcription *in vitro* by eukaryotic DNA-dependent RNA polymerase II (Sassone-Corsi *et al.*, 1981) the enzyme which is evidently also involved in viroid replication (Mühlbach & Sänger, 1979; Rackwitz *et al.*, 1981). Thus, one could postulate that certain plants may have acquired specific DNA sequences from prokaryotic plant pathogens either by integration or in form of extrachromosomal and autonomously replicating elements. Alternatively, viroid RNA might already have existed as such in a bacterial or fungal plant pathogen prior to the uptake by the susceptible plant. The finding that, in the presence of an appropriate primer, viroid-complementary DNA synthesis is catalysed by *E. coli* DNA polymerase I *in vitro* might be taken as an indication that viroid-specific or viroid-like nucleic acids might be found in non-plant systems, provided that such elements would be replicated and escape the regulation through the corresponding hosts once they are introduced.

Without any doubt, all these hypotheses will catalyse new experimental approaches, not only in viroid research but also in the field of mRNA processing and regulation.

CONCLUSIONS

Viroids were the first pathogens for which the complete molecular structure was elucidated in detail. After their discovery in 1971 scepticism and even rejection prevailed for many years until their existence as a rather unusual reality became well established and removed a number of psychological blocks in our thinking. Viroids have extended our concept of microbiological disease agents into the realm of small RNA molecules. Along that way several unwritten dogmas have broken down. Infectious nucleic acids were previously considered to exist and survive only in an encapsidated form, the virus particle. Viroids showed, however, that 'naked' infectious nucleic acid molecules can also exist, because they developed a unique structure which confers the necessary stability. Furthermore, the minimal size of infectious replicating nucleic acids is no longer restricted to molecules of 10^6 molecular weight, which corresponds to a chain length of about 3600 nucleotides. Viroids are only one tenth of that size and yet infectious and replicating.

Infectious viral nucleic acids exert some of their pathogenic functions through the virus-specific proteins they code for. The lack of translation potential *in vitro*, together with the absence of viroid-specific proteins in infected plants, indicates that this concept may not apply to viroid pathogenesis. Viroid RNA has the ability to act *in vitro* as an efficient template for DNA-dependent RNA polymerase II and III, to form a binary complex with these polymerases, to compete with DNA for the template binding site on these enzymes and to strongly inhibit DNA-directed RNA synthesis. If viroid RNA is, in fact, also capable of directing DNA-dependent RNA polymerase II and III to perform its 'selfish' replication *in vivo* then viroid pathogenicity could result from the direct interference of viroid replication with the messenger and nuclear RNA synthesis of the host. The presence in different viroids of strictly conserved sequences (some of which exhibit surprising homologies and complementarity with the known consensus sequences of exon–intron junctions, the 5′-end of the nuclear U1 RNA and the nucleolar U3 RNA) furthermore suggested that viroid RNA might also interfere with messenger and nucleolar RNA maturation, in particular with the mechanisms of splicing and ligation. Thus, viroids could function like some of the tentative regulatory RNAs which have been repeatedly postulated in the past. However, our own most recent data suggest that the mechanism of viroid pathogenesis is in some way related to the interaction of the viroid RNA molecule with the nucleolar DNA-dependent RNA polymerase I which is known to transcribe precursors of the 25S, 18S and 5.8S cytoplasmic ribosomal RNAs.

Viroids differ from all presently known nucleic acids by the combination of circularity, unusual secondary structure and the specific dynamics of highly cooperative structural transitions. In addition, viroids represent RNA molecules with thermodynamic and kinetic properties of a DNA of the same size and G : C content. However, these features do not result from a DNA-like secondary structure but are the consequence of the serial arrangement of double-helical segments and internal loops which are only guaranteed by a very specific nucleotide sequence. Random sequences of the same number of nucleotides and composition do not yield the high degree of base-pairing and the extraordinary cooperativity observed for viroids. There is little doubt that the combination of these unique features is of functional importance for replication, ligation, pathogenicity and 'survival'. After the elucidation of viroid structure, the first light has been shed on the possible mechanisms of viroid replication and pathogenesis. One can safely assume already that both mechanisms must differ fundamentally from the corresponding ones of conventional viruses. The study of the more dynamic aspects of the viroid–host-plant interaction at the cellular level will progress rapidly because several technical problems inherent in plant systems have been overcome and suitable plant cells and protoplasts are now available for such investigations. Nevertheless, one should remember that in higher plants certain pathogenic responses are based on functions of the intact tissue or even the entire plant.

One of the most interesting questions centres around the possible existence of viroids or viroid-like pathogens as causative agents of certain unconventional diseases in humans and animals. It must be emphasized that so far viroids have been found only in higher plants and that there is no direct evidence yet that they also exist in other organisms. Experience has shown, however, that viruses, mycoplasma, bacteria and fungi are found as pathogens in all organisms. Therefore, it would be a rather unique situation if viroids were confined to higher plants. Despite the recent developments in the elucidation of the unusual nature of the scrapie agent and its fundamental differences from viroids (Prusiner, 1982; Diener *et al.*, 1982) it is well justified to continue the search for viroids or viroid-like molecules in all those transmissible diseases where the involvement of conventional viruses has been ruled out by the classical methods of virology.

In conclusion, viroids represent a completely new class of pathogenic RNAs in plants, with unique structural features and a still enigmatic mechanism of replication and pathogenesis. Regarding the relation between structure and function, viroids represent an optimal compromise between structural stability and functional flexibility, which combines maximal self-protection with efficient replication. Their RNA genome seems to contain only the information for its own unique structure. In terms of their interaction with the host plant, viroids may be considered as replicating competitive inhibitors of mRNA and nuclear RNA synthesis and maturation. Their stable yet flexible DNA-like structure confers on them a selective advantage over other cellular nucleic acids by guaranteeing their replication by host enzymes and their 'survival'.

Because of their small size, their known structure, their relative stability and their intimate relationship to central functions of the host cell metabolism, viroids are not only an excellent system for investigating the relationship between molecular structure and biological function, but also the molecular biology and biochemistry of higher plants in general. However, due to the increasing complexity of the unresolved problems, the future pace of progress will certainly be slower than in the past. Nevertheless, all these features render viroids fascinating models for investigation by various scientific disciplines.

I wish to thank all my associates identified in the respective publications for their continual collaboration and excellent efforts. The generous support by the Deutsche Forschungsgemeinschaft through Sonderforschungsbereich 47 (Virologie), personal grants and stipends especially in the critical early stages of our work is gratefully acknowledged.

REFERENCES

BIEBRICHER, C. K. & ORGEL, L. E. (1973). An RNA that multiplies indefinitely with DNA-dependent RNA polymerase: selection from a random copolymer. *Proceedings of the National Academy of Sciences, USA*, **70**, 934–8.

BOEGE, F., ROHDE, W. & SÄNGER, H. L. (1982). *In vitro* transcription of viroid RNA into full length copies by RNA-dependent RNA polymerase from healthy tomato leaf tissue. *Bioscience Reports*, **2**, 185–94.

BOEGE, F. & SÄNGER, H. L. (1980). RNA-dependent RNA polymerase from healthy tomato leaf tissue. *FEBS Letters*, **121**, 91–6.

BRALANT, C., KROL, A., EBEL, J.-P., LAZAR, E., GALLINARO, H., JACOB, M., SRI-WIDADA, J. & JEANTEUR, P. (1980). Nucleotide sequences of nuclear U1A RNAs from chicken, rat and man. *Nucleic Acids Research*, **8**, 4143–54.

BRANCH, A. D. & DICKSON, E. (1980). Tomato DNA contains no detectable regions complementary to potato spindle tuber viroid as assayed by Southern hybridization. *Virology*, **104**, 10–26.

BRANCH, A. D., ROBERTSON, H. D. & DICKSON, E. (1981). Longer-than-unit-length viroid minus strands are present in RNA from infected plants. *Proceedings of the National Academy of Sciences, USA*, **78**, 6381–5.

BRANCH, A. D., ROBERTSON, A. D., GREER, C., GEGENHEIMER, P., PEEBLES, C. & ABELSON, J. (1982). Cell-free circularization of viroid progeny RNA by an RNA ligase from wheat germ. *Science*, **217**, 1147–9.

BROWN, F. & MARTIN, S. J. (1965). A new model for virus ribonucleic acid replication. *Nature*, **208**, 861–3.

BUSCH, H., REDDY, R., ROTHBLUM, L. & CHOI, Y. C. (1982). SnRNAs, SnRNPs, and RNA processing. *Annual Review of Biochemistry*, **51**, 617–54.

CALAVAN, E. C., WEATHERS, L. G. & CHRISTIANSEN, D. W. (1968). Effect of exocortis on production and growth of Valencia orange trees on trifoliate orange rootstock. In *Proceedings of the 4th Conference of the International Organization of Citrus Virologists*, ed. J. F. L. Childs, pp. 101–4. Gainsville: University of Florida Press.

CAMACHO-HENRIQUEZ, A. & SÄNGER, H. L. (1982*a*). Gelelectrophoretic analysis of phenol-extractable leaf proteins from different viroid/host combinations. *Archives of Virology*, **74**, 167–80.

CAMACHO-HENRIQUEZ, A. & SÄNGER, H. L. (1982*b*). Analysis of acid-extractable tomato leaf proteins after infection with a viroid, two viruses and a fungus and partial purification of the 'pathogenesis-related' protein p. 14. *Arch. Virol.*, **74**, 181–93.

CHEN WEI, TIEN PO, ZHU YU XIANG & LIU YONG. (1983). Viroid-like RNAs associated with Burdock stunt disease. Journal of General Virology, **64**, 409–14.

CHIFFLOT, S., SOMMER, P., HARTMANN, D., STUSSI-GARAUD, C. & HIRTH, L. (1980). Replication of alfalfa mosaic virus RNA: evidence for a soluble replicase in healthy and infected tobacco-leaves. *Virology*, **100**, 91–100.

CRICK, F. (1979). Split genes and RNA splicing. *Science*, **204**, 264–71.

DALE, J. L. & ALLEN, R. N. (1979). Avocado affected by sunblotch disease contains low molecular weight ribonucleic acid. *Australian Plant Pathology Society Newsletter*, **8**, 3–4.

DAVIES, J. W., KAESBERG, P. & DIENER, T. O. (1974). Potato spindle tuber viroid. XII. An investigation of viroid RNA as a messenger for protein synthesis. *Virology*, **61**, 281–6.

DE BOKX, J. A. & PIRON, P. G. M. (1981). Transmission of potato spindle tuber viroid by aphids. *Netherlands Journal of Plant Pathology*, **87**, 31–4.

DICKSON, E. (1979). Viroids: infectious RNA in plants. In *Nucleic Acids in Plants*, vol. 2, ed. T. C. Hall & J. W. Davies, pp. 153–93.

DICKSON, E. (1981). A model for the involvement of viroids in RNA splicing. *Virology*, **115**, 216–21.

DICKSON, E., DIENER, T. O. & ROBERTSON, H. D. (1978). Potato spindle tuber and citrus exocortis viroids undergo no major sequence changes during replication in two different hosts. *Proceedings of the National Academy of Sciences, USA*, **75**, 951–4.

DICKSON, E. & ROBERTSON, H. D. (1976). Potential regulatory roles for RNA in cellular development. *Cancer Research*, **36**, 3387–93.

DICKSON, E., ROBERTSON, H. D., NIBLETT, C. L., HORST, R. K. & ZAITLIN, M. (1979). Minor differences between nucleotide sequences of mild and severe strains of potato spindle tuber viroid. *Nature*, **277**, 60–2.

DIENER, T. O. (1971a). Potato spindle tuber virus: a plant virus with properties of a free nucleic acid. III. Subcellular location of PSTV-RNA and the question of whether virions exist in extracts or *in situ*. *Virology*, **43**, 75–89.

DIENER, T. O. (1971b). Potato spindle tuber 'virus'. IV. A replicating, low molecular weight RNA. *Virology*, **45**, 411–28.

DIENER, T. O. (1974). Viroids: the smallest known agents of infectious disease. *Annual Reviews of Microbiology*, **28**, 23–39.

DIENER, T. O. (1978). Are viroids autoinducing regulatory RNAs? In *Persistent Viruses*, ed. J. G. Stevens, G. J. Todaro & C. F. Fox, pp. 297–309. New York: Academic Press.

DIENER, T. O. (1979a). Viroids: structure and function. *Science*, **205**, 859–66.

DIENER, T. O. (1979b). *Viroids and Viroid Diseases*. New York: John Wiley & Sons.

DIENER, T. O. (1981a). Are viroids escaped introns? *Proceedings of the National Academy of Sciences USA*, **78**, 5014–15.

DIENER, T. O. (1981b). Viroids: abnormal products of plant metabolism. *Annual Reviews of Plant Physiology*, **32**, 313–25.

DIENER, T. O. (1980). Viroids: minimal biological systems. *Bioscience*, **32**, 38–44.

DIENER, T. O. & LAWSON, R. H. (1973). *Virology*, **51**, 94–101.

DIENER, T. O., McKINLEY, M. P., PRUSINER, S. B. (1982). Viroids and prions. *Proceedings of the National Academy of Sciences, USA*, **79**, 5220–4.

DUDA, C. T. (1976). Plant RNA polymerases. *Annual Review of Plant Physiology*, **27**, 119–32.

DUDA, C. T., ZAITLIN, M. & SIEGEL, A. (1973). *In vitro* synthesis of double stranded RNA by an enzyme system isolated from tobacco leaves. *Biochimica et Biophysica acta*, **319**, 62–71.

FRAENKEL-CONRAT, H. (1979). RNA-dependent RNA polymerases of plants. *Trends in Biochemical Science*, **4**, 184–6.

GALINDO, J., SMITH, D. R. & DIENER, T. O. (1982). Etiology of planta macho, a viroid disease of tomato. *Phytopathology*, **72**, 49–54.

GARNSEY, S. M. & WHIDDEN, R. (1973). Efficiency of mechanical inoculation procedures for citrus exocortis virus. *Plant Disease Reports*, **57**, 886–9.

GILBERT, W. & DRESSLER, D. (1968). DNA-replication: The rolling circle model. *Cold Spring Harbor Symposia of Quantitative Biology*, **33**, 473–84.

GILLILAND, J. M. & SYMONS, R. H. (1968). Properties of a plant virus-induced RNA polymerase in cucumbers infected with cucumber mosaic virus. *Virology*, **36**, 232–40.

GOLDMANN, D. & MERRIL, C. R. (1982). Silver staining of DNA in polyacrylamide gels: linearity and effect of fragment size. *Electrophoresis*, **3**, 24–6.

GRABOWSKI, P. J., ZANG, A. J. & CECH, T. R. (1981). The intervening sequence of the ribosomal RNA precursor is converted to a circular RNA in isolated nuclei of Tetrahymena. *Cell*, **23**, 447–76.

GRILL, L. K. & SEMANCIK, J. S. (1978). RNA sequences complementary to citrus exocortis viroid in nucleic acid preparations from infected Gynura aurantiaca. *Proceedings of the National Academy of Sciences, USA*, **75**, 896–900.

GRILL, L. K. & SEMANCIK, J. S. (1980). Viroid synthesis: the question of inhibition by actinomycin D. *Nature*, **283**, 399–400.

GROSS, H. J., DOMDEY, H., LOSSOW, C., JANK, P., RABA, M., ALBERTY, H. & SÄNGER, H. L. (1978). Nucleotide sequence and secondary structure of potato spindle tuber viroid. *Nature*, **272**, 203–8.

GROSS, H. J., DOMDEY, H. & SÄNGER, H. L. (1977). Comparative oligonucleotide fingerprints of three plant viroids, *Nucleic Acids Research*, **4**, 2021–8.

GROSS, H. J., KRUPP, G., DOMDEY, H., RABA, M., ALBERTY, H., LOSSOW, C. H., RAMM, K. & SÄNGER, H. L. (1982). Nucleotide sequence and secondary structure of citrus exocortis and chrysanthemum stunt viroid. *European Journal of Biochemistry*, **121**, 249–57.

GROSS, H. J., LIEBL, U., ALBERTY, H., KRUPP, G., DOMDEY, H. RAMM, K. & SÄNGER, H. L. (1981). A severe and a mild potato spindle tuber viroid isolate differ in 3 nucleotide exchanges only . *Bioscience Reports*, **1**, 235–41.

GRUMMT, I. (1982). Nucleotide sequence requirements for specific initiation of transcription by RNA polymerase I. *Proceedings of the National Academy of Sciences, USA*, **79**, 6908–11.

HADIDI, A., JONES, D. M., GILLESPIE, D. H., WONG-STAAL, F. & DIENER, T. O. (1976). Hybridization of potato spindle tuber viroid to cellular DNA of normal plants. *Proceedings of the National Academy of Sciences, USA*, **73**, 2453–7.

HALBREICH, A., PAJOT, P., FOUCHER, M., GRANDCHAMP, C. & SLONIMSKY, P. (1980). A pathway of cytochrome b mRNA processing in yeast mitochondria: specific splicing steps and an intron-derived circular RNA. *Cell*, **19**, 321–9.

HALL, T. C., WEPPRICH, R. K., DAVIES, J. W., WEATHERS, L. G. & SEMANCIK, J. S. (1974). Functional distinctions between the ribonucleic acids from citrus exocortis viroid and plant viruses: cell-free translation and aminoacylation reactions. *Virology*, **61**, 486–92.

HARDY, S. F., GERMAN, T. L., LOESCH-FRIES, L. S. & HALL, T. C. (1979). Highly active template-specific RNA-dependent RNA polymerase from barley leaves infected with brome mosaic virus. *Proceedings of the National Academy of Sciences, USA*, **76**, 4956–60.

HASELOFF, J. & SYMONS, R. H. (1981). Chrysanthemum stunt viroid-primary sequence and secondary structure. *Nucleic Acids Research*, **9**, 2741–52.

HASELOFF, J., MOHAMED, N. A. & SYMONS, R. H. (1982). Viroid RNA of cadang-cadang disease of coconuts. *Nature*, **299**, 316–21.

HOLLINGS, M. (1965). Disease control though virus-free stock. *Annual Review of Phytopathology*, **3**, 367–96.

HOLLINGS, M. & STONE, O. M. (1970). Attempts to eliminate chrysanthemum stunt from chrysanthemum by meristem-tip culture after heat-treatment. *Annals of Applied Biology*, **65**, 311–15.

HOLLINGS, M. & STONE, O. M. (1973). *Annals of Applied Biology*, **74**, 333–348.

HORST, R. K. & ROMAINE, C. P. (1975). Chrysanthemum chlorotic mottle: a viroid disease. *New York's Food and Life Sciences Quarterly*, **8**, 11–14.

IKEGAMI, M. & FRAENKEL-CONRAT, H. (1978). RNA-dependent RNA polymerase of tobacco plants. *Proceedings of the National Academy of Sciences, USA*, **75**, 2122–4.

IKEGAMI, M. & FRAENKEL-CONRAT, H. (1979a). Characterization of the RNA-dependent RNA-polymerase of tobacco leaves. *Journal of Biological Chemistry*, **254**, 149–54.

IKEGAMI, M. AND FRAENKEL-CONRAT, H. (1979b). Characterization of double-stranded ribonucleic acid in tobacco leaves. *Proceedings of the National Academy of Sciences, USA*, **76**, 3637–46.

IMPERIAL, J. S., RODRIGUEZ, J. B., RANDLES, J. M. (1981). Variation in the viroid-like RNA associated with cadang-cadang disease: evidence for an increase in molecular-weight with disease progress. *Journal of General Virology*, **56**, 77–85.

INGRAM, D. S. (1976). Growth of biotrophic parasites in tissue culture. In *Physiological Plant Pathology*, ed. R. Heitefuss & P. H. Williams, pp. 743–59. Berlin, Heidelberg, New York: Springer.

JENSEN, E. O., POLUDAN, K., HYLDIG-NIELSON, J. J., JORGENSEN, P. & MARCKER, K. A. (1982). The structure of a chromosomal leghaemoglobin gene from soybean. *Nature*, **291**, 677–9.

KAMEN, R. I. (1975). Structure and function of the Qβ replicase. In *RNA Phages*, ed. N. D. Zinder, pp. 203–34. Cold Spring Harbor: Cold Spring Harbor Laboratories.

KELLER, J. R. (1951). Report on indicator plants for chrysanthemum stunt virus and on a previously unreported chrysanthemum virus. *Phytopathology*, **41**, 947–9.

KELLER, J. R. (1953). Investigations on chrysanthemum stunt virus and chrysanthemum virus Q. *Cornell University Agricultural Experimental Station Memoir*, **324**, 40 pp.

KIKUCHI, Y., TYC, K., FILIPOWICZ, W., SÄNGER, H. L. & GROSS, H. J. (1982). Circularization of linear viroid RNA via 2′-phosphomonoester, 3′, 5′-phosphodiester bonds by a novel type of RNA ligase from wheat germ and *Chlamydomonas*. *Nucleic Acids Research*, **10**, 7521–9.

KLOTZ, G. & SÄNGER, H. L. (1981). Electron microscopic evidence for viroid conformers. *European Journal of Cell Biology*, **25**, 5–7.

KONARSKA, M. FILIPOWICA, W., DOMDEY, H. & GROSS, H. J. (1981). Formation of a 2′-phosphomonoester, 3′, 5′-phosphodiester linkage by a novel RNA ligase in wheat germ. *Nature*, **293**, 112–16.

LAZAR, E., WALTER, B., STUSSI-GARAUD, C. & HIRTH, L. (1979). RNA-dependent RNA polymerases from healthy and tobacco necrosis virus-infected Phaseolus aurus: assay of localization in fractions of cellular homogenates. *Virology*, **96**, 553–63.

LEARY, J. J., BRIGATI, D. J. & WARD, D. C. (1983). Rapid and sensitive colorimetric method for visualizing biotin-labeled DNA probes hybridized to DNA or RNA immobilized on nitrocellulose: bio-blots. *Proceedings of the National Academy of Sciences, USA*, **80**, 4045–9.

LERNER, M. R., BOYLE, J. A., MOUNT, S. M., WOLIN, S. L. & STEITZ, J. A. (1980). Are snRNPs involved in splicing? *Nature*, **283**, 220–4.

LINDELL, T. J., WEINBERG, F., MORRIS, P. W., ROEDER, R. G. & RUTTER, W. J. (1970). Specific inhibition of nuclear RNA polymerase II by α-amanitin. *Science*, **170**, 447–9.

LIZARAGA, R. E., SALAZAR, L. F., ROCA, W. M. & SCHILDE-RENTSCHLER, L. (1980). Elimination of potato spindle tuber viroid by low temperature and meristem culture. *Phytopathology*, **70**, 754–5.

MCCLEAN, A. P. D. (1931). Bunchy top disease of tomato. *South African Department of Agricultural Science Bulletin*, **100**, 36 p.

MACQUAIR, G., MONSION, M., BACHELIE, J. C., FAYDI, C. & DUNEZ, J. (1981). The slab gel-electrophoretic assay for detection and investigation of chrysanthemum chlorotic mottle viroid (CCHMV) in infected plants. *Agronomie Tropicale*, **1**, 99–103.

MAY, J. T., GILLILAND, J. M. & SYMONS, R. H. (1969). Plant virus-induced RNA polymerase: properties of the enzyme partly purified from cucumber cotyledons infected with cucumber mosaic virus. *Virology*, **39**, 54–65.

MAY, J. T. & SYMONS, R. H. (1971). Specificity of the cucumber mosaic virus-induced RNA polymerase for RNA and polynucleotide template. *Virology*, **44**, 517–26.

MORRIS, T. J. & SMITH, E. M. (1977). Potato spindle tuber disease: procedures for the detection of viroid RNA and certification of disease-free potato tubers. *Phytopathology*, **67**, 145–50.

MORRIS, T. J. & WRIGHT, N. S. (1975). Detection on polyacrylamide gel of a diagnostic nucleic acid from tissue infected with potato spindle tuber viroid. *American Potato Journal*, **52**, 57–63.

MOUNT, S. M., PETTERSON, I., HINTERBERGER, M., KARMAS, A. & STEITZ, J. (1983). The U1 small nuclear-RNA–protein complex selectively binds a 5' splice site *in vitro*. *Cell*, **33**, 509–18.

MÜHLBACH, H.-P. (1982). Plant cell cultures and protoplasts in plant virus research. *Current Topics in Microbiology and Immunology*, **99**, 81–129.

MÜHLBACH, H.-P., CAMACHO-HENRIQUEZ, A. & SÄNGER, H. L (1977). Isolation and properties of protoplasts from leaves of healthy and viroid-infected tomato plants. *Plant Science Letters*, **8**, 183–89.

MÜHLBACH, H.-P., FAUSTMANN, O., & SÄNGER, H. L. (1983). Conditions for optimal growth of a PSTV-infected potato cell suspension and detection of viroid-complementary, longer-than-unit-length RNA in these cells. *Plant Molecular Biology*, (in press).

MÜHLBACH, H.-P., & SÄNGER, H. L. (1977). Multiplication of cucumber pale fruit viroid in inoculated tomato leaf protoplasts. *Journal of General Virology*, **35**, 377–86.

MÜHLBACH, H.-P. & SÄNGER, H. L. (1979). Viroid replication is inhibited by α-amanitin. *Nature*, **278**, 185–8.

MÜHLBACH, H.-P. & SÄNGER, H. L. (1981). Continuous replication of potato spindle tuber viroid (PSTV) in permanent cell-cultures of potato and tomato. *Bioscience Reports*, **1**, 79–87.

OHNO, T., TAKAMATSU, N. MESHI, T. & OKADA, Y. (1983). Hop stunt viroid: molecular cloning and nucleotide sequence of the complete cDNA copy. *Nucleic Acids Research*, **11**, 509–18.

OWENS, R. A. (1978). *In vitro* synthesis and characterization of DNA complementary to potato spindle tuber viroid. *Virology*, **89**, 380–7.

OWENS, R. A. & CRESS, D. E. (1980). Molecular cloning and characterization of potato spindle tuber viroid cDNA sequences. *Proceedings of the National Academy of Sciences, USA*, **77**, 5302–6.

OWENS, R. A. & DIENER, T. O. (1981). Sensitive and rapid diagnosis of potato spindle tuber viroid disease by nucleic-acid hybridization. *Science*, **213**, 670–2.

OWENS, R. A. & DIENER, T. O. (1982). RNA intermediates in potato spindle tuber viroid replication. *Proceedings of the National Academy of Sciences, USA*, **79**, 113–17.

PALUKAITIS, P. & SYMONS, R. H. (1978). Synthesis and characterization of complementary DNA probe for chrysanthemum stunt viroid. *FEBS Letters*, **92**, 268–72.

PALUKAITIS, P. & SYMONS, R. H. (1980). Purification and characterization of the circular form of chrysanthemum stunt viroid. *Journal of General Virology*, **46**, 477–89.

PFANNENSTIEL, M. A., SLACK, S. A. & LANE, L. C. (1980). Detection of PSTV in field-grown potatoes by an improved electrophoretic assay. *Phytopathology*, **70**, 1015–18.

PRUSINER, S. B. (1982). Novel proteinaceous infectious particles cause scrapie. *Science*, **216**, 136–44.

RACKWITZ, H. R., ROHDE, W. & SÄNGER, H. L. (1981). DNA-dependent RNA polymerase II of plant-origin transcribes viroid RNA into full-length copies. *Nature*, **291**, 297–301.

RANDLES, J. W. (1975). Association of two ribonucleic acid species with cadang-cadang disease of coconut palm. *Phytopathology*, **65**, 163–7.

RANDLES, J. W., RILLO, E. P. & DIENER, T. O. (1976). The viroid-like structure and cellular location of anomalous RNA associated with the cadang-cadang disease. *Virology*, **74**, 128–39.

RANDLES, J. W., STEGER, G. & RIESNER, D. (1982). Structural transitions in viroid-like RNAs associated with cadang-cadang disease, velvet tobacco virus, and *Solanum nodiflorum* mottle virus. *Nucleic Acids Research*, **10**, 5569–86.

REANNEY, D. C. (1975). A regulatory role for viral RNA in eucaryotes. *Journal of Theoretical Biology*, **49**, 461–92.

REDDY, R., RO-CHOI, T. S., HENNING, D. & BUSCH, H. (1974). Primary sequence of U-1 nuclear ribonucleic acid of Novikoff hepatoma ascites cells. *Journal of Biological Chemistry*, **249**, 6486–94.

RIESNER, D., HENCO, K., ROKOHL, U., KLOTZ, G., KLEINSCHMIDT, A. K., GROSS, H. J., DOMDEY, H. SÄNGER, H. L. (1979). Structure and structure formation of viroids. *Journal of Molecular Biology*, **133**, 85–115.

RIESNER, D., KAPER, J. M. & RANDLES, J. W. (1982). Stiffness of viroids and viroid-like RNA in solution. *Nucleic Acids Research*, **10**, 5587–98.

RIESNER, D., STEGER, G., SCHUMACHER, J., GROSS, H. J. & SÄNGER, H. L. (1981). Structure and function of viroids. *Biophysical and Structural Mechanics*, **7**, 240–1.

RIESNER, D., STEGER, G., SCHUMACHER, J., GROSS, H. J., RANDLES, J. W. & SÄNGER, H. L. (1983). Structure and function of viroids. *Biophysical and Structural Mechanics*, **9**, 145–70.

ROBERTS, R. J. (1978). Intervening sequences excised *in vitro*. *Nature*, **274**, 530.

ROBERTS, R. (1980). Small RNAs and splicing. *Nature*, **283**, 132–3.

ROBERTSON, H. D. & DICKSON, E. (1974). E. RNA processing and the control of gene expression. In *Processing of RNA*, *Brookhaven Symp. Biol*, **26**, 240–66.

ROGERS, J. & WALL, R. (1980). A mechanism for RNA splicing. *Proceedings of the National Academy of Sciences, USA*, **77**, 1877–9.

ROHDE, W. & SÄNGER, H. L. (1981). Detection of complementary RNA intermediates of viroid replication by northern blot hybridization. *Bioscience Reports*, **1**, 327–36.

ROHDE, W., SCHNÖLZER, M., RACKWITZ, H. R., HAAS, B., SELIGER, H. & SÄNGER, H. L. (1981*a*). Specifically primed synthesis in vitro of full-length DNA complementary to potato-spindle-tuber viroid. *European Journal of Biochemistry*, **118**, 151–7.

ROHDE, W., SCHNÖLZER, M. & SÄNGER, H. L. (1981*b*). Sequence-specific priming of the *in vitro* synthesis of DNA complementary to citrus exocortis viroid. *FEBS Letters*, **130**, 208–12.

ROHDE, W., RACKWITZ, H.-R., BOEGE, F. & SÄNGER H. L. (1982). Viroid RNA is accepted as a template for *in vitro* transcription by DNA-dependent DNA polymerase I from *Escherichia coli*. *Bioscience Reports*, **2**, 929–39.

ROISTACHER, C. N., CALAVAN, E. C. & BLUE, R. L. (1969). Citrus exocortis virus – chemical inactivation on tools, tolerance to heat and separation of isolates. *Plant Disease Reporter*, **53**, 333–6.

ROMAINE, C. P. & HORST, R. K. (1975). Suggested viroid etiology for chrysanthemum chlorotic mottle disease. *Virology*, **64**, 86–95.

ROMAINE, C. P. & ZAITLIN, M. (1978). RNA-dependent RNA polymerases in uninfected and tobacco mosaic virus-infected tobacco leaves: viral-induced stimulation of a host polymerase activity. *Virology*, **86**, 241–53.

SAMMONS, D. W., ADAMS, L. D. & NISHIZAWA, E. (1981). Ultrasensitive silver-based color staining of polypeptides in polyacrylamide gels. *Electrophoresis*, **2**, 135–41.

SANGER, F., AIR, G. M., BARRELL, B. G., BROWN, N. L., COULSON, A., FIDDES, J. C., HUTCHINSON III, C. A., SLOCOMBE, P. M. & SMITH, M. (1977). Nucleotide sequence of bacteriophage ϕX174 DNA. *Nature*, **265**, 687–95.

SÄNGER, H. L. (1972). An infectious and replicating RNA of low molecular weight: the agent of the exocortis disease of citrus. *Advances in Bioscience*, **8**, 103–16.

SÄNGER, H. L. (1980). Structure and possible functions of viroids. *Annals of the New York Academy of Sciences*, **354**, 251–78.

SÄNGER, H. L. (1982). Biology, structure, functions and possible origin of viroids. In *Encyclopedia of Plant Physiology, New Series*, vol. 14B, ed. B. Parthier & D. Boulter, pp. 368–454. Berlin, Heidelberg, New York: Springer.

SÄNGER, H. L., KLOTZ, G., RIESNER, D., GROSS, H. J. & KLEINSCHMIDT, A. K. (1976). Viroids are single-stranded covalently closed circular RNA molecules existing as highly base-paired rod-like structures. *Proceedings of the National Academy of Sciences, USA*, **73**, 3852–6.

SÄNGER, H. L. & RAMM, K. (1975). Radioactive labelling of viroid RNA. In *Modification of Information Content of Plant Cells*, ed. R. Markham, D. R. Davies, D. A. Hopwood & R. W. Horne, pp. 229–52. Amsterdam: North Holland/American Elsevier Publishing Company.

SÄNGER, H. L., RAMM, K., DOMDEY, H., GROSS, H. J., HENCO, K. & RIESNER, D. (1979). Conversion of circular viroid molecules to linear strands. *FEBS Letters*, **99**, 117–22.

SASAKI, R., GOTO, H., ARIMA, K. & SASAKI, Y. (1974a). Effect of polyribonucleotides on eukaryotic DNA-dependent RNA polymerase. *Biochimica et biophysica acta*, **366**, 435–42.

SASAKI, Y., GOTO, H., WAKE, T. & SASAKI, R. (1974b). Purine ribonucleotide homopolymer formation activity of RNA polymerase from cauliflower. *Biochimica et biophysica acta*, **366**, 443–53.

SASAKI, Y., GOTO, H., OHTA, H. & KAMIKUBO, R. (1976). Template activity of synthetic deoxyribonucleotide polymers in the eukaryotic DNA-dependent RNA polymerase reaction. *European Journal of Biochemistry*, **70**, 369–75.

SASAKI, M. & SHIKATA, E. (1977a). Studies on the host range of hop stunt disease in Japan. *Proceedings of the Japan Academy*, **53B**, 103–8.

SASAKI, M. & SHIKATA, E. (1977b). On some properties of hop stunt disease agent, a viroid. *Proceedings of the Japan Academy*, **53B**, 109–12.

SASSONE-CORSI, P., CORDEN, J., KEDINGER, C. & CHAMBON, P. (1981). Promotion of specific *in vitro* transcription by excised 'TATA' box sequences inserted in a foreign nucleotide environment. *Nucleic Acids Research*, **9**, 3941–58.

SCHULTZ, E. S. & FOLSOM, D. (1923a). D. Transmission, variation, and control of certain degeneration diseases of Irish potatoes. *Journal of Agricultural Research*, **25**, 43–117.

SCHULTZ, E. S. & FOLSOM, D. (1923b). A 'spindling-tuber disease' of Irish potatoes. *Science*, **57**, 149.

SCHUMACHER, J., RANDLES, J. W. & RIESNER, D. (1983b). A two-dimensional electrophoretic technique for the detection of circular viroids and virusoids. *Analytical Biochemistry*, (in press).

SCHUMACHER, J., SÄNGER, H. L. & RIESNER, D. (1983a). Subcellular localization of viroids in highly purified nuclei from tomato leaf tissue. *The EMBO Journal*, **2**, 1549–55.

SEMANCIK, J. S., CONJERO, V. & GERHART, J. (1977). Citrus exocortis viroid: survey of protein synthesis in *Xenopus laevis* oocytes following addition of viroid RNA. *Virology*, **80**, 218–21.

SEMANCIK, J. S., TSURUDA, D., ZANER, L., GEELEN, J. L. M. C. & WEATHERS, J. G. (1976). Exocortis disease: subcellular distribution of pathogenic (viroid) RNA. *Virology*, **69**, 669–76.

SEMANCIK, J. S. & WEATHERS, L. G. (1972a). Exocortis disease: evidence for a new species of 'infectious' low molecular weight RNA in plants, *Nature New Biology*, **237**, 242–4.

SEMANCIK, J. S. & WEATHERS, L. G. (1972b). Pathogenic 10S RNA from exocortis disease recovered from tomato bunchy-top plants similar to potato spindle tuber virus infection. *Virology*, **49**, 622–5.

SHARP, P. A. (1981). Speculations on RNA splicing. *Cell*, **23**, 643–6.

SINGH, R. P. (1973). Experimental host range of the potato spindle tuber 'virus'. *American Potato Journal*, **50**, 111–23.

SINGH, R. P. (1977). Piperonyl butoxyde as a protectant against potato spindle tuber viroid infection. *Phytopathology*, **67**, 933–5.

SINGH, R. P. & CLARK, M. C. (1971). Infectious low-molecular-weight ribonucleic acid. *Biochemical and Biophysical Research Communications*, **44**, 1077–82.

SINGH, R. P., FINNIE, R. E. & BAGNALL, R. H. (1971). Losses due to the potato spindle tuber virus. *American Potato Journal*, **48**, 262–7.

SINGH, R. P., MICHNIEWICZ, J. J. & NARANG, S. A. (1975). Piperonyl butoxyde, a potent inhibitor of potato spindle tuber viroid in Scopolia sinensis. *Canadian Journal of Biochemistry*, **53**, 1130–2.

SOMMERVILLE, L. L. & WANG, K. (1981). The ultrasensitive silver 'protein' stain also detects nanograms of nucleic acids. *Biochemical and Biophysical Research Communications*, **102**, 53–8.

SPIESMACHER, E., MÜHLBACH, H.-P., SCHNÖLZER, M., HAAS, B. & SÄNGER, H. L. (1983). Oligomeric forms of potato spindle tuber viroid (PSTV) and of its complementary RNA are present in nuclei isolated from viroid-infected potato cells. *Bioscience Reports*, **3**, 767–74.

STACE-SMITH, R. & MELLOR, F. C. (1970). Eradication of potato spindle tuber virus by thermotherapy and axillary bud culture. *Phytopathology*, **60**, 1857–8.

SUN, S. M., SLIGHTOM, J. L. & HALL, T. C. (1981). Intervening sequences in a plant gene – comparison of the partial sequence of cDNA genomic DNA of French bean phaseolin. *Nature*, **289**, 37–41.

SYMONS, R. H. (1981). Avocado sunblotch viroid – primary sequence and proposed secondary structure. *Nucleic Acids Research*, **9**, 6527–37.

TAKAHASHI, T. & DIENER, T. O. (1975). Potato spindle tuber viroid. XIV. Replication in nuclei isolated from infected leaves. *Virology*, **64**, 106–14.

TAKANAMI, Y. & FRAENKEL-CONRAT, H. (1982). Nonviral gene is able to elicit RNA-dependent RNA polymerase in cucumber mosaic virus-infected cucumber cotyledons. *Virology*, **116**, 372–4.

THOMAS, W. & MOHAMED, N. A. (1979). Avocado sunblotch – a viroid disease? *Australian Plant Pathology Society Newsletter*, **8**, 1–3.

TRAPNELL, B. C., TOLSTOSHEV, P. & CRYSTAL, R. G. (1980). Secondary structures for splice junctions in eucaryotic and viral messenger RNA precursors. *Nucleic Acids Research*, **8**, 3659–72.

VAN DORST, H. J. M. & PETERS, D. (1974). Some biological observations on pale fruit, a viroid-incited disease of cucumber. *Netherlands Journal of Plant Pathology*, **80**, 85–96.

VAN MONTAGU, M. & SCHELL, J. (1982). The Ti plasmids of agrobacterium. *Current Topics in Microbiology and Immunology*, **96**, 236–54.

VISVADER, J. E., GOULD, A. R., BRUENING, G. E. & SYMONS, R. H. (1982). Citrus exocortis viroid-nucleotide-sequence and secondary structure of an Australian isolate. *FEBS Letters*, **137**, 288–92.

WALTER, B. (1981). Un viroide de la Tomate en Afrique de l'Ouest: identité avec le viroide du 'potato spindle tuber'? *Comptes rendus hebdomadaire des scéances de l'Academie des sciences*, **292**, Serie III, 537–42.

WEINMANN, R. & ROEDER, R. G. (1974). Role of DNA-dependent RNA polymerase III in the transcription of the tRNA and 5S RNA genes. *Proceedings of the National Academy of Sciences, USA*, **71**, 1790–5.

WIELAND, TH. & FAULSTICH, H. (1978). Amatoxins, phallotoxins, phallolysin and antamanide: the biologically active components of poisonous *Amanita* mushrooms. *Critical Reviews in Biochemistry*, **5**, 185–260.

WILLMITZER, L., SCHMALENBACH, W. & SCHELL, J. (1981). Transcription of T DNA in octopine and nopaline crown gall tumours is inhibited by low concentrations of α-amanitin. *Nucleic Acid Research*, **9**, 4801–12.

WOLLGIEHN, R. (1982). RNA polymerase and regulation of transcription. In Encyclopedia of Plant Physiology, New Series, vol. 14B, ed. B. Parthier & D. Boulter, pp. 125–59. Berlin, Heidelberg, New York: Springer.

ZAITLIN, M., NIBLETT, C. L., DICKSON, E. & GOLDBERG, R. B. (1980). Tomato DNA contains no detectable regions complementary to potato spindle tuber viroid as assayed by solution and filter hybridization. *Virology*, **104**, 1–9.

ZAUG, J. A., GRABOWSKI, P. J. & CECH, T. R. (1983). Autocatalytic cyclization of an excised intervening sequence RNA is a cleavage-ligation reaction. *Nature*, **301**, 578–83.

ZELCER, A., VANADELS, J., LEONHARD, D. A. ZAITLIN, M. (1981). Plant-suspension cultures sustain longterm replication of potato spindle tuber viroid. *Virology*, **109**, 314–22.

ZELCER, A., ZAITLIN, M., ROBERTSON, H. D. & DICKSON, E. (1982). Potato spindle tuber viroid-infected tissues contain RNA complementary to the entire viroid. *Journal of General Virology*, **59**, 139–48.

ZIEVE, G. W. (1981). Two groups of small stable RNAs. *Cell*, **35**, 296–7.

ZIMMERN, D. (1982). Do viroids and RNA viruses derive from a system that exchanges genetic information between eukaryotic cells? *Trends in Biochemical Science*, **8**, 205–7.

ZYLBER, E. A. & PENMAN, S. (1971). Products of RNA polymerases in HeLa cell nuclei. *Proceedings of the National Academy of Sciences, USA*, **68**, 2861–5.

INDEX

Italic page numbers refer to tables.

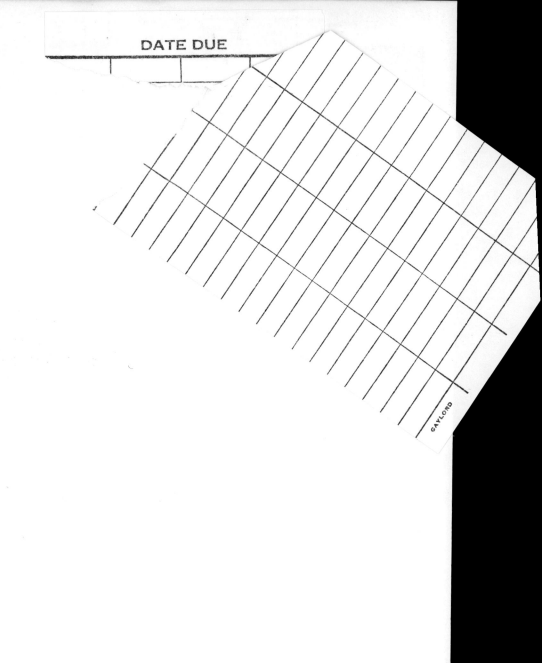